Delay and Disruption Tolerant Networks

Interplanetary and Earth-Bound –
Architecture, Protocols,
and Applications

Delay and Disruption Tolerant Networks

Interplanetary and Earth-Bound –
Architecture, Protocols,
and Applications

Edited by

Aloizio Pereira da Silva

Scott Burleigh

Katia Obraczka

CRC Press
Taylor & Francis Group
Boca Raton London New York

CRC Press is an imprint of the
Taylor & Francis Group, an **informa** business

CRC Press
Taylor & Francis Group
6000 Broken Sound Parkway NW, Suite 300
Boca Raton, FL 33487-2742

© 2019 by Taylor & Francis Group, LLC
CRC Press is an imprint of Taylor & Francis Group, an Informa business

No claim to original U.S. Government works

Printed on acid-free paper
Version Date: 20180728

International Standard Book Number-13: 978-1-1381-9806-7 (Hardback)

Visit the Taylor & Francis Web site at
http://www.taylorandfrancis.com

and the CRC Press Web site at
http://www.crcpress.com

Contents

Preface

Delay and Disruption Tolerant Networks: Interplanetary and Earth-Bound – architecture, protocols and applications

Vinton G Cerf
Vice President and Chief Internet Evangelist for Google

Every once in a while, an idea which is patently timely comes along that clearly anticipates a future need. Such is the case with Delay and Disruption Tolerant Networks. In the late 1990s, a small team at the Jet Propulsion Laboratory began work on an "Interplanetary Internet". The premise was that in 25 years, such a system would be needed to support manned and robotic missions to near and especially deep space. It is now nearly 25 years since that auspicious beginning. We are starting to see startling progress in commercial space missions such as those launched by SpaceX and Blue Origins that are intended to place humans on Mars within the next decade, to mine the asteroids, to return to the Moon and to continue scientific exploration of our solar system and perhaps, also, a mission to Alpha Centauri.

We live in a time when the Internet has become ubiquitous for about 50% of the world's population and when networking is a critical component of daily life. It is natural to incorporate digital networking into ongoing plans for space exploration. Astronauts will want to stay connected to Earth, however tenuously, and will need local communication as they begin exploration of Mars and the other planets in tandem with a flotilla of surface and orbital resources with which they will need to coordinate. Robust and reliable communication is needed to manage the spacecraft and onboard instruments and to return data to Earth for further processing.

The Jet Propulsion Laboratory team has expanded over the course of two decades to include most of the NASA Laboratories and sponsors at NASA Headquarters. Moreover the effort rapidly took on international character as participation of the

European Space Agency (ESA), the Japanese Space Agency (JAXA), and the Korean national space agency (KARI) ramped up through the Consultative Committee on Space Data Communication (CCSDS). Through the efforts undertaken with the sponsorship of CCSDS, standards have evolved and more are coming by way of the Internet Engineering Task Force (IETF). These are all welcomed developments as they move these important protocols to standard practice in delay and disruption prone environments.

These are not limited to deep space. We see similar conditions in terrestrial mobile telecommunications and in tactical military communication. It is thus extremely timely and very welcomed to have in hand a definitive volume devoted to the design and implementation of the Bundle Protocol (an instance of DTN communication architecture) and the Licklider Transport Protocol, among other important elements including security, congestion control and management. A considerable amount of operational experience has been gained through the use of these protocols and their predecessors on the Mars rovers and orbiters as well as the International Space Station, Earth-based resources and special test missions such as EPOXI. The time is ripe to deploy these new capabilities in support of current and planned space missions as well as emerging terrestrial applications.

Contributors

Nicholas Ansell
Independent Researcher
England, United Kingdom

Edward Birrane
The Johns Hopkins Applied Physics
    Laboratory (APL)
Laurel, Maryland

Carlo Caini
University of Bologna
Italy

Vinton G Cerf
Vice President and Chief Internet
    Evangelist for Google

Tomaso de Cola
German Aerospace Center
Germany

Marius Feldman
TU Dresden
Germany

Jorge M. Finochietto
Universidad Nacional de Córdoba -
    CONICET
Argentina

Juan A. Fraire
Universidad Nacional de Córdoba -
    CONICET
Argentina

Gianluigi Liva
German Aerospace Center
Germany

Daniel Oberhaus
Science Journalist in The Atlantic,
    Popular Mechanics
Phoenix, Arizona

Keith Scott
The MITRE Corporation
Virginia, USA

Felix Walter
TU Dresden
Germany

Chapter 1

Introduction

Daniel Oberhaus

Science Journalist in The Atlantic, Popular Mechanics, Nautilus, Vice

CONTENTS

1.1   Introduction

In July of 2010, a computer scientist named Stephen Farrell found himself in a remote Swedish village just north of the Arctic Circle. Outside of his teepee a few dozen villagers were busy herding thousands of reindeer just as their ancestors had done for centuries. The spectacle must've been an odd one for Farrell, but the Dubliner had not journeyed hundreds of kilometers north to look at reindeer. Instead, he was tasked with figuring out how to connect these quasi-nomadic reindeer herders to the internet when the nearest power grid was nearly 20 kilometers away.

As someone writing this from my home in Brooklyn, New York, it's difficult to imagine living in a place so remote that internet connectivity is impossible. These days, connectivity is nearly constant—whether you are a mountaineer standing on the summit of Everest or an astronaut docking at the International Space Station, it's possible to log on to social media, send an email, or video chat with friends from around the world.

But as with the reindeer herders, there are still cases where connections are intermittent at best and non-existent at worst. Whether this is the result of harsh environmental conditions, lack of infrastructure, or a lifestyle predicated on constant mobility, there is no shortage of situations that preclude the connectivity that makes the internet as we know it possible.

The assumption of end-to-end connectivity is a defining feature of the internet protocol suite that determines how data is transmitted over the internet. At its core, the suite is comprised of two packet switching protocols known as the Transmission Control Protocol (TCP) and the Internet Protocol (IP). Whereas IP is responsible for packaging and delivering the data sent across networks, TCP is responsible for correctly organizing these data packets so that they are sent and delivered as intended. Without TCP, a number of internet applications we use every day—such as the World Wide Web, email, music streaming platforms, or peer-to-peer file sharing—would be impossible. TCP and IP complement each other, but unlike IP, TCP is connection-oriented meaning that it requires establishing a connection between the hosts before the data can be delivered. The upside of TCP is that it is reliable—the connection ensures that the data will be sent and received in the intended order. The downside, of course, is that any of the previously mentioned applications that rely on ordered data transmission won't work in situations where establishing a connection between two hosts is impossible at the time of transmission.

Depending on the situation, the inability to connect to the internet can be a minor nuisance or life-threatening. For the reindeer herders, it meant having to fly in helicopters back to a city just to do simple tasks like checking their bank account. For a military unit in a combat zone or the first settlers on Mars, however, having a connection to headquarters can mean the difference between life and death.

1.1.1   What Is DTN?

Delay- or Disruption-tolerant networking is an approach to networking architecture that is specifically designed for environments with intermittent connectivity. Like the internet suite, DTN has its own protocols for routing data through a network whose nodes are constantly coming in and out of contact with one another. The main difference between delay tolerant networks and the 'normal' internet is that DTNs have a memory component and nodes are able to store data for an arbitrary amount of time, whereas nodes in the internet only relay data and all memory resources are concentrated at the end points of a connection.

The origins of DTN can be traced back to 1973, when DARPA funded a research project at the Stanford Research Institute in California whose aim was to establish a radio network capable of transmitting packets of digital data. The project was to be a proof of concept for implementing the packet switching protocols we are all familiar with today—TCP/IP—on ARPANET, the network that prefigured the global internet.

The packet radio network developed at the Stanford Research Institute—known as PRNET—was building on the work of the first wireless packet network, ALO-

HAnet, which was developed in 1971. Whereas ALOHAnet was using only fixed terminals to wirelessly transmit data, the goal of PRNET was to create a network comprised of fixed and mobile terminals. In this sense, the project was successful: PRNET was able to exchange data with ARPANET in one of the first instances of internetworking.

More importantly, however, PRNET developed the technologies that would make wireless ad hoc networking (decentralized networks that don't rely on pre-existing or fixed infrastructure) possible. A key technology created for PRNET was a repeater capable of performing a store-and-forward function, meaning that it was able to store received data packets until a mobile terminal came within range. This store-and-forward approach to networking pioneered by PRNET is the keystone of the delay tolerant networks being built today.

Fast forward to 1998. At this point the World Wide Web has been around for about seven years and IEEE 802.11, the protocol standard for wireless local area networking (colloquially known as WiFi) isn't even a year old. The advent of 802.11 coupled with the decreasing size and price of personal computers created a research boom in wireless ad hoc networking, with a particular focus on mobile ad hoc networks. These nets were designed such that each node in the network is able to move independently, changing its links to other nodes in an opportunistic fashion depending on its movement. In other words, the network could be sustained even if all the network nodes weren't always available.

These wireless ad hoc networks allowed for highly dynamic networks that were able to perform well in degraded communication conditions, as might be found in a war zone or disaster area, but they still required a connection to be established between nodes in the network before data could be sent. In situations where maintaining this connection would be impossible, a wireless ad hoc network could be as good as having no connection at all. There needed to be a way to pass information along a network in increments, where data packets would hop from node to node on the basis of their availability and eventually reach their destination—an approach to routing now known as store-carry-and-forward. The store-and-forward approach to routing had been pioneered by PRNET in the 70s, but the 'carry' element wasn't added until 1998. This was the year that Vint Cerf, better known as one of the 'fathers of the internet' for his role in designing TCP/IP, would meet with Adrian Hooke, a senior scientist at NASA's Jet Propulsion Laboratory. Their task was to figure out a networking scheme that would connect nodes moving at thousands of kilometers per hour and millions of kilometers apart—an interplanetary internet.

1.1.2 What Is IPN?

The interplanetary internet (IPN) began when DARPA funded a small group of scientists at the Jet Propulsion Laboratory to figure out how to apply a store-and-forward approach to networking in outer space. The result was the Bundle Protocol (BP), which has become synonymous with delay-tolerant networking since it was designed by Cerf and Hooke in the late 1990s.

The aptly-named Bundle Protocol works by tying together lower-layer internet protocols (i.e., the transport, data link, network and physical layers) in self-contained bundles. These bundles are then forwarded from node to node, temporarily stored at each point until a connection becomes available that allows the bundle to progress toward its final destination. You might think of the bundle of data as an airline passenger on a multi-stop flight—each airport is the node, and the passenger must hang out at the node until their connecting flight is ready to take off. What constitutes a node in the interplanetary internet depends on the situation: it might be a ground-based satellite dish that is part of NASA's Deep Space Network, the International Space Station, a satellite in Low Earth Orbit, or a rover on the surface of Mars.

The bundle protocol was designed with the challenging conditions of space in mind. Radio waves propagate in a straight line, but satellites and planets are constantly moving which means the intended recipient of a message isn't always going to be in line of sight. Thus the bundle protocol had to ensure that a node on the network is able to store information for an arbitrary amount of time until a connection with another node becomes available. Moreover, the bundle protocol also had to be able to account for massive communication delays. Every location on Earth is only a fraction of a light second away from every other location, so protocols like TCP/IP that depend on constant connectivity between two nodes isn't a problem—errors can be detected and corrected almost immediately. In space, however, radio and optical transmissions take minutes or hours to travel the gulf between Earth and the message's destination. The bundle protocol couldn't rely on the instantaneous feedback mechanisms integral to the terrestrial internet if it was going to work in space.

Imagine that one of the rovers on Mars needs to send some photos and data back to a NASA ground station. The rover could technically just directly beam this data back to Earth from the Martian surface, but because of the way the rovers were built and the conditions of the Martian atmosphere, this technique would slow down the data transmission by 4000 times. It makes much more sense for the rover to wait until a satellite in orbit around Mars passes overhead and transmit data to that satellite. Once that satellite has received this data, it can store it until the next node in the interplanetary internet is able to establish a connection—perhaps this is a satellite in Low Earth Orbit. On the other hand, if the data transmitted by the rover is corrupted, the orbiter can reestablish a connection with the rover in a matter of seconds, as opposed to dozens of minutes it would take to transmit directly to Earth and wait for a reply.

Although the Bundle Protocol was first described in the late 90s as a specific solution for an interplanetary internet, it wasn't deployed in space until almost a decade later. A major step in this direction was the standardization of a DTN architecture specifically for space applications called the Interplanetary Overlay Network (ION), which was first proposed by JPL researchers in 2007. Shortly thereafter, in 2008, the Bundle Protocol was used by a satellite in the UK's Disaster Monitoring Constellation to send a large image file (150 megabytes) to a ground station. In this case, the satellite was on a 100 minute polar orbit, which means that it was available to any particular ground station for about 8-14 minutes at a time. These short data transmission windows and long delays between them meant that the image had to be segmented into two smaller bundles which were transmitted to ground stations and combined again after the fact. This was successfully accomplished in August 2008 when an image of the Cape of Good Hope became the first to be transmitted to Earth from space using the Bundle Protocol. That same year, the bundle protocol was uploaded to the Deep Impact spacecraft 15 million miles from Earth, which had launched in 2005 on a mission to crash into comet. It was NASA's first experiment with ION and scientists at JPL sent the Deep Impact spacecraft photos which were then relayed back to Earth without data loss. Although this was only an experimental demonstration of the bundle protocol, Deep Impact was technically the first node of the interplanetary internet in deep space—albeit temporarily.

In the decade since these pioneering DTN experiments, researchers at NASA and other institutions have been hard at work figuring out how to leverage DTN protocols, and ION specifically, to create a solar system internet (SSI). The goal is to link planetary internets (e.g., on Earth and Mars) to one another via an interplanetary backbone of long-haul wireless links consisting of dedicated telecommunications orbiters or retrofitted spacecraft. The Mars Telecommunications Orbiter was meant to be the first permanent node of the interplanetary internet and would have functioned as a permanent deep space relay link between rovers on the Martian surface and mission control on Earth. A 2005 budget crunch killed plans for the orbiter, but the dream lives on in plans for the Mars 2022 telecommunications satellite. Still, dedicated IPN relay nodes are a tough sell for cash-strapped space agencies, which has made the notion of repurposing spacecraft whose primary missions have ended as IPN nodes a more attractive option.

In the meantime, the bundle protocol continues to improve. One of the latest developments is the bundle streaming service, which would allow for data streaming in an environment characterized by delay and disruption. This is done by implementing a library on the user's end which is responsible for storing and reordering all the data as it comes streaming into the user. So if a Martian is trying to stream video being relayed from Earth, due to the high error rate and significant delays inherent in interplanetary communication some of this video stream will be corrupted as the user is watching it. On the backend of the application, however, the library will be storing and reordering the data as it is received. If part of the stream was corrupted the first time, when that part of the bundle is retransmitted, the library will be responsible for incorporating it into the appropriate place in the stream. That way, if the Martian wants to play back the video later, they can watch the whole file as it was intended without the gaps experienced during the original transmission.

Although space applications for DTN have been deployed for over a decade in various experimental and operational capacities, it is still early days for delay tolerant networking in the final frontier. There are a number of pressing problems in need of solutions, ranging from securing data as it propagates through the SSI, which may be composed of nodes belonging to several different space agencies, to standardizing the most effective routing protocols. These challenges, and proposed solutions, will be considered throughout this text.

1.1.3 DTN Applications

Although the bulk of research into delay-tolerant networking over the last decade has been carried out in the context of interplanetary networking, deep space is certainly not the only application for delay tolerant networks.

Consider Farrell and the reindeer herders in the Arctic. During Farrell's sojourn to the land of perpetual twilight, he was testing out the bundle protocol for a terrestrial application. He and his colleagues set up solar-powered WiFi hotspots at different villages that were several kilometers distant from one another. Over the course of eight weeks, they acted as data mules ferrying packets of data from one WiFi hotspot to another, giving the villagers access to email by way of the bundle protocol.

Outside of connecting reindeer herders to the global internet, delay-tolerant networking has been successfully applied in dozens of different scenarios. For example, various projects have adopted delay tolerant networking protocols to provide internet access to remote villages around the world. In one example considered in this textbook, a service called Bytewalla turns Android phones into DTN nodes that can ferry data from point to point in rural Africa. Another interesting application of delay-tolerant networking was to monitor water quality in European lakes. In this instance, a ship would cruise around the lake, picking up and storing data from specialized nodes within the lake and then ferry this information as a data mule back to the dock, where a regular internet connection was available and then transmit this data from the lake to users on the global internet. Other uses have considered turning vehicles with regular routes (such as city busses or mail trucks) into information centers that would allow for efficient and scheduled data transfers. One of the biggest emerging DTN applications is the Internet of Things, a global network of embedded systems that range from everyday home appliances to industrial machines. Although many of these embedded systems are stationary or narrowly mobile, there is an increasing demand for extremely mobile networked objects that can withstand disruptions in connectivity when the object moves out of range of its nearest neighbor. This book will consider the unique challenges that come with various terrestrial applications of DTN.

1.1.4 Outline of the Rest of This Book

These ideas and many others will be explored throughout this book. In the second chapter, you will first be acquainted with the basic principles of delay-tolerant networking before jumping into the nitty gritty of DTN architecture and routing protocols. The third chapter provides a comprehensive overview of the diversity of DTN platforms and details how various package protocols and network operation concepts can be tailored to specific DTN use cases. The fourth chapter is a deep exploration of delay-tolerant networking as applied to the interplanetary network. Here you will learn about the satellites and rovers that make up the backbone of the interplanetary 'net' as well as specific implementations of the Bundle Protocol, such as the Interplanetary Overlay Network.

Chapter 5 focuses on the different methods of routing data packets through a delay tolerant network, such as forwarding strategies, flooding strategies, as well as Epidemic and Prophet routing. The sixth chapter of this text is dedicated to the art and science of coding for delay-tolerant networks. The seventh chapter builds on the knowledge base in Chapter 4 and explains how DTN is implemented on spacecraft. The eighth chapter will focus on the field of security, one of the most important aspects of taking DTN from an experimental networking architecture to mainstream use. While DTN attackers are limited in the same ways as legitimate users (e.g., few nodes, latency, etc.), delay tolerant networks are also more vulnerable to certain kinds of attacks. For example, most defenses against Man-in-the-middle attacks, in which an attacker occupies a spot in the network between a sender and receiver to distort or disrupt the data being sent, are not applicable in a DTN because they rely on session-establishment or timing analysis.

Chapter 9 discusses emerging applications of DTN in light of the burgeoning Internet of Things and the various technical issues associated with networking the world of objects. As the authors note, so far disruptions in IoT networks have received little to no attention from researchers. Although mobile embedded devices have been well studied, these systems never consider cases of extreme mobility, where connected devices move out of range of one another for arbitrary periods of time. Moreover, IoT devices are specifically designed to offload computationally intensive tasks to the end points of the network, but DTN requires nodes to have a decent amount of memory to function. The challenges of adapting DTN to the IoT context will be considered in detail.

The tenth chapter focuses on congestion control, which is closely related to DTN security. Although delay tolerant networks are still largely experimental, as they become more common they will need to be able to handle heavy network loads. When there is too much data on the network, DTN nodes must decide how to mitigate the situation by selectively dropping packets of data. This is arguably one of the least understood facets of delay tolerant networking and current research paradigms are discussed. Chapter 11 is all about the Licklider Transmission Protocol, which is a way to communicate between two nodes in deep space. Finally, Chapter 12 considers DTNperf 3, a tool to evaluate the performance of delay tolerant networks. Unlike TCP/IP networks, evaluating the performance of a DTN is more complicated due to the number of possible confounding factors, such as long delays and intermittent links. DTNperf 3 was developed as a way to take these factors into consideration when evaluating the performance of a challenged network.

1.1.5 Summary

Delay-tolerant networking is one of the most exciting fields in networking today, and the text you have before you is the most robust guide to this burgeoning field of research. The relative novelty of delay-tolerant networking as a discipline means that ground-breaking DTN innovations and applications are happening every year. In our increasingly networked world, DTN is sure to become an integral aspect of the global terrestrial internet in ways that have yet to be conceived. At the same time, DTN has a clear utility in space exploration and has already been heavily deployed for this purpose. Indeed, humanity's extraterrestrial future depends on improved communications and delay tolerant networks are sure to play an integral role in our exploration of the cosmos. This book is meant to be both a comprehensive introductory text for those unacquainted with DTN, as well as a source of inspiration for researchers already working on the DTN applications of the future.

Chapter 2

Delay and Disruption Tolerant Network Architecture

Aloizio P. Silva

University of Bristol

Scott Burleigh

JPL-NASA/Caltech

Katia Obraczka

University of California Santa Cruz

CONTENTS

2.1 Introduction

The history of DTN dates from early 1998 when Vint Cerf and a group of scientists from NASA's Jet Propulsion Laboratory (JPL) and industry came up with the idea of an "interplanetary backbone" to provide Internet communication for space missions. With the vision of space networking communication in mind they started developing an Interplanetary Internet (IPN) architecture which is discussed in more detail in Chapter 4. To support this work, the Internet Research Task Force (IRTF) formed the Interplanetary Networking Research Group (IPNRG) [148]. IPNRG started by exploring the possibility of using the TCP/IP architecture for interplanetary communication: since it worked on Earth it was expected also to work on Mars. However, the main question was, "Would TCP/IP work to support deep space communication?"

And the answer is "No." The two principal reasons for this answer are:

■ The speed of light is low relative to distances in the solar system. A one-way radio signal from Earth takes between 4 and 20 minutes to reach Mars.

■ Planetary motion is the second reason. If there is communication with an entity on the surface of a planet, it breaks as the planet moves. In this case it is necessary to wait until the planet moves back around again to reestablish communication.

In other words, in deep space communication, both long delays and connectivity disruptions (intermittent connectivity) must be taken into account and TCP/IP does not handle lengthy signal propagation delays and arbitrarily frequent- and long-lived disconnections well. Moreover, low and highly asymmetric bandwidth as well as high bit-error rates (BERs) distinguish interplanetary network communication from terrestrial networking scenarios. Another important consideration is that the TCP/IP protocols do not retain data in memory long enough while waiting for interrupted network connectivity to be resumed.

In summary, the communication model of the Internet is based on several assumptions, namely: the existence of a durable end-to-end path between message origin and destination entities; short round-trip delays; symmetric data rates; bidirectional end-to-end path between two network nodes; and low error rates. These assumptions do not necessarily hold in so-called "extreme" environments. Consequently, one of IPNRG's main conclusions was that extending the Internet's network architecture to operate end-to-end in extreme, challenged environments such as deep space communication would not work. The IPNRG then decided to explore network models able to withstand arbitrarily long delays and disconnections.

In 2002, the IPNRG transitioned to a new research working group, named Delay Tolerant Network Research Group (DTNRG) [147], that aimed to specify the architectural and protocol design principles to support communication in extreme environments. As such, interplanetary internetworking began to be considered an example application scenario within the emerging field of Delay- and Disruption-Tolerant Networking (DTN).

In 2003, the DTNRG published the first DTN architecture specification as RFC 4838 [75]. Since then the DTN architecture has been extended, adapted, and improved. The DTN communication model is a departure from the Internet's "end-to-end" paradigm and is based on a different set of assumptions. For example, as part of the proposed DTN architecture a new *Bundle Protocol* [244] was developed. DTN "bundles" are functionally similar to Internet packets in the sense that they are discrete chunks of information to be transmitted by the network. "Bundles" may be arbitrarily large and are sent through the DTN using *store-carry-and-forward* procedures, an extension traditional packet switching's *store-carry-and-forward* mechanism. Outbound bundles are stored as long as necessary while awaiting a transmission "opportunity", or "contact", since connectivity disprusions are part of DTN's typical modus operandi. Bundles may be stored within the network for minutes, even hours. And, the declining cost of memory makes this architecture increasingly more viable. Bundles do not need to be retransmitted from end to end; instead, retransmissions may be performed at intermediate points in the path.

In 2008 a delay- and disruption-tolerant point-to-point protocol specification, named Licklider Transmission Protocol (LTP) [55], was published by the DTNRG. The main idea behind LTP is to achieve reliability by enabling retransmission over links characterized by very long propagation delays. More details regarding LTP are provided in Chapter 11.

Even though the DTNRG concluded its activities as of May 2016, it had been quite active since its creation. It addressed a wide range of DTN applications including both deep-space as well as terrestrial scenarios, such as vehicular networks, military and civilian applications, bridging the digital divide in remote regions and developing parts of the world, sensor networks, just to name a few.

This chapter presents the DTN architecture as proposed by RFC 4838. It describes the DTN architecture in a general way addressing both interplanetary and earth-bound scenarios. It also presents the Bundle Protocol as defined in RFC 5050. Throughout the chapter, we also introduce DTN basic concepts and terminology. We start with a brief description of the Internet architecture.

2.2 The Internet Architecture

The Internet consists of collection of interconnected networks providing global connectivity to users and devices around the world. The Internet's network architecture is based on the layered system design principles and uses the TCP/IP protocol stack illustrated in Figure 2.1. Like any layered system, the Internet structures the functions needed to perform end-to-end data communication in layers, where each layer builds upon functionality offered by the layer below through well-defined interfaces.

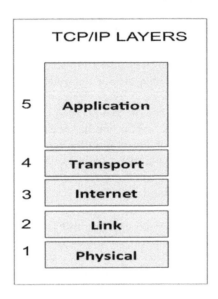

Figure 2.1: The Internet's TCP/IP protocol stack.

The functions provided at each layer of the TCP/IP stack can be summarized as follows:

1. **Layer 1 - Physical Layer (or PHY):** The PHY is responsible for transmitting and receiving raw bits of information over physical links (wired or wireless) connecting nodes to one another.

2. **Layer 2 - Data Link Layer (or DLL):** The DLL controls communication between 2 directly connected network nodes. It encapsulates a packet from the network layer into a frame with layer 2 source and destination addresses. Depending on the Layer 2 protocol used, it can also perform hop-by-hop reliable delivery as well as flow control.

3. **Layer 3 - Internet/Network Layer:** Routing and forwarding are two of Layer 3's main functions. Forwarding uses paths computed by routing to decide how

to forward a packet at each hop on its way from source to destination. The Internet's network layer is considered the "glue" of the Internet: through the IP protocol it provides a common addressing scheme (IP addresses) as well as transmission data unit (packet) that enables networks based on different Layer 2 and Layer 1 technologies to interconnect.

4. **Layer 4 - Transport Layer:** Also known as host-to-host layer, it is responsible for providing end-to-end communication. The Internet's two most widely deployed transport protocols are:

   - Transmission Control Protocol (TCP): TCP provides connection-oriented, reliable end-to-end communication service which ensures data is delivered in order and with no duplicates. TCP also performs flow- and congestion control to avoid senders from overunning receivers as well as not congesting the network.

   - User Datagram Protocol (UDP): Unlike TCP, UDP provides connection-less, unreliable communications service. UDP is typically used for short-lived request-response communication for which the overhead of establishing a TCP connection is not justified or when the applications or upper layer protocols provide their own reliable delivery.

5. **Layer 5 - Application Layer:** It is the layer closest to the user and is where network applications run generating and consuming data. Example of Internet applications and their protocols include:

   - The Hypertext Transfer Protocol (HTTP) is the protocol used by the World Wide Web (WWW) for communication between Web clients (browsers) amd Web servers.

   - The File Transfer Protocol (FTP) is used for interactive file transfer.

   - The Simple Mail Transfer Protocol (SMTP) is the standard Internet protocol for electronic mail (e-mail) transmission.

   - The Domain Name System (DNS) is the Internet's directory service used to resolve names (e.g., host names) to IP addresses.

   - The Simple Network Management Protocol (SNMP) is an Internet standard for collecting and exchanging network management and monitoring information. It also allows remote management actions such as configuration changes through modifications of network management variables and parameters.

As previously discussed, the Internet was designed to operate under assumptions that cannot be guaranteed in DTN scenarios. These assumptions are summarized below and illustrated in in Figure 2.2.

- An end-to-end path between any two communicating nodes exists and is available throughout a communication session between them.

- Most functions are implemented at the edges leaving a minimalist network core.

- Consequently, a timely and stable way to receive end-to-end feedback to provide error-, flow-, and congestion control must exist.

- Transmission losses are relatively small.

- Interoperability is achieved through a packet switching mechanism.

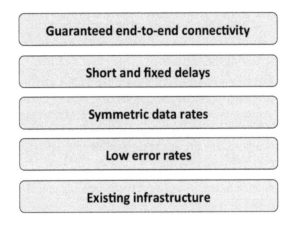

Figure 2.2: TCP/IP traditional characteristics and assumptions.

2.3   DTN Characteristics

The Internet's design principles do not hold in extreme or challenged networking environments, which are characterized by:

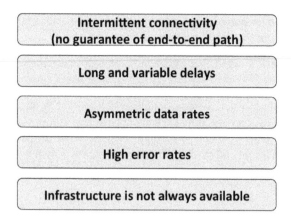

Figure 2.3: DTN's main features.

Intermittent Connectivity: DTNs cannot rely on the existence of an end-to-end path between source and destination as connectivity disruptions may result in network partitions.

Long and Variable Delays: Due to potentially long distances between nodes (e.g., IPNs which are characterized by "astronomical distances"), episodic network connectivity, or physical characteristics of the propagation medium (e.g., underwater), communication between nodes may be subject to long and/or highly variable signal propagation delays.

Asymmetric Data Rates: Source-destinaton path characterisitics may be significantly different from reverse path characteristics, e.g. either one may be unavailable, as well as data rates, routes, and bit error rates may exhibit significant mismatch.

High Error Rates: DTN links may also exhibit high bit error rates (BERs). For example, communication links in space have BERs between 10^{-9} to 10^{-7}.

2.4 DTN Architecture

The main entity in the DTN ecosystem is the "DTN node" also called "Bundle node". It can send and receive protocol data units (PDUs), named "bundles", which are generated according to the DTN Bundle protocol. Each bundle is formed by a sequence protocol data block as described in the next sections. The DTN node is composed of three components:

1. A **bundle protocol agent** that offers bundle protocol services,

2. A set of **convergence layer adapters (CLAs)** that enable bundle transmission and reception,

3. An **application agent** that uses bundle protocol services for communication purposes.

Figure 2.4 illustrates the main components of a DTN node. Administrative Records are application data units exchanged between Administrative Elements of DTN nodes' Application Agents to accomplish some administrative task. RFC 5050 defines the only administrative record: the status report which is presented in Section 2.12.2.

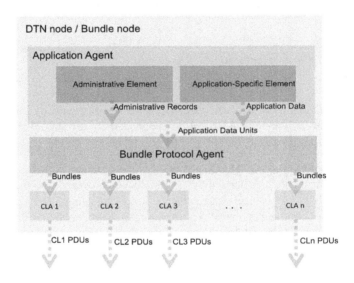

Figure 2.4: The main components of a DTN node.

2.4.1 Store-Carry-and-Forward

The DTN architecture introduced the *store-carry-and-forward* paradigm, an extension of the Internet's *store-and-forward*. In *store-carry-and-forward*, the DTN node uses persistent memory to store and carry data until a *contact* opportunity arises. At that time, the node may decide to forward data. Data is thus forwarded from node to node whenever links (or contacts) are available until the data reaches the destination. Figure 2.5 illustrates the store-carry-and-forward paradigm: a source DTN node has a message (bundle) to send to a destination DTN node. The source DTN node moves and carries the bundle stored in its persistent memory. A contact opportunity arises with an intermediate DTN node *A* and the source DTN node forwards the bundle to *A*. The intermediate DTN node *A* stores the bundle in its persistent memory for future forwarding. DTN node *A* encounters DTN node *B* and *A* forwards the bundle to *B*. *B* then stores and carries the bundle until it finally encounters the destination DTN node and delivers the bundle to it.

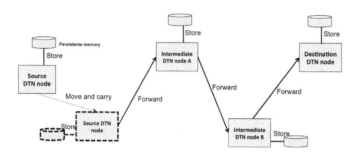

Figure 2.5: DTN's store-carry-and-forward paradigm.

DTN uses the store-carry-and-forward mechanism to withstand intermittent connectivity by overlaying a new protocol layer named "bundle layer" which enables the DTN node to transmit messages which are called "bundles".

2.4.2 Network of Networks

DTNs may consist of a collection of interconnected networks. The DTN architecture is a network of Internets. Similarly to the Internet's Network layer, the DTN Bundle layer functions as the "convergence" layer enabling interconnectivity amongst different DTN "regional" networks each of which may employ different network technologies, network protocols, addressing schemes, etc. Each regional network is identified by an unique ID and forwarding messages/bundles between regional networks is done through DTN gateways.

Just like network prefixes, regional network IDs enable bundles to be routed across different regional networks. "Entity IDs" uniquely identify a DTN node within a region network, so that, once in the destination regional network, the bundle will be delivered to the destination DTN node identified by the entity ID. In other words, a globally unique ID in the DTN architecture is represented by the tuple (*<region.id; entity.id >*).

Both regional network ID and entity ID can be named in the Internet's Domain Name System (DNS). In this case a name-to-address binding function may exist to support routing. The DTN architecture defines late binding where the entity ID of a tuple is only interpreted inside of the applicable regional network. This approach

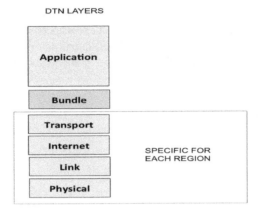

DTN LAYERS

Figure 2.6: The DTN stack.

gives much autonomy to the regional network and avoids having an universal name-to-address binding space.

2.4.3 The Bundle Layer

As discussed in Section 2.4.2, the DTN architecture introduces a new "convergence" layer, named "Bundle", located under the the application layer and typically above the transport layer. Figure 2.6 illustrates the DTN stack.

The main functions associated with the Bundle layer are:

■ Tolerate disruptions/intermittent connectivity.

■ Support late binding across heterogeneous networks acting as an overlay.

■ Provide reliable delivery since DTN nodes hold messages as "custodians" (DTN Custody transfers are explained in Section 2.5 below).

■ Support different types of connections or contacts (e.g., scheduled, predicted, and opportunistic).

The unit of information exchanged in a DTN is named "bundle". And each DTN node in the network is defined as an entity that implements the bundle layer as well as the bundle protocol defined by RFC 5050 [244]. A DTN node can assume different roles. It can be a router, a host, and a gateway.

2.5 Bundle Protocol

The DTN architecture implements through the bundle layer a new transmission protocol, called Bundle Protocol (BP). The BP is for DTNs what IP is for the Internet, i.e., it allows application programs to operate across networks that may be using a variety of lower layer protocol stacks while being able to withstand arbitrarily long lived delays and disconnections. It runs at the application layer of the Internet stack and interfaces with lower layer stacks the *convergence layer adapter (CLA)* as illustrated in Figure 2.7.

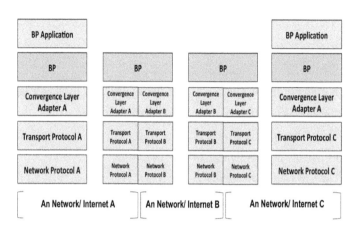

Figure 2.7: The Bundle protocol and convergence layer adapter.

The BP uses the following terminology as defined in RFC 5050.

Bundle: As previously presented, bundle is a protocol data unit in DTN. Each bundle comprises a sequence of two or more "blocks" of protocol data, which serve various purposes. Bundles contains a primary block and more other blocks of data. The primary block contains basic information, for instance destination ID. The other blocks contain information that arises in the header or/and payload section of the protocol data units in other protocol architectures.

Endpoint ID: is an uniform resource identifier (URI) in the format $<scheme\_name >: <scheme\_specific\_part >$ where:

- $<scheme\_name >$ is "dtn" or "ipn" for terrestrial and interplanetary networks, respectively.
- $<scheme\_specific\_part >$ is application-specific or administrative. Generally used when forwarding bundles node from node to node. For example, "dtn://madridServer/interfaces" is application-specific and "dtn://madridServer/" is administrative.

Bundle Payload: is the application data that should be transmitted to a specific destination.

Fragment: is part of a bundle that has been fragmented in multiple small bundles during transmission. Any number of fragments can be generated. These fragments can be reassembled anywhere in the network, reconstituting the original bundle.

Bundle Protocol Agent (BPA): is a component in the DTN node that makes the BP services available. The BPA can be implemented in hardware or software. The BPA is responsible for providing the following services to the AA:

- Register a node in an endpoint;
- Summarize a registration;
- Perform the transition of a registration between Active and Passive states;
- Transmit a bundle to an identified bundle endpoint;
- Cancel a transmission;
- Poll a registration that is in the passive state;
- Deliver a received bundle.

Application Agent(AA): is a DTN node component that accesses the BP services provided via BPA. Basically, the AA is composed of two elements:

- Administrative Element: is responsible for constructing and requesting transmission of administrative records (status reports and custody signals), and it accepts delivery of and processes any custody signals that the node receives.
- Application-Specific Element: is responsible for constructing, requesting transmission of, accepting delivery of, and processing application specific application data units.

Bundle Endpoint: is a set of one or more DTN nodes that identify themselves to the BP through a single string, named "bundle endpoint ID" (EID). As previously described, source DTN nodes and destination DTN nodes are identified via EID.

Registration: is a state machine that characterizes a node's membership in a specific bundle endpoint. The node's membership can be in one of two states: active or passive. The node's membership registration is in the active state when the reception of a bundle that is deliverable subject to this registration must cause the bundle to be delivered automatically. The node's membership registration is in the passive state when the reception of a bundle that is deliverable subject to this registration must cause delivery of the bundle to be abandoned or deferred as mandated by the registration's current delivery failure action.

Custody: A custody transfer enables a DTN node to be assigned the responsibility to deliver a specific bundle to its next hop on the way to the bundle's ultimate destination. The DTN node that currently stores the bundle is named the "custodian" for that bundle. Once the next DTN node accepts custody of the bundle, it becomes the bundle's new "custodian" and the previous custodian may delete the bundle from its buffer. Custody transfers require that the bundle pass from one custodian to the other, and each custodian stores the bundle until it has confirmation that the bundle was accepted in custody by another custodian. Custody can be released as follows.

■ After receiving a notification that another node has accepted custody.

■ After receiving a notification that the bundle has been received in its destination.

■ When the bundle's TTL has expired and the bundle is discarded.

Figure 2.8 shows an example of custody transfer. In this example a "source DTN node A" needs to send a bundle to the "destination DTN node E". Note that a DTN node can accept or refuse a custody request. If a DTN node refuses the custody request it still can forward the bundle to the next hop.

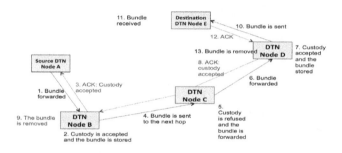

Figure 2.8: Example of custody transfer.

Hop-by-hop retransmission is supported by BP using custody transfer. The custody transfer is an agreement between the BPAs of intermediate nodes and the source application where the custody process was started.

2.6   Bundle Format

Bundles are formed by a sequence of two or more "blocks", where the first block, or *primary block* is unique block. One of the subsequent blocks is the payload block. All blocks after the primary block in the sequence must have the "last block" flag (block processing control flags) set to 0 except the last one that must have it set to 1. Figure 2.9 shows the bundle control flags specification.

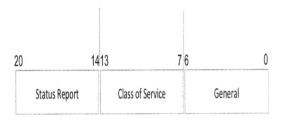

Figure 2.9: Bundle processing control flag specification.

The seven least-significant bits (bits 0-6) are "general" flags and are defined as follows:

0. Bundle is a fragment.

1. Application data unit is an administrative record.

2. Bundle must not be fragmented.

3. Custody transfer is requested.

4. Destination endpoint is a singleton.

5. Acknowledgment by application is requested.

6. Reserved for future use.

The bits from 7 to 13 are used to indicate the bundle's class of service. The bits 8 and 7 constitute a two-bit priority field indicating the bundle's priority, with higher values being of higher priority:

■ 00 : bulk

■ 01 : normal

■ 10 : expedited

■ 11 : reserved for future use

The bits from 9 to 13 are reserved for future use. The bits from 14 to 20 define the status report request flags as follows.

14. Request reporting of bundle reception.

15. Request reporting of custody acceptance.

16. Request reporting of bundle forwarding.

17. Request reporting of bundle delivery.

18. Request reporting of bundle deletion.

19. Reserved for future use.

20. Reserved for future use.

2.6.1 Block Processing Control Flags

Every block in the sequence after the primary block is a Self-Delimiting Numeric Value (SDNV)[1]. The value stored in this SDNV is used to select the block processing control features. The block processing control flags are defined in bits as follows.

0. Block must be replicated in every fragment.

1. Transmit status report if block cannot be processed.

2. Delete bundle if block cannot be processed.

3. Last block.

4. Discard block if it cannot be processed.

5. Block was forwarded without being processed.

6. Block contains an EID-reference field.

2.6.2 Primary and Payload Block Format

Figure 2.10 shows the primary block and payload block for a bundle.
The primary bundle block is composed of the following fields.

Version: is an unsigned integer value indicating the version of the bundle protocol that constructed the current block.

Bundle Processing Control Flags: It has been described before.

Block Length: is the length of all remaining fields of the current block.

Destination Scheme Offset: contains the offset within the dictionary byte array of the scheme name of the endpoint ID of the bundle's destination.

Destination SSP Offset: contains the offset within the dictionary byte array of the scheme-specific part of the endpoint ID of the bundle's destination.

---

[1] An SDNV is a numeric value encoded in N octets, the last of which has its most significant bit (MSB) set to zero; the MSB of every other octet in the SDNV must be set to 1. The value encoded in an SDNV is the unsigned binary number obtained by concatenating into a single bit string the 7 least significant bits of each octet of the SDNV.

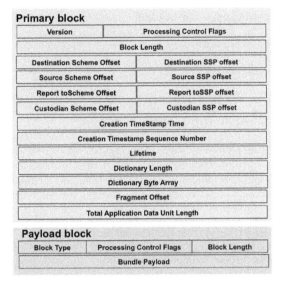

Figure 2.10: Bundle block formats (Primary block and Payload block).

Source Scheme Offset: contains the offset within the dictionary byte array of the scheme name of the endpoint ID of the bundle's nominal source.

Source SSP Offset: contains the offset within the dictionary byte array of the scheme-specific part of the endpoint ID of the bundle's nominal source.

Report-to Scheme Offset: contains the offset within the dictionary byte array of the scheme name of the ID of the endpoint to which status reports pertaining to the forwarding and delivery of this bundle are to be transmitted.

Report-to SSP Offset: contains the offset within the dictionary byte array of the scheme-specific part of the ID of the endpoint to which status reports pertaining to the forwarding and delivery of this bundle are to be transmitted.

Custodian Scheme Offset: contains the offset within the dictionary byte array of the scheme name of the current custodian endpoint ID. The current custodian endpoint ID of a primary bundle block identifies an endpoint whose membership includes the node that most recently accepted custody of the bundle upon forwarding this bundle.

Custodian SSP Offset: contains the offset within the dictionary byte array of the scheme-specific part of the current custodian endpoint ID.

Creation Timestamp: contains two unsigned integers that, together with the source node ID and (if the bundle is a fragment) the fragment offset and payload length, serve to identify the bundle. The two unsigned integers are described as follows.

1. Bundle's creation time which is expressed in seconds since the start of the year 2000, on the Coordinated Universal Time (UTC) scale [UTC] - at which the transmission request was received that resulted in the creation of the bundle

2. Bundle's creation timestamp sequence number is the latest value (as of the time at which that transmission request was received) of a monotonically increasing positive integer counter managed by the source node's bundle protocol agent that may be reset to zero whenever the current time advances by one second.

Lifetime: consists of an unsigned integer that indicates the time at which the bundle's payload will no longer be useful, encoded as a number of microseconds past the creation time. When a bundle's lifetime expires, the bundle node can delete it from the network.

Dictionary Length: contains the length of the dictionary byte array.

Dictionary Byte Array: consists of an array of bytes formed by concatenating the null-terminated scheme names and SSPs of all endpoint IDs referenced by any fields in this Primary Block together with, potentially, other endpoint IDs referenced by fields in other blocks.

Fragment offset: consists of an unsigned integer indicating the offset from the start of the original application data unit at which the bytes comprising the payload of this bundle were located. This field must not be empty if the bundle processing control flags of the current primary block indicates that the bundle is a fragment.

Total Application Data Unit Length: If the current bundle is a fragment, this field is a SDNV indicating the total length of the original application data unit of which this bundle's payload is a part. Otherwise, the total application data unit length field is omitted from the current block.

The payload block fields are described below.

Block type code: is 1 byte field that indicates the type of the block. For the bundle payload block, this field contains the value 1.

Block processing control flags: was previously discussed.

Block data length: contains an unsigned integer that represents the length of the bundle's payload.

Bundle Payload: contains the application data unit, or some contiguous extent thereof.

2.6.3 Canonical Bundle Block Format

Canonical bundle blocks are all bundle blocks except the primary block. Each canonical bundle block is composed of the following fields.

Block type code: contains an unsigned integer that can have the following values.

1. indicates that the block is a bundle payload block.
2. reserved (see Section 2.6.4).
3. reserved (see Section 2.6.4).
4. reserved (see Section 2.6.4).
5. reserved (see Section 2.6.4).
6. reserved (see Section 2.6.4).
7. reserved (see Section 2.6.4).
8. reserved (see Section 2.6.4).
9. reserved (see Section 2.6.4).
10. reserved (see Section 2.6.4).

Block type codes 192 through 255 are not reserved and are available for private and/or experimental use. All other block type code values are reserved for future use.

Block number: contains an unsigned integer that identifies the block within the bundle, enabling blocks to explicitly reference other blocks in the same bundle. Block numbers do not need to be a continuous sequence, and blocks do not need to appear in block number sequence in the bundle. The block number of the payload block is always zero.

Block processing control flags: was previously described.

Block EID Reference Count: is optional. If and only if the block references EID elements in the primary block's dictionary, the flag "block contains an EID reference" is set to 1 and the block contains an EID reference field that consist of a count of EID references followed by the EID references.

Block data length: contains an unsigned integer that represents the aggregate length of all remaining fields of the block.

Block-type-specific: defines a specific format and order and the aggregate length in octets is the value of the block data length field. For the Payload Block in particular (block type 1), there shall be exactly one block- type-specific data field, called "payload".

2.6.4 Extension Block

The extension blocks are all blocks other than the primary and payload blocks. A node may not be able to process an extension block since some extension blocks are not defined in the BP specification. In this case, the values of the block processing control flags can indicate the action to be taken by the bundle protocol agent when the extension block cannot be processed.

When a bundle is forwarded and it contains extension blocks that could not be processed, the flag "block was forwarded without being processed" is set to 1. If the next hop it is able to process the extension block this flag can be set to zero.

2.7 Bundle Processing

The bundle processing is performed in the BPA and AA (see Figure 2.4). The transmission of a bundle starts when a DTN node's AA needs to send a bundle. If this bundle requires an administrative record the BPA through the administrative element generates a new bundle that includes the record and transmits it.

Given a bundle transmission request, the bundle processing follows the steps below.

Step 1: Transmission of the bundle is initiated. An outbound bundle is created per the parameters of the bundle transmission request, with the retention constraint "Dispatch pending". The source node ID of the bundle is either a null endpoint ID indicating that the source of the bundle is anonymous), or a singleton endpoint whose only member is the node of which the BPA is a component.

Step 2: This is the starting point of forwarding process. The retention constraint "Forward pending" is added to the bundle, and the bundle's "Dispatch pending" retention constraint is removed.

Step 3: The BPA determines whether or not the bundle forwarding is constrained for any of the following reasons which are included in the bundle status report.

- No additional information.
- Lifetime expired.
- Forward over unidirectional link.
- Tranmission canceled.
- Storage overflow.
- Destination endpoint ID unintelligible.
- No known route to destination from here.
- No timely contact with next on route.
- Block unintelligible.
- Hop limit exceeded.

Step 4: If the bundle cannot be forwarded due to any reasons cited above the BPA will determine whether or not to declare failure in forwarding the bundle. If forwarding failure is declared, then the forwarding failed procedure is performed:

> **Step 4.1:** BPA may forward the bundle back to the node that sent it, as identified by the previous node block, if present.

> **Step 4.2:** If the bundle's destination endpoint is an endpoint of which the node is a member, then the bundle's "Forward pending" retention constraint is removed. Otherwise, the bundle is deleted as follows.

> > **Step 4.2.1:** If the "request reporting of bundle deletion" flag in the bundle's status report request field is set to 1, and if status reporting is enabled, then a bundle deletion status report citing the reason for deletion is generated, destined for the bundle's report-to endpoint ID.

> > **Step 4.2.2:** All of the bundle's retention constraints are removed.

Otherwise, at some point in the future if the forwarding of this bundle ceases to be constrained, processing proceeds.

Step 5: For each node selected for forwarding, BPA invokes the services of the selected CLAs in order to effect the sending of the bundle to that node.

Step 6: After all CLAs have informed the BPA that the data sending procedure is concluded for the current bundle, the following steps might take place.

> **Step 6.1:** If the "request reporting of bundle forwarding" flag in the bundle's status report request field is set to 1, and status reporting is enabled, then a bundle forwarding status report is generated, destined for the bundle's report-to endpoint ID. The reason code on this bundle forwarding status report must be "no additional information".

> **Step 6.2:** If any applicable bundle protocol extensions mandate generation of status reports upon conclusion of convergence-layer data sending procedures, all such status reports are generated with extension mandated reason codes.

> **Step 6.3:** The bundle's "Forward pending" retention constraint is removed.

Note that the bundle can expire at any point during its processing. When this happens, the BPA must delete the bundle for the reason "lifetime expired" and the steps 4.2.1 and 4.2.2 should take place.

2.8 Bundle Reception

When a DTN node receives a bundle the following steps are executed.

Step 1: The retention constraint "Dispatch pending" is added to the bundle.

Step 2: If the "request reporting of bundle reception" flag in the bundle's status report request field is set to 1, and status reporting is enabled, then a bundle reception status report with reason code "No additional information" is generated, destined for the bundle's report-to endpoint ID.

Step 3: For each block in the bundle that is an extension block that the bundle protocol agent cannot process:

- If the block processing flags in that block indicate that a status report is requested in this event, and status reporting is enabled, then a bundle reception status report with reason code "Block unintelligible" is generated, destined for the bundle's report-to endpoint ID.

- If the block processing flags in that block indicate that the bundle must be deleted in this event, then the bundle protocol agent deletes the bundle for the reason "Block unintelligible" and all remaining steps of the bundle reception procedure are skipped.

- If the block processing flags in that block do not indicate that the bundle must be deleted in this event but do indicate that the block must be discarded, then the BPA removes this block from the bundle.

- If the block processing flags in that block indicate neither that the bundle must be deleted nor that the block must be discarded, then processing continues with the next extension block that the bundle protocol agent cannot process.

Step 4: The bundle dispatching procedure is executed as follows. If the bundle's destination endpoint is an endpoint of which the node is a member, the local bundle delivery procedure described in Section 2.9 is executed and for the purposes of all subsequent processing of this bundle at this node the node's membership in the bundle's destination endpoint is rejected.

Step 5: The forwarding procedure should follow the Step 4 in Section 2.7.

2.9   Local Bundle Delivery

This section describes the steps that take place when the bundle being delivered is destined for an endpoint of which the current node is a member.

Step 1: If the received bundle is a fragment, the application data unit reassembly procedure should be followed (see Section 2.11). If this procedure results in reassembly of the entire original application data unit, processing of this bundle proceeds from Step 2; otherwise, the retention constraint "Reassembly pending" is added to the bundle and all remaining steps of this procedure are skipped.

Step 2: Delivery depends on the state of the registration whose endpoint ID matches that of the destination of the bundle:

■ If the registration is in the Active state, then the bundle is delivered automatically as soon as it is the next bundle that is due for delivery according to the BPA's bundle delivery scheduling policy, an implementation matter.

■ If the registration is in the Passive state, or if delivery of the bundle fails for some implementation specific reason, then the registration's delivery failure action is taken. Delivery failure action can be one of the two:

1. defer delivery of the bundle subject to this registration until this bundle is the least recently received of all bundles currently deliverable subject to this registration and either the registration is polled or else the registration is in the Active state, and also perform any additional delivery deferral procedure associated with the registration;

2. abandon delivery of the bundle subject to this registration.

Step 3: Once the bundle has been delivered, if the "request reporting of bundle delivery" flag in the bundle's status report request field is set to 1 and bundle status reporting is enabled, then a bundle delivery status report is generated, destined for the bundle's report-to endpoint ID.

2.10 Bundle Fragmentation

The Bundle Protocol permits both proactive and reactive bundle fragmentation. Proactive fragmentation is performed to meet constraints on maximum bundle size on intermittent DTN hops. Reactive fragmentation aims to avoid retransmitting already acknowledged data during disconnection. This functionality enables different regions to handle different maximum sizes. Basically, the bundle fragmentation process breaks bundles into smaller pieces (fragments), so that bundles may be formed that can pass through a link with a smaller maximum transmission unit (MTU) than the original bundle size. Fragmentation is also helpful if a node to which a bundle is to be forwarded is accessible only via intermittent contacts and no upcoming contact is long enough to enable the forwarding of the entire bundle. The fragments are reassembled by the receiving host.

The fragmentation process reduces the bundle's size. For example, to fragment a bundle whose payload is of size M we replace it with two "fragments" which are new bundles with the same source node ID and creation timestamp as the original bundle. The payloads of the two fragmentary bundles are the first N and the last $(M - N)$ bytes of the original bundle's payload, where $0 < N < M$. Note that fragments may themselves be fragmented, so fragmentation may in effect replace the original bundle with more than two fragments.

Any bundle whose primary block's bundle processing flags do not indicate that it must not be fragmented may be fragmented at any time, for any purpose, at the discretion of the bundle protocol agent.

The fragmentation of a bundle can be limited as follows.

■ The concatenation of the payloads of all fragments produced by fragmentation must always be identical to the payload of the fragmented bundle. Note that the payloads of fragments resulting from different fragmentation episodes, in different parts of the network, may be overlapping subsets of the fragmented bundle's payload.

■ The primary block of each fragment must differ from that of the fragmented bundle, in that the bundle processing flags of the fragment indicates that the bundle is a fragment and both fragment offset and total application data unit length must be provided.

■ The payload blocks of fragments will differ from that of the fragmented bundle as follows.

   ■ If the fragmented bundle is not a fragment or is the fragment with offset zero, then all extension blocks of the fragmented bundle are replicated in the fragment whose offset is zero.

   ■ Each of the fragmented bundle's extension blocks whose "Block must be replicated in every fragment" flag is set to 1 is replicated in every fragment.

2.11   Fragment Reassembly

Bundles can be fragmented during transmission into multiple small bundles (fragments) which, when concatenated, reconstitute the original bundle. These fragments can also be further fragmented and reassembled anywhere in the network.

If the concatenation as informed by fragment offsets and payload lengths of the payloads of all previously received fragments with the same source node ID and creation timestamp as this fragment, together with the payload of this fragment, forms a byte array whose length is equal to the total application data unit length in the fragment's primary block, then:

■ This byte array replaces the payload of this fragment.

■ The "Reassembly pending" retention constraint is removed from every other fragment whose payload is a subset of the reassembled application data unit.

2.12 Administrative Record Processing

2.12.1 Administrative Records

Administrative records are standard application data units that are used in providing some of the BP's features. RFC 5050 specifies one type of administrative record, named "bundle status reports". Each administrative record is represented as a tuple with the following fields:

1. Record type code which is an unsigned integer. Record type code equal 1 defines a bundle status report. Other type codes are reserved for future use.

2. Record content in type-specific format.

2.12.2 Bundle Status Reports

The bundle status reports provide information about how bundles are progressing through the system, including notices of receipt, forwarding, final delivery, and deletion. They are transmitted to the Report-to endpoints of bundles. The transmission of "bundle status reports" under specified conditions is an option that can be invoked when transmission of a bundle is requested.

The first item of the bundle status report is the bundle status information represented as an array of at least 4 elements. The first four items of the bundle status information array shall provide information on the following four status assertions, in this order:

1. Reporting node received bundle.

2. Reporting node forwarded the bundle.

3. Reporting node delivered the bundle.

4. Reporting node deleted the bundle.

Each item of the bundle status information array is a bundle status item represented as an array. The size of this array is 2 if the value of the first item of this bundle status item is 1 and the "Report status time" flag was set to 1 in the bundle processing flags of the bundle whose status is being reported. Otherwise the size is 1. The first item of the bundle status item array is a status indicator (a boolean value) that indicates whether or not the corresponding bundle status is asserted. The second item of the bundle status item array, if present, indicates the time at which the indicated status was asserted for this bundle.

The second item of the bundle status report array is the bundle status report reason code explaining the value of the status indicator. Valid status report reason codes are listed in Table 2.1.

The third item of the bundle status report array is the source node ID identifying the source of the bundle whose status is being reported.

The fourth item of the bundle status report array is the creation timestamp of the

Table 2.1: Status report Reason codes.

Value	Description
0	No additional information.
1	Lifetime expired.
2	Forwarded over unidirectional link.
3	Transmission canceled.
4	Storage overflow.
5	Destination endpoint ID unintelligible.
6	No known route to destination from here.
7	No timely contact with next on route.
8	Block unintelligible.
9	Reserved for future use.

bundle whose status is being reported. The creation timestamp is a tuple where the first value is a DTN time[2] and the second one is the creation timestamp's sequence number.

The fifth item of the bundle status report array exists only if the bundle whose status is being reported contains a fragment offset. In this case it is the subject bundle's fragment offset.

The sixth item of the bundle status report array exists only if the bundle whose status is being reported contains a fragment offset. If present, it is the length of the subject bundle's payload.

2.12.3 Generation of Administrative Records

The application agent's administrative element is directed by the BPA to generate an administrative record with reference to some bundle. The procedure to generate the record includes the following steps.

Step 1: The administrative record is constructed. If the administrative record references a bundle and the referenced bundle is a fragment, the administrative record must contain the fragment offset and fragment length.

Step 2: A request for transmission of a bundle whose payload is this administrative record must be sent to BPA.

---

[2]A DTN time is an unsigned integer indicating a count of seconds since the start of the year 2000 on the Coordinated Universal Time (UTC) scale.

2.13 Convergence Layers Service

The BP end-to-end operation depends on the operation of underlying protocols, named "convergence layers"; these protocols provide communication between nodes. A wide variety of protocols may serve this purpose, so long as each convergence layer protocol adapter provides a defined minimal set of services to the bundle protocol agent.

Each convergence layer protocol adapter is expected to provide the following services to the bundle protocol agent:

■ sending a bundle to a bundle node that is reachable via the convergence layer protocol;

■ delivering to the bundle protocol agent a bundle that was sent by a bundle node via the convergence layer protocol.

2.14 Bundle Protocol Security

The bundle protocol security architecture is described in more details in the Bundle Security Protocol specification [BPSEC] [268]. This section gives a brief overview of BP security.

The BPSEC extensions to Bundle Protocol enable each block of a bundle to be individually authenticated by a signature block (Block Integrity Block, or BIB) and also enable each block of a bundle other than the primary block to be individually encrypted by a block based codebook (BCB).

Because the security mechanisms are extension blocks that are themselves inserted into the bundle, the integrity and confidentiality of bundle blocks are protected while the bundle is at rest, awaiting transmission at the next forwarding opportunity, as well as in transit.

Additionally, convergence-layer protocols that ensure authenticity of communication between adjacent nodes in BP network topology could be used where available, to minimize the ability of unauthenticated nodes to introduce inauthentic traffic into the network.

Note that, while the primary block must remain in the clear for routing purposes, the bundle protocol can be protected against traffic analysis to some extent by using bundle-in-bundle encapsulation to tunnel bundles to a safe forward distribution point: the encapsulated bundle forms the payload of an encapsulating bundle, and that payload block may be encrypted by a BCB.

The BPSEC extensions accommodate an open-ended range of ciphersuites; different ciphersuites may be utilized to protect different blocks. One possible variation is to sign and/or encrypt blocks in symmetric keys securely formed by Diffie-Hellman procedures (such as EKDH) using the public and private keys of the sending and receiving nodes. For this purpose, the key distribution problem reduces to the problem of trustworthy delay-tolerant distribution of public keys.

Inclusion of the bundle security protocol in any bundle protocol implementation is recommended and the usage of the bundle security protocol in bundle protocol operations is optional, subject to the following consideration:

■ Every block that is not a BPSEC extension block of every bundle is authenticated by a BIB citing the ID of the node that inserted that block. BIB authentication can be omitted on any initial end-to-end path segments on which it would impose unacceptable overhead, provided that satisfactory authentication is ensured at the convergence layer and that BIB authentication is asserted on the first path segment on which the resulting overhead is acceptable and on all subsequent path segments.

■ If any segment of the end-to-end path of a bundle will traverse the Internet or any other potentially insecure communication environment, then the payload block is encrypted by a BCB on this path segment and all subsequent segments of the end-to-end path.

2.15   Summary

The DTN architecture originated from the work on the Interplanetary Internet and it is initially specified in RFC 4838 [75]. This architecture includes the bundle protocol specification that is described in RFC 5050 [244]. The use of the new layer, named bundle layer, is guided not only by its own design principles, but also by a few application design principles as follows.

■ Applications should minimize the number of round-trip exchanges.

■ Applications should cope with restarts after failure while network transactions remain pending.

■ Applications should inform the network of the useful life and relative importance of data to be delivered.

The DTN architecture addresses many of the problems of heterogeneous networks that must operate in environments subject to long delays and discontinuous end-to-end connectivity. It is based on asynchronous messaging and uses postal mail as a model of service classes and delivery semantics. It accommodates many different forms of connectivity, including scheduled, predicted, and opportunistically connected delivery paths. It introduces a novel approach to end-to-end reliability across frequently partitioned and unreliable networks. It also proposes a model for securing the network infrastructure against unauthorized access.

Although DTN was developed with space applications in mind, the benefits hold true for terrestrial applications where frequent disruptions and high-error rates are common. Some examples include disaster response and wireless sensor networks, satellite networks with moderate delays and periodic connectivity, and underwater acoustic networks with moderate delays and frequent interruptions due to environmental factors.

Chapter 3

DTN Platforms

Aloizio P. Silva

University of Bristol

Scott Burleigh

JPL-NASA/Caltech

CONTENTS

3.1 Introduction

A variety of delay and disruption tolerant systems and applications have been engineered in recent years.

The initial standardization effort came out of the Internet Research Task Force (IRTF)'s Delay Tolerant Networking Research Group (DTNRG). The first specifications were RFC 4838 for DTN architecture and RFC 5050 for the bundle protocol. The DTN reference implementation of these two RFCs is written in C^{++}. This reference implementation, named DTN2 [3], is mainly implemented to demonstrate basic functionalities and provide a vehicle for research.

This chapter provides an overview of the available DTN implementations, which follow the reference implementation from the DTNRG.

3.2   DTN2 : Reference Implementation

The DTN reference implementation was originally developed by Mike Demmer at the University of California, Berkeley, but more recently it has mainly been maintained at Trinity College Dublin. It was primarily developed in C^{++} and includes a simulator for DTN prototyping. It can operate in Linux, MAC, Win and FreeBSD operating systems. DTN2 is available from a SourceForge repository under an open-source license.

The DTN2 Bundle Protocol Agent is implemented as a user space daemon called "dtnd". The daemon has a configuration and control interface which can be run remotely over a TCP connection when the daemon is running. The DTN applications interface to "dtnd" is through the application API, which is a simple Sun RPC mechanism.

In addition to the basic C language binding for the API, there are also Perl, Python and TCL bindings. DTN2 provides a flexible experimental platform and it can be configured through a TCL console and configuration files. It provides extensions for routing, storage and convergence layers attached through XML interfaces. In particular, table-based routing, ProPHET, DTLSR, Bonjour and epidemic routing are supported.

3.2.1   DTN2 Architecture

The DTN2 architecture consists of eight components that are described as follows.

- **Bundle Router** The component of DTN2 architecture that is mainly responsible for routing decisions is the bundle router. The bundle router includes route selection and scheduling policies that support the routing procedure. The bundle router works together with the bundle forwarder which executes the routing decisions.

- **Bundle Forwarder** The forwarder component takes the responsibility to execute the routing decisions and any encoded instructions received from the bundle router. During the processing of a routing decision the bundle forwarder interacts with three other DTN2 architecture components: Convergence Layers, Registration, and Persistent Store.

- **Convergence Layers**

  The convergence layers component is responsible for performing basic data plane functions that include, for instance, transmitting and receiving bundles. More specifically, the convergence layers are adapters that enable the DTN bundle protocol to talk to other underlying transport protocols. The following convergence layers are included in DTN2:

  1. Transmission Control Protocol (TCP): TCP Convergence Layer enables to connect DTN nodes via a TCP channel.

     User Datagram Protocol (UDP): UDP Convergence Layer enables

DTN nodes to be connected via a UDP channel. The bundles are sent inside UDP datagrams. Bundles which exceed 64000 bytes are not transmitted or received by the UDP convergence layer.

2. Ethernet (ETH): Ethernet Convergence Layer enables access through ethernet interfaces to support neighbor discovery.

3. Bluetooth (BT): Bluetooth Convergence Layer is based on piconets with master and slave architecture.

4. Serial: In the serial convergence layer a connection is either a receiver or a sender. First it tries to resolve the next hop address and if it is able to do that then a teletypewriter (tty) connection is open.

5. Nack-Oriented Reliable Multicast (NORM): In NORM convergence layer a selective, negative acknowledgment (NACK) mechanism for transport reliability is used and additional protocol mechanisms enable reliable multicast sessions.

6. External (EXTCL): External convergence layer operates as a gateway between DTN2 and convergence layer adapters (CLAs) running outside of DTN2. EXTCLs connect to DTN2 through a socket and exchange XML messages.

7. NULL: The Null convergence layer simply discards all bundles.

■ **Persistent Store**

DTNs are based on a store-carry-and-forward paradigm that requires a persistent storage component. The persistent storage enables the DTN nodes to store bundles for a long period of time until, for example, a contact opportunity arises.

■ **Fragmentation Module** The fragmentation module is the component responsible for fragmenting and reassembling bundle fragments. It also notifies the bundle router when all the fragments of a subject bundle have been received.

■ **Contact Manager** The Contact Manager component is responsible for providing contact information that includes, for instance, contacts currently available and future contacts as well as historic information about previous contacts. One of the core tasks of this component is to use the information about previous contacts to create abstract contact descriptions that can be used by the bundle router.

■ **Management Interface** The management interface component is an interprocess communication interface that enables the user to notify the bundle router about policy constraints that may affect its data routing decisions.

■ **Console Module** The console module components provide a command line interface for debugging as well as a structured method to set initial configuration options.

■ **Registration Module** The interactions between the DTN application and the bundle router are performed through libraries that allow interprocess communications. This interaction is enabled through the registration module component.

The DTN2 architecture provides a set of API functions as follows:

■ Create a DTN registration

■ Remove a DTN registration

■ Checking for an existing registration

■ Modify an existing registration

■ Associate a registration with the current IPC channel

■ Send a bundle

■ Cancel a bundle transmission

■ Blocking receive for a bundle

■ Remove an association from the current IPC handle

■ Open a new connection to the router

■ Close an open DTN handle

■ Set the error associated with the given handle

■ Build a local endpoint

■ Blocking query for new subscribers on a session

■ Return a file descriptor for a given handle

■ Start a polling period for incoming bundles

DTN2 also provides support for Bundle Security Protocol (BSP) as defined in RFC 6257 [268]; its implementation is based on OpenSSL [6].

3.3   IBR-DTN : Embedded Systems and Mobile Nodes

IBR-DTN [242] is a light implementation of DTN that is based on DTN2 and designed to run OpenWRT [7], a Linux distribution that can be run on embedded devices. IBR-DTN was developed on a Mikrotik router board 532 at Technical University Braunschweig.

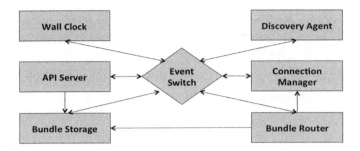

Figure 3.1: IBR-DTN Architecture [242].

3.3.1 IBR-DTN Architecture

IBR-DTN is composed of six main building blocks as shown in Figure 3.1. It also supports BPS extensions.

1. **Event Switch:** the core building block of the IBR-DTN architecture is the event switch. It is responsible for dispatching events to other building blocks.

2. **Discovery Agent:** is responsible for the node discovery procedure. The DTN IP Neighbor Discovery and IP Discovery are included in the discovery agent implementation.

3. **Convergence Layer:** the convergence layers, which are responsible for providing an interface to transfer bundles, are implemented as modules in IBR-DTN. Each module is managed by the Connection Manager building block. IBR-DTN implements the following four convergence layers.

    (a) TCPCL [99]: this convergence layer is based on TCP. It includes a handshake mechanism between DTN daemons as well as a bundle segmentation procedure.

    (b) UDPCL [167]: the UDPCL convergence layer enables transmission of either bundles or LTP blocks over UDP.

    (c) HTTPCL: this convergence layer is based on using Hypertext Transfer Protocol (HTTP) server to send and to receive bundles.

    (d) LowPANCL: This convergence layer splits up bundles into segments and sends them over IEEE 802.15.4.

4. **Bundle Storage:** again, the DTN architecture is based on a store-carry-and-forward paradigm. A DTN node may be required to store bundles for a long period of time until a contact opportunity arises. The bundle storage building block enables the DTN node to store and retrieve bundles. IBR-DTN supports the following storage media:

    ■ Memory: non-persistent memory by default.

- File: file based storage depends on a configuration setting by which a specified file is used for persistent storage.

- SQLite: based on the SQLite database, which is a software library that implements a self-contained, serverless, zero-configuration, transactional SQL database engine.

5. **Base Router:** the base router building block implements routing schemes. Each routing scheme contacts the connection manager to request the proper convergence layer to transfer the bundle to the next hop. IBR-DTN implements the following routing schemes:

  - Epidemic: in epidemic routing IBR-DTN does not use a summary vector; instead it uses a Bloom-Filter mechanism. A Bloom-Filter is a memory-efficient data structure designed to enable rapid determination of whether or not an element is present in a set.

  - Static: all the information about the route is provided a priori through the configuration file.

  - Neighbor: bundles are forwarded to nodes discovered by the discovery agent.

  - Retransmission: if an error occurs during bundle transmission this bundle is queued for retransmission later.

  - PRoPHET: the Probabilistic Routing Protocol using History of Encounters and Transitivity (PRoPHET) system uses the delivery predictability of node encounters and transitivity to select and forward bundles to neighbor nodes based on their distance.

6. **Wall Clock:** the wall clock building block provides the current time in the DTN by evaluating the local host's clock. It also provides a time tick event that is provided every second to the other building blocks.

7. **API Server:** IBR-DTN provides a socket API interface, which can be either a TCP socket or a Unix Domain Socket that operates locally.

The IBR-DTN distribution also provides some basic shell tools as follows:

- **dtnsend** and **dtnrecv** which enable the node to send and to receive files, respectively.

- **dtnping** that enables the node to send bundles to another DTN node and waits for a bundle with the same payload that serves as a reply. After all ping exchanges have been completed, the minimum time needed to send the bundle and receive a response is provided.

- **dtntracepath** shows the route of a bundle that is discovered through "bundle forwarded" status reports from each hop.

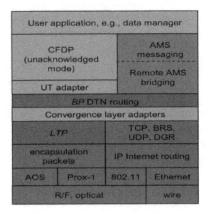

Figure 3.2: DTN protocol stack.

- **dtntunnel** enables the DTN node to act as an IP network router for TCP and UDP protocols. The TCP or UDP traffic is redirected to the dtntunnel with the help of IPTables and TProxy.

- **dtninbox** and **dtnoutbox** enables transfer of files between two folders in different computers.

3.4 ION : Space Communication

The Interplanetary Overlay Network (ION) software distribution is an implementation of Delay-Tolerant Networking (DTN) architecture as described in Internet RFC 4838. It is designed to enable inexpensive insertion of DTN functionality into embedded systems such as robotic spacecraft. The intent of ION deployment in space flight mission systems is to reduce cost and risk in mission communications by simplifying the construction and operation of automated digital data communication networks spanning space links, planetary surface links, and terrestrial links.

The DTN architecture is similar to the architecture of the Internet, except that it is one layer higher in the familiar ISO protocol "stack". The DTN analog to the Internet Protocol (IP), called "Bundle Protocol" (BP), is designed to function as an "overlay" network protocol that interconnects "internets" - including both Internet-structured networks and also data paths that utilize only space communication links as defined by the Consultative Committee for Space Data Systems (CCSDS) - in much the same way that IP interconnects "subnets" such as those built on Ethernet, SONET, etc. By implementing the DTN architecture, ION provides communication software configured as a protocol stack that looks like Figure 3.2.

Data traversing a DTN are conveyed in DTN bundles - which are functionally analogous to IP packets - between BP endpoints which are functionally analogous to sockets. Multiple BP endpoints may reside on the same computer - termed a node - just as multiple sockets may reside on the same computer (host or router) in the Internet.

BP endpoints are identified by Universal Record Identifiers (URIs), which are ASCII text strings of the general form: *scheme_name:scheme_specific_part*

For example: *dtn://topquark.caltech.edu/mail*

But for space flight communications this general textual representation might impose more transmission overhead than missions can afford. For this reason, ION is optimized for networks of endpoints whose IDs conform more narrowly to the following scheme: *ipn:node_number.service_number*

This enables them to be abbreviated to pairs of unsigned binary integers via a technique called Compressed Bundle Header Encoding (CBHE) [58]. CBHE-conformant BP endpoint IDs (EIDs) are not only functionally similar to Internet socket addresses but also structurally similar: node numbers are roughly analogous to Internet node numbers (IP addresses), in that they typically identify the flight or ground data system computers on which network software executes, and service numbers are roughly analogous to TCP and UDP port numbers.

More generally, the node numbers in CBHE-conformant BP endpoint IDs are one manifestation of the fundamental ION notion of network node number: in the ION architecture there is a natural one-to-one mapping not only between node numbers and BP endpoint node numbers but also between node numbers and:

- LTP and BSSP engine IDs

- AMS continuum numbers

- CFDP entity numbers

Starting with version 3.1 of ION, this endpoint naming rule was experimentally extended to accommodate bundle multicast, i.e., the delivery of copies of a single transmitted bundle to multiple nodes at which interest in that bundle's payload has been expressed. Multicast in ION - "Interplanetary Multicast" (IMC) - is accomplished by simply issuing a bundle whose destination endpoint ID conforms to the following scheme: *imc:group_number.service_number*

A copy of the bundle will automatically be delivered at every node that has registered in the destination endpoint. (Note: for now, the operational significance of a given group number must be privately negotiated among ION users. If this multicast mechanism proves useful, IANA may at some point establish a registry for IMC group numbers.)

3.4.1 Structure and Function

The ION distribution comprises the following software packages:

ici (Interplanetary Communication Infrastructure), a set of general-purpose libraries providing common functionality to the other packages. The ici package includes a security policy component that supports the implementation of security mechanisms at multiple layers of the protocol stack.

ltp (Licklider Transmission Protocol), a core DTN protocol that provides transmission reliability based on delay-tolerant acknowledgments, timeouts, and retransmissions. The LTP specification is defined in Internet RFC 5326.

bp (Bundle Protocol), a core DTN protocol that provides delay-tolerant forwarding of data through a network in which continuous end-to-end connectivity is never assured, including support for delay-tolerant dynamic routing. The BP specification is defined in Internet RFC 5050.

dgr (Datagram Retransmission), an alternative implementation of LTP that is designed for use in the Internet. Equipped with algorithms for TCP-like congestion control, DGR enables data to be transmitted via UDP with reliability comparable to that provided by TCP. The dgr system is provided primarily for the conveyance of Meta-AMS (see below) protocol traffic in an Internet-like environment.

ams (Asynchronous Message Service), an application-layer service that is not part of the DTN architecture but utilizes underlying DTN protocols. AMS comprises three protocols supporting the distribution of brief messages within a network:

- The core AAMS (Application AMS) protocol, which does message distribution on both the publish/subscribe model and the client/server model, as required by the application.

- The MAMS (Meta-AMS) protocol, which distributes control information enabling the operation of the Application AMS protocol.

- The RAMS (Remote AMS) protocol, which performs aggregated message distribution to end nodes that may be numerous and/or accessible only over very expensive links, using an aggregation tree structure similar to the distribution trees used by Internet multicast technologies.

cfdp (CCSDS File Delivery Protocol), another application-layer service that is not part of the DTN architecture but utilizes underlying DTN protocols. CFDP performs the segmentation, transmission, reception, reassembly, and delivery of files in a delay-tolerant manner. ION's implementation of CFDP conforms to the "class 1" definition of the protocol in the CFDP standard, utilizing DTN (BP, nominally over LTP) as its "unitdata transport" layer.

bss (Bundle Streaming Service), a system for efficient data streaming over a delay-tolerant network. The bss package includes:

1. a convergence-layer protocol (bssp) that preserves in-order arrival of all data that were never lost en route, yet ensures that all data arrive at the destination eventually, and

2. a library for building delay-tolerant streaming applications, which enables low-latency presentation of streamed data received in real time while offering rewind/playback capability for the entire stream including late-arriving retransmitted data.

Taken together, the packages included in the ION software distribution constitute a communication capability characterized by the following operational features:

■ Reliable conveyance of data over a delay-tolerant network (dtnet), i.e., a network in which it might never be possible for any node to have reliable information about the detailed current state of any other node.

■ Built on this capability, reliable data streaming, reliable file delivery, and reliable distribution of short messages to multiple recipients (subscribers) residing in such a network.

■ Management of traffic through such a network, taking into consideration:

　■ requirements for data security

　■ scheduled times and durations of communication opportunities

　■ fluctuating limits on data storage and transmission resources

　■ data rate asymmetry

　■ the sizes of application data units

　■ and user-specified final destination, priority, and useful lifetime for those data units.

■ Facilities for monitoring the performance of the network.

■ Robustness against node failure.

■ Portability across heterogeneous computing platforms.

■ High speed with low overhead.

■ Easy integration with heterogeneous underlying communication infrastructure, ranging from Internet to dedicated spacecraft communication links.

3.4.2   *Constraints on the Design*

A DTN implementation intended to function in an interplanetary network environment - specifically, aboard interplanetary research spacecraft separated from Earth and from one another by vast distances - must operate successfully within two general classes of design constraints: link constraints and processor constraints.

3.4.2.1 Link Constraints

All communications among interplanetary spacecraft are, obviously, wireless. Less obviously, those wireless links are generally slow and are usually asymmetric.

The electrical power provided to on-board radios is limited and antennas are relatively small, so signals are weak. This limits the speed at which data can be transmitted intelligibly from an interplanetary spacecraft to Earth, usually to some rate on the order of 256 *Kbps* to 6 *Mbps*.

The electrical power provided to transmitters on Earth is certainly much greater, but the sensitivity of receivers on spacecraft is again constrained by limited power and antenna mass allowances. Because historically the volume of command traffic that had to be sent to spacecraft was far less than the volume of telemetry the spacecraft were expected to return, spacecraft receivers have historically been engineered for even lower data rates from Earth to the spacecraft, on the order of 1 to 2 *Kbps*.

As a result, the cost per octet of data transmission or reception is high and the links are heavily subscribed. Economical use of transmission and reception opportunities is therefore important, and transmission is designed to enable useful information to be obtained from brief communication opportunities: units of transmission are typically small, and the immediate delivery of even a small part (carefully delimited) of a large data object may be preferable to deferring delivery of the entire object until all parts have been acquired.

3.4.2.2 Processor Constraints

The computing capability aboard a robotic interplanetary spacecraft is typically quite different from that provided by an engineering workstation on Earth. In part this is due, again, to the limited available electrical power and limited mass allowance within which a flight computer must operate. But these factors are exacerbated by the often intense radiation environment of deep space. In order to minimize errors in computation and storage, flight processors must be radiation-hardened and both dynamic memory and non-volatile storage (typically flash memory) must be radiation-tolerant. The additional engineering required for these adaptations takes time and is not inexpensive, and the market for radiation-hardened spacecraft computers is relatively small; for these reasons, the latest advances in processing technology are typically not available for use on interplanetary spacecraft, so flight computers are invariably slower than their Earth-bound counterparts. As a result, the cost per processing cycle is high and processors are heavily subscribed; economical use of processing resources is very important.

The nature of interplanetary spacecraft operations imposes a further constraint. These spacecraft are wholly robotic and are far beyond the reach of mission technicians; hands-on repairs are out of the question. Therefore the processing performed by the flight computer must be highly reliable, which in turn generally means that it must be highly predictable. Flight software is typically required to meet "hard" real-time processing deadlines, for which purpose it must be run within a hard real-time operating system (RTOS).

One other implication of the requirement for high reliability in flight software is

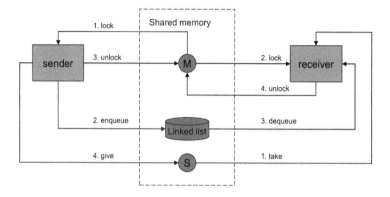

Figure 3.3: ION inter-task communication.

that the dynamic allocation of system memory may be prohibited except in certain well-understood states, such as at system start-up. Unrestrained dynamic allocation of system memory introduces a degree of unpredictability into the overall flight system that can threaten the reliability of the computing environment and jeopardize the health of the vehicle.

3.4.3 Design Principles

The design of the ION software distribution reflects several core principles that are intended to address these constraints.

3.4.3.1 Shared Memory

Since ION must run on flight processors, it had to be designed to function successfully within an RTOS. Many real-time operating systems improve processing determinism by omitting the support for protected-memory models that is provided by Unix-like operating systems: all tasks have direct access to all regions of system memory. (In effect, all tasks operate in kernel mode rather than in user mode.) ION therefore had to be designed with no expectation of memory protection.

But universally shared access to all memory can be viewed not only as a hazard but also as an opportunity. Placing a data object in shared memory is an extremely efficient means of passing data from one software task to another.

ION is designed to exploit this opportunity as fully as possible. In particular, virtually all inter-task data interchange in ION follows the model shown in Figure 3.3:

■ The sending task takes a mutual exclusion semaphore (mutex) protecting a linked list in shared memory (either DRAM or non-volatile memory), appends a data item to the list, releases the mutex, and gives a "signal" semaphore associated with the list to announce that the list is now non-empty.

■ The receiving task, which is already pended on the linked list's associated signal semaphore, resumes execution when the semaphore is given. It takes the associated mutex, extracts the next data item from the list, releases the mutex, and proceeds to operate on the data item from the sending task.

Semaphore operations are typically extremely fast, as is the storage and retrieval of data in memory, so this inter-task data interchange model is suitably efficient for flight software.

3.4.3.2   Zero-Copy Procedures

Given ION's orientation toward the shared memory model, a further strategy for processing efficiency offers itself: if the data item appended to a linked list is merely a pointer to a large data object, rather than a copy, then we can further reduce processing overhead by eliminating the cost of byte-for-byte copying of large objects.

Moreover, in the event that multiple software elements need to access the same large object at the same time, we can provide each such software element with a pointer to the object rather than its own copy (maintaining a count of references to assure that the object is not destroyed until all elements have relinquished their pointers). This serves to reduce somewhat the amount of memory needed for ION operations.

3.4.3.3   Highly Distributed Processing

The efficiency of inter-task communications based on shared memory makes it practical to distribute ION processing among multiple relatively simple pipelined tasks rather than localize it in a single, somewhat more complex daemon. This strategy has a number of advantages:

■ The simplicity of each task reduces the sizes of the software modules, making them easier to understand and maintain, and thus it can somewhat reduce the incidence of errors.

■ The scope of the ION operating stack can be adjusted incrementally at run time, by spawning or terminating instances of configurable software elements, without increasing the size or complexity of any single task and without requiring that the stack as a whole be halted and restarted in a new configuration. In theory, a module could even be upgraded with new functionality and integrated into the stack without interrupting operations.

■ The clear interfaces between tasks simplify the implementation of flow control measures to prevent uncontrolled resource consumption.

3.4.3.4   Portability

Designs based on these kinds of principles are foreign to many software developers, who may be far more comfortable in development environments supported by

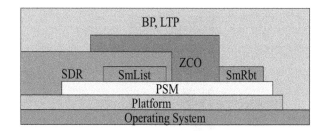

BP, LTP	Bundle Protocol and Licklider Transmission Protocol libraries and daemons
ZCO	Zero-copy objects capability: minimize data copying up and down the stack
SDR	Spacecraft Data Recorder: persistent object database in shared memory, using PSM and SmList
SmList	linked lists in shared memory using PSM
SmRbt	red-black trees in shared memory using PSM
PSM	Personal Space Management: memory management within a pre-allocated memory partition
Platform	common access to O.S.: shared memory, system time, IPC mechanisms
Operating System	POSIX thread spawn/destroy, file system, time

Figure 3.4: ION software functional dependencies.

protected memory. It is typically much easier, for example, to develop software in a Linux environment than in VxWorks 5.4 [101]. However, the Linux environment is not the only one in which ION software must ultimately run.

Consequently, ION has been designed for easy portability. POSIX^{TM} [207] API functions are widely used, and differences in operating system support that are not concealed within the POSIX abstractions are mostly encapsulated in two small modules of platform-sensitive ION code. The bulk of the ION software runs, without any source code modification whatsoever, equally well in Linux^{TM} (Red Hat$^®$, Fedora^{TM}, and Ubuntu^{TM}, so far), FreeBSD$^®$, Solaris$^®$ 9, Microsoft Windows (the MinGW environment), OS/X$^®$, VxWorks$^®$ 5.4, and RTEMS^{TM}, on both 32-bit and 64-bit processors. Developers may compile and test ION modules in whatever environment they find most convenient.

3.4.4 Organizational Overview

Two broad overviews of the organization of ION may be helpful at this point. Figure 3.4 shows a summary view of the main functional dependencies among ION software elements.

That is, BP and LTP invoke functions provided by the sdr, zco, psm, and platform elements of the ici package, in addition to functions provided by the operating system itself; the zco functions themselves also invoke sdr, psm, and platform functions; and so on.

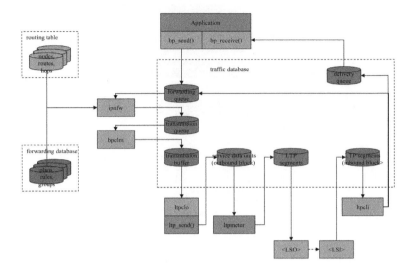

Figure 3.5: Main line of ION data flow.

Figure 3.5 shows a summary view of the main line of data flow in ION's DTN protocol implementations.

Note that data objects residing in shared memory, many of them in a nominally non-volatile SDR data store, constitute the central organizing principle of the design. Here as in other diagrams showing data flow in this chapter:

- Ordered collections of data objects are shown as cylinders.

- Darker greyscale data entities indicate data that are managed in the SDR data store, while lighter greyscale data entities indicate data that are managed in volatile DRAM to improve performance.

- Rectangles indicate processing elements (tasks, processes, threads), sometimes with library references specifically identified.

A few notes on this main line data flow:

- For simplicity, the data flow depicted here is a "loopback" flow in which a single BP "node" is shown sending data to itself (a useful configuration for test purposes). To depict typical operations over a network we would need two instances of this node diagram, such that the < *LSO* > task of one node is shown sending data to the < *LSI* > task of the other and vice versa.

- A BP application or application service (such as Remote AMS [116]) that has access to the local BP node, for our purposes, the "sender" invokes the bp_send function to send a unit of application data to a remote counterpart.

The destination of the application data unit is expressed as a BP endpoint ID (EID). The application data unit is encapsulated in a bundle and is queued for forwarding.

■ The forwarder task identified by the "scheme" portion of the bundle's destination EID removes the bundle from the forwarding queue and computes a route to the destination EID. The first node on the route is termed the "proximate node" for the computed route. The forwarder appends the bundle to the transmission queue for the convergence-layer manager (CLM) daemon that is responsible for transmission to the proximate node.

■ The CLM daemon removes the bundle from the transmission queue and imposes rate control, fragments the bundle as necessary, and appends the bundle to the transmission buffer for some underlying "convergence layer" (CL) protocol interface to the proximate node, termed an outduct. In the event that multiple outducts are available for transmission to that node (e.g., multiple radio frequency bands), the CLM invokes mission-supplied code to select the appropriate duct. Each outduct is serviced by some CL-specific output task that communicates with the proximate node - in this case, the LTP output task ltpclo. (Other CL protocols supported by ION include TCP and UDP.)

■ The output task for LTP transmission to the selected proximate node removes the bundle from the transmission buffer and invokes the ltp_send function to append it to a block that is being assembled for transmission to the proximate node. (Because LTP acknowledgment traffic is issued on a per-block basis, we can limit the amount of acknowledgment traffic on the network by aggregating multiple bundles into a single block rather than transmitting each bundle in its own block.)

■ The ltpmeter task for the selected proximate node divides the aggregated block into multiple segments and enqueues them for transmission by underlying link-layer transmission software, such as an implementation of the CCSDS AOS protocol.

■ Underlying link-layer software at the sending node transmits the segments to its counterpart at the proximate node (the receiver), where they are used to reassemble the transmission block.

■ The receiving node's input task for LTP reception extracts the bundles from the reassembled block and dispatches them: each bundle whose final destination is some other node is queued for forwarding, just like bundles created by local applications, while each bundle whose final destination is the local node is queued for delivery to whatever application "opens" the BP endpoint identified by the bundle's final destination endpoint ID. (Note that a multicast bundle may be both queued for forwarding, possibly to multiple neighboring nodes, and also queued for delivery.)

■ The destination application or application service at the receiving node opens the appropriate BP endpoint and invokes the bp_receive function to remove the bundle from the associated delivery queue and extract the original application data unit, which it can then process.

Finally, note that the data flow shown here represents the sustained operational configuration of a node that has been successfully instantiated on a suitable computer. The sequence of operations performed to reach this configuration is not shown. That startup sequence will necessarily vary depending on the nature of the computing platform and the supporting link services. Broadly, the first step normally is to run the ionadmin utility program to initialize the data management infrastructure required by all elements of ION. Following this initialization, the next steps normally are:

1. any necessary initialization of link service protocols,

2. any necessary initialization of convergence-layer protocols (e.g., LTP - the lt-padmin utility program),

3. and finally initialization of the Bundle Protocol by means of the bpadmin utility program. BP applications should not try to commence operation until BP has been initialized.

3.4.5 Resource Management in ION

Successful Delay-Tolerant Networking relies on retention of bundle protocol agent state information - including protocol traffic that is awaiting a transmission opportunity - for potentially lengthy intervals. The nature of that state information will fluctuate rapidly as the protocol agent passes through different phases of operation, so efficient management of the storage resources allocated to state information is a key consideration in the design of ION.

Two general classes of storage resources are managed by ION: volatile "working memory" and non-volatile "heap".

3.4.5.1 Working Memory

ION's "working memory" is a fixed-size pool of shared memory (dynamic RAM) that is allocated from system RAM at the time the bundle protocol agent commences operation. Working memory is used by ION tasks to store temporary data of all kinds: linked lists, red-black trees, transient buffers, volatile databases, etc. All intermediate data products and temporary data structures that ought not to be retained in the event of a system power cycle are written to working memory.

Data structures residing in working memory may be shared among ION tasks or may be created and managed privately by individual ION tasks. The dynamic allocation of working memory to ION tasks is accomplished by the Personal Space Management (PSM) service, described later. All of the working memory for any single ION bundle protocol agent is managed as a single PSM "partition". The size

SDR heap

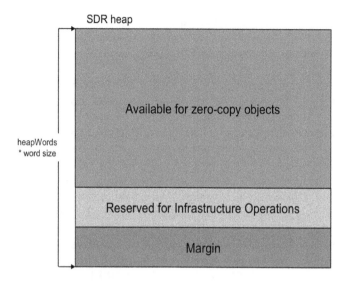

heapWords
* word size

Available for zero-copy objects

Reserved for Infrastructure Operations

Margin

Figure 3.6: ION heap space use.

of the partition is specified in the wmSize parameter of the ionconfig file supplied at the time ION is initialized.

3.4.5.2 Heap

ION's "heap" is a fixed-size pool of notionally non-volatile storage that is likewise allocated at the time the bundle protocol agent commences operation. This notionally non-volatile space may occupy a fixed-size pool of shared memory (dynamic RAM, which might or might not be battery-backed), or it may occupy only a single fixed-size file in the file system, or it may occupy both. In the latter case, all heap data are written both to memory and to the file but are read only from memory; this configuration offers the reliable non-volatility of file storage coupled with the high performance of retrieval from dynamic RAM.

We characterize ION's heap storage as "notionally" non-volatile because the heap may be configured to reside only in memory (or, for that matter, in a file that resides in the file system of a RAM disk). When the heap resides only in memory, its contents are truly non-volatile only if that memory is battery-backed. Otherwise heap storage is in reality as volatile as working memory: heap contents will be lost upon a system power cycle (which may in fact be the preferred behavior for any given deployment of ION). However, the heap should not be thought of as "memory" even when it in fact resides only in DRAM, just as a disk device should not be thought of as "memory" even when it is in fact a RAM disk.

The ION heap is used for storage of data that (in at least some deployments) would have to be retained in the event of a system power cycle to ensure the correct continued operation of the node. For example, all queues of bundles awaiting route computation, transmission, or delivery reside in the node's heap. So do the non-volatile databases for all of the protocols implemented within ION, together with all of the node's persistent configuration parameters.

The dynamic allocation of heap space to ION tasks is accomplished by the Simple Data Recorder (SDR) service, described later. The entire heap for any single ION bundle protocol agent is managed as a single SDR "data store".

Space within the ION heap is apportioned as shown in Figure 3.6. The total number of bytes of storage space in the heap is computed as the product of the size of a "word" on the deployment platform (normally the size of a pointer) multiplied by the value of the heapWords parameter of the ionconfig file supplied at the time ION is initialized. Of this total, 20% is normally reserved as margin and another 40% is normally reserved for various infrastructure operations. (Both of these percentages are macros that may be overridden at compile time.) The remainder is available for storage of protocol state data in the form of "zero-copy objects", described later. At any given moment, the data encapsulated in a zero-copy object may "belong" to any one of the protocols in the ION stack (AMS, CFDP, BP, LTP), depending on processing state; the available heap space is a single common resource to which all of the protocols share concurrent access.

Because the heap is used to store queues of bundles awaiting processing, blocks of LTP data awaiting transmission or reassembly, etc., the heap for any single ION node must be large enough to contain the maximum volume of such data that the node will be required to retain during operations. Demand for heap space is substantially mitigated if most of the application data units passed to ION for transmission are file-resident, as the file contents themselves need not be copied into the heap. In general, however, computing the optimum ION heap size for a given deployment remains a research topic.

3.4.6 Package Overviews

3.4.6.1 Interplanetary Communication Infrastructure (ICI)

The ICI package in ION provides a number of core services that, from ION's point of view, implement what amounts to an extended POSIX-based operating system. ICI services include the following:

1. **Platform** The platform system contains operating-system-sensitive code that enables ICI to present a single, consistent programming interface to those common operating system services that multiple ION modules utilize. For example, the platform system implements a standard semaphore abstraction that may invisibly be mapped to underlying POSIX semaphores, SVR4 IPC semaphores, Windows Events, or VxWorks semaphores, depending on which operating system the package is compiled for. The platform system also implements a standard shared-memory abstraction, enabling software running on operating systems both with and without memory protection to participate readily in ION's shared-memory-based computing environment.

2. **Personal Space Management (PSM)** Although sound flight software design may prohibit the uncontrolled dynamic management of system memory, private management of assigned, fixed blocks of system memory is standard practice. Often that private management amounts to merely controlling the reuse of fixed-size rows in static tables, but such techniques can be awkward and may not make the most efficient use of available memory. The ICI package provides an alternative, called PSM, which performs high-speed dynamic allocation and recovery of variable-size memory objects within an assigned memory block of fixed size. A given PSM-managed memory block may be either private or shared memory.

3. **Memmgr** The static allocation of privately-managed blocks of system memory for different purposes implies the need for multiple memory management regimes, and in some cases a program that interacts with multiple software elements may need to participate in the private shared-memory management regimes of each. ICI's memmgr system enables multiple memory managers - for multiple privately-managed blocks of system memory - to coexist within ION and be concurrently available to ION software elements.

4. **Lyst** The lyst system is a comprehensive, powerful, and efficient system for managing doubly-linked lists in private memory. It is the model for a number of other list management systems supported by ICI; as noted earlier, linked lists are heavily used in ION inter-task communication.

5. **Llcv** The llcv (Linked-List Condition Variables) system is an inter-thread communication abstraction that integrates POSIX thread condition variables (vice semaphores) with doubly-linked lists in private memory.

6. **Smlist** Smlist is another doubly-linked list management service. It differs from lyst in that the lists it manages reside in shared (rather than private) DRAM, so operations on them must be semaphore-protected to prevent race conditions.

7. **SmRbt** The SmRbt service provides mechanisms for populating and navigating "red/black trees" (RBTs) residing in shared DRAM. RBTs offer an alternative to linked lists: like linked lists they can be navigated as queues, but locating a single element of an RBT by its "key" value can be much quicker than the equivalent search through an ordered linked list.

8. **Simple Data Recorder (SDR)** SDR is a system for managing non-volatile storage, built on exactly the same model as PSM. Put another way, SDR is a small and simple "persistent object" system or "object database" management system. It enables straightforward management of linked lists (and other data structures of arbitrary complexity) in non-volatile storage, notionally within a single file whose size is pre-defined and fixed.

   SDR includes a transaction mechanism that protects database integrity by ensuring that the failure of any database operation will cause all other operations undertaken within the same transaction to be backed out. The intent of the system is to assure retention of coherent protocol engine state even in the event of an unplanned flight computer reboot in the midst of communication activity.

9. **Sptrace** The sptrace system is an embedded diagnostic facility that monitors the performance of the PSM and SDR space management systems. It can be used, for example, to detect memory "leaks" and other memory management errors.

10. **Zco** ION's zco (zero-copy objects) system leverages the SDR system's storage flexibility to enable user application data to be encapsulated in any number of layers of protocol without copying the successively augmented protocol data unit from one layer to the next. It also implements a reference counting system that enables protocol data to be processed safely by multiple software elements concurrently - e.g., a bundle may be both delivered to a local endpoint and, at the same time, queued for forwarding to another node - without requiring that distinct copies of the data be provided to each element.

11. **Rfx** The ION rfx (R/F Contacts) system manages lists of scheduled communication opportunities in support of a number of LTP and BP functions.

12. **Ionsec** The IONSEC (ION security) system manages information that supports the implementation of security mechanisms in the other packages: security policy rules and computation keys.

3.4.6.2   Licklider Transmission Protocol (LTP)

The ION implementation of LTP conforms fully to RFC 5326, but it also provides two additional features that enhance functionality without affecting interoperability with other implementations:

■ The service data units - nominally bundles - passed to LTP for transmission may be aggregated into larger blocks before segmentation. By controlling block size we can control the volume of acknowledgment traffic generated as blocks are received, for improved accommodation of highly asynchronous data rates.

■ The maximum number of transmission sessions that may be concurrently managed by LTP (a protocol control parameter) constitutes a transmission "window" - the basis for a delay-tolerant, non-conversational flow control service over interplanetary links.

In the ION stack, LTP serves effectively the same role that is performed by an LLC protocol (such as IEEE 802.2) in the Internet architecture, providing flow control and retransmission-based reliability between topologically adjacent bundle protocol agents.

All LTP session state is safely retained in the ION heap for rapid recovery from a spacecraft or software fault. Chapter 11 provides more details about LTP.

3.4.6.3 Bundle Protocol (BP)

The ION implementation of BP conforms fully to RFC 5050, including support for the following standard capabilities:

- Prioritization of data flows

- Proactive bundle fragmentation

- Bundle reassembly from fragments

- Flexible status reporting

- Custody transfer, including re-forwarding of custodial bundles upon timeout interval expiration or failure of nominally reliable convergence-layer transmission.

The system also provides three additional features that enhance functionality without affecting interoperability with other implementations:

- Rate control provides support for congestion forecasting and avoidance.

- Bundle headers are encoded into compressed form (CBHE, as noted earlier) before issuance, to reduce protocol overhead and improve link utilization.

- Bundles may be "multicast" to all nodes that have registered within a given multicast group endpoint.

In addition, ION BP includes a system for computing dynamic routes through time-varying network topology assembled from scheduled, bounded communication opportunities. This system, called "Contact Graph Routing," is described later in Chapter 5.

In short, BP serves effectively the same role that is performed by IP in the Internet architecture, providing route computation, forwarding, congestion avoidance, and control over quality of service.

All bundle transmission state is safely retained in the ION heap for rapid recovery from a spacecraft or software fault.

3.4.6.4 Asynchronous Message Service (AMS)

The ION implementation of the CCSDS AMS [116] standard conforms fully to CCSDS 735.0-B-1. AMS is a data system communications architecture under which the modules of mission systems may be designed as if they were to operate in isolation, each one producing and consuming mission information without explicit awareness of which other modules are currently operating. Communication relationships among such modules are self-configuring; this tends to minimize complexity in the development and operations of modular data systems.

A system built on this model is a "society" of generally autonomous interoperating modules that may fluctuate freely over time in response to changing mission objectives, modules' functional upgrades, and recovery from individual module failure. The purpose of AMS, then, is to reduce mission cost and risk by providing standard, reusable infrastructure for the exchange of information among data system modules in a manner that is simple to use, highly automated, flexible, robust, scalable, and efficient.

A detailed discussion of AMS is beyond the scope of this book. For more information, please see the AMS Programmer's Guide.

3.4.6.5 Datagram Retransmission (DGR)

The DGR package in ION is an alternative implementation of LTP that is designed to operate responsibly - i.e., with built-in congestion control - in the Internet or other IP-based networks. It is provided as a candidate "primary transfer service" in support of AMS operations in an Internet-like (non-delay-tolerant) environment. The DGR design combines LTP's concept of concurrent transmission transactions with congestion control and timeout interval computation algorithms adapted from TCP.

3.4.6.6 CCSDS File Delivery Protocol (CFDP)

The ION implementation of CFDP [73] conforms fully to Service Class 1 (Unreliable Transfer) of CCSDS 727.0-B-4, including support for the following standard capabilities:

- Segmentation of files on user-specified record boundaries.

- Transmission of file segments in protocol data units that are conveyed by an underlying Unitdata Transfer service, in this case the DTN protocol stack. File data segments may optionally be protected by CRCs. When the DTN protocol stack is configured for reliable data delivery (i.e., with BP custody transfer running over a reliable convergence-layer protocol such as LTP), file delivery is reliable; CFDP need not perform retransmission of lost data itself.

- Reassembly of files from received segments, possibly arriving over a variety of routes through the delay-tolerant network. The integrity of the delivered files is protected by checksums.

- User-specified fault handling procedures.

- Operations (e.g., directory creation, file renaming) on remote file systems.

All CFDP transaction state is safely retained in the ION heap for rapid recovery from a spacecraft or software fault.

3.4.6.7 Bundle Streaming Service (BSS)

The BSS service provided in ION enables a stream of video, audio, or other continuously generated application data units, transmitted over a delay-tolerant network, to be presented to a destination application in two useful modes concurrently:

■ In the order in which the data units were generated, with the least possible end-to-end delivery latency, but possibly with some gaps due to transient data loss or corruption.

■ In the order in which the data units were generated, without gaps (i.e., including lost or corrupt data units which were omitted from the real-time presentation but were subsequently retransmitted), but in a non-real-time "playback" mode.

3.4.7 Network Operation Concepts

A small number of network operation design elements - fragmentation and reassembly, bandwidth management, and delivery assurance (retransmission) - can potentially be addressed at multiple layers of the protocol stack, possibly in different ways for different reasons. In stack design it's important to allocate this functionality carefully so that the effects at lower layers complement, rather than subvert, the effects imposed at higher layers of the stack. This allocation of functionality is discussed below, together with a discussion of several related key concepts in the ION design.

3.4.7.1 Fragmentation and Reassembly

To minimize transmission overhead and accommodate asymmetric links (i.e., limited "uplink" data rate from a ground data system to a spacecraft) in an interplanetary network, we ideally want to send "downlink" data in the largest possible aggregations - coarse-grained transmission.

But to minimize head-of-line blocking (i.e., delay in transmission of a newly presented high-priority item) and minimize data delivery latency by using parallel paths (i.e., to provide fine-grained partial data delivery, and to minimize the impact of unexpected link termination), we want to send "downlink" data in the smallest possible aggregations - fine-grained transmission.

We reconcile these impulses by doing both, but at different layers of the ION protocol stack.

First, at the application service layer (AMS and CFDP) we present relatively small application data units (ADUs) - on the order of 64 KB - to BP for encapsulation in bundles. This establishes an upper bound on head-of-line blocking when bundles are de-queued for transmission, and it provides perforations in the data stream at which forwarding can readily be switched from one link (route) to another, enabling partial data delivery at relatively fine, application-appropriate granularity.

(Alternatively, large application data units may be presented to BP and the resulting large bundles may be proactively fragmented at the time they are presented to the convergence-layer manager. This capability is meant to accommodate environments in which the convergence-layer manager has better information than the application as to the optimal bundle size, such as when the residual capacity of a contact is known to be less than the size of the bundle.)

Then, at the BP/LTP convergence layer adapter lower in the stack, we aggregate these small bundles into blocks for presentation to LTP:

> Any continuous sequence of bundles that are to be shipped to the same LTP engine and that all require assured delivery may be aggregated into a single block, to reduce overhead and minimize report traffic.
>
> However, this aggregation is constrained by an aggregation size limit rule: aggregation must stop and the block must be transmitted as soon as the sum of the sizes of all bundles aggregated into the block exceeds the block aggregation threshhold value declared for the applicable span (the relationship between the local node and the receiving LTP engine) during LTP protocol configuration via ltpadmin.

Given a preferred block acknowledgment period - e.g., a preferred acknowledgment traffic rate of one report per second - the nominal block aggregation threshhold is notionally computed as the amount of data that can be sent over the link to the receiving LTP engine in a single block acknowledgment period at the planned outbound data rate to that engine.

Taken together, application-level fragmentation (or BP proactive fragmentation) and LTP aggregation place an upper limit on the amount of data that would need to be re-transmitted over a given link at next contact in the event of an unexpected link termination that caused delivery of an entire block to fail. For example, if the data rate is 1 *Mbps* and the nominal block size is 128 *KB* (equivalent to 1 second of transmission time), we would prefer to avoid the risk of having wasted five minutes of downlink in sending a 37.5 *MB* file that fails on transmission of the last kilobyte, forcing retransmission of the entire 37.5 *MB*. We therefore divide the file into, say, 1200 *bundles* of 32 *KB* each which are aggregated into blocks of 128 *KB* each: only a single block failed, so only that block (containing just 4 bundles) needs to be retransmitted. The cost of this retransmission is only 1 second of link time rather than 5 minutes. By controlling the cost of convergence-layer protocol failure in this way, we avoid the overhead and complexity of "reactive fragmentation" in the BP implementation.

Finally, within LTP itself we fragment the block as necessary to accommodate the Maximum Transfer Unit (MTU) size of the underlying link service, typically the transfer frame size of the applicable CCSDS link protocol.

3.4.7.2 Bandwidth Management

The allocation of bandwidth (transmission opportunity) to application data is requested by the application task that's passing data to DTN, but it is necessarily accomplished only at the lowest layer of the stack at which bandwidth allocation decisions can be made - and then always in the context of node policy decisions that have global effect.

The transmission queue interface to a given neighbor in the network is actually three queues of outbound bundles rather than one: one queue for each of the defined levels of priority ("class of service") supported by BP. When an application presents an ADU to BP for encapsulation in a bundle, it indicates its own assessment of the

ADU's priority. Upon selection of a proximate forwarding destination node for that bundle, the bundle is appended to whichever of the queues corresponds to the ADU's priority.

Normally the convergence-layer manager (CLM) task servicing a given proximate node extracts bundles in strict priority order from the heads of the three queues. That is, the bundle at the head of the highest-priority non-empty queue is always extracted.

However, if the ION_BANDWIDTH_RESERVED compiler option is selected at the time ION is built, the convergence-layer manager task servicing a given proximate node extracts bundles in interleaved fashion from the heads of the node's three queues:

■ Whenever the priority-2 ("express") queue is non-empty, the bundle at the head of that queue is the next one extracted.

■ At all other times, bundles from both the priority-1 queue and the priority-0 queue are extracted, but over a given period of time twice as many bytes of priority-1 bundles will be extracted as bytes of priority-0 bundles.

Following insertion of the extracted bundles into transmission buffers, CLO tasks other than ltpclo simply segment the buffered bundles as necessary and transmit them using the underlying convergence-layer protocols. In the case of ltpclo, the output task aggregates the buffered bundles into blocks as described earlier and a second daemon task named ltpmeter waits for aggregated blocks to be completed; ltpmeter, rather than the CLO task itself, segments each completed block as necessary and passes the segments to the link service protocol that underlies LTP. Either way, the transmission ordering requested by application tasks is preserved.

3.4.7.3 Contact Plans

In the Internet, protocol operations can be largely driven by currently effective information that is discovered opportunistically and immediately, at the time it is needed, because the latency in communicating this information over the network is negligible: distances between communicating entities are small and connectivity is continuous. In a DTN-based network, however, ad-hoc information discovery would in many cases take so much time that it could not be completed before the information lost currency and effectiveness. Instead, protocol operations must be largely driven by information that is pre-placed at the network nodes and tagged with the dates and times at which it becomes effective. This information takes the form of contact plans that are managed by the R/F Contacts (rfx) services of ION's ici package.

The structure of ION's RFX (contact plan) database, the rfx system elements that populate and use that data, and affected portions of the BP and LTP protocol state databases are shown in Figure 3.7.

To clarify the notation of this diagram, which is also used in other database structure diagrams in this chapter:

■ Data objects of defined structure are shown as circles. Dark greyscale indicates

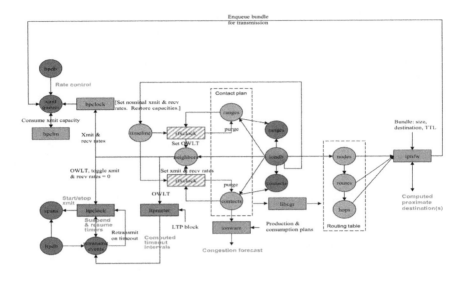

Figure 3.7: RFX services in ION.

notionally non-volatile data retained in "heap" storage, while lighter greyscale indicates volatile data retained in dynamic random access memory.

■ Solid arrows connecting circles indicate one-to-many cardinality.

■ A dashed arrow between circles indicates a potentially many-to-one reference mapping.

■ Arrows from processing elements (rectangles) to data entities indicate data production, while arrows from data entities to processing elements indicate data retrieval.

A contact is here defined as an interval during which it is expected that data will be transmitted by DTN node A (the contact's transmitting node) and most or all of the transmitted data will be received by node B (the contact's receiving node). Implicitly, the transmitting node will utilize some "convergence-layer" protocol underneath the Bundle. Protocol to effect this direct transmission of data to the receiving node. Each contact is characterized by its start time, its end time, the identities of the transmitting and receiving nodes, and the rate at which data are expected to be transmitted by the transmitting node throughout the indicated time period.

Note that a contact is specifically not an episode of activity on a link. Episodes of activity on different links - e.g., different radio transponders operating on the same spacecraft - may well overlap, but contacts by definition cannot; they are bounded time intervals and as such are innately "tiled". For example, suppose transmission on link X from node A to node B, at data rate RX, begins at time $T1$ and ends at time $T2$; also, transmission on link Y from node A to node B, at data rate RY begins at time $T3$ and ends at time $T4$. If $T1 = T3$ and $T2 = T4$, then there is a single contact from time T1 to time T2 at data rate $RX + RY$. If $T1 < T3$ and $T2 = T4$, then there are two contiguous contacts: one from $T1$ to $T3$ at data rate RX, then one from $T3$ to $T2$ at data rate $RX + RY$. If $T1 < T3$ and $T3 < T2 < T4$, then there are three contiguous contacts: one from $T1$ to $T3$ at data rate RX, then one from $T3$ to $T2$ at data rate $RX + RY$, then one from $T2$ to $T4$ at data rate RY. And so on.

A range interval is a period of time during which the displacement between two nodes A and B is expected to vary by less than 1 light second from a stated anticipated distance. (We expect this information to be readily computable from the known orbital elements of all nodes.) Each range interval is characterized by its start time, its end time, the identities of the two nodes to which it pertains, and the anticipated approximate distance between those nodes throughout the indicated time period, to the nearest light second.

The topology timeline at each node in the network is a time-ordered list of scheduled or anticipated changes in the topology of the network. Entries in this list are of two types:

1. Contact entries characterize scheduled contacts.

2. Range entries characterize anticipated range intervals.

Each node to which, according to the RFX database, the local node transmits data directly via some convergence-layer protocol at some time is termed a neighbor of the local node. Each neighbor is associated with one or more outduct for the applicable BP convergence-layer (CL) protocol adapter(s), so bundles that are to be transmitted directly to this neighbor can simply be queued for transmission by outduct (as discussed in the Bandwidth Management notes above).

At startup, and at any time while the system is running, ionadmin inserts and removes Contact and Range entries in the topology timeline of the RFX database. Inserting or removing a Contact or Range entry will cause routing tables to be recomputed for the destination nodes of all subsequently forwarded bundles, as described in the discussion of Contact Graph Routing below.

Once per second, the rfxclock task (which appears in multiple locations on the diagram to simplify the geometry) applies all topology timeline events (Contact and Range start, stop, purge) with effective time in the past. Applying a Contact event that cites a neighboring node revises the transmission or reception data rate between the local node and that Neighbor. Applying a Range event that cites a neighboring node revises the One-Way Light Time (OWLT) between the local node and that neighbor. Setting data rate or OWLT for a node with which the local node will at some time be in direct communication may entail creation of a Neighbor object.

3.4.7.4 Route Computation

ION's computation of a route for a given bundle with a given destination endpoint is accomplished by one of several methods, depending on the destination. In every case, the result of successful routing is the insertion of the bundle into an outbound transmission queue (selected according to the bundle's priority) for one or more neighboring nodes.

But before discussing these methods it will be helpful to establish some terminology:

Egress plans ION can only forward bundles to a neighboring node by queuing them on some explicitly specified transmission queue. Specifications that associate neighboring nodes with outducts are termed egress plans. They are retained in ION's unicast forwarding database.

Static routes ION can be configured to forward to some specified node all bundles that are destined for a given node to which no dynamic route can be discovered from an examination of the contact graph, as described later. Static routing is implemented by means of the "exit" mechanism described below.

Unicast When the destination of a bundle is a single node that is registered within a known "singleton endpoint" (that is, an endpoint that is known to have exactly one member), then transmission of that bundle is termed unicast. For this purpose, the destination endpoint ID must be a URI formed in either the "dtn" scheme (e.g., *dtn://bobsmac/mail*) or the "ipn" scheme (e.g., *ipn:913.11*).

Exits When unicast routes must be computed to nodes for which no contact plan information is known (e.g., the size of the network makes it impractical to distribute all Contact and Range information for all nodes to every node, or the destination nodes do not participate in Contact Graph Routing at all), the job of computing routes to all nodes may be partitioned among multiple exit nodes. Each exit is responsible for managing routing information (for example, a comprehensive contact graph) for some subset of the total network population - a group comprising all nodes whose node numbers fall within the range of node numbers assigned to the gateway. A bundle destined for a node for which no dynamic route can be computed from the local node's contact graph may be routed to the exit node for the group within whose range the destination's node number falls. Exits are defined in ION's unicast forwarding database. (Note that the exit mechanism implements static routing in ION in addition to improving scalability.)

Multicast When the destination of a bundle is all nodes that are registered within a known "multicast endpoint" (that is, an endpoint that is not known to have exactly one member), then transmission of that bundle is termed multicast. For this purpose (in ION), the destination endpoint ID must be a URI formed in the "imc" scheme (e.g., *imc:913.11*).

Multicast Groups A multicast group is the set of all nodes in the network that are members of a given multicast endpoint. Forwarding a bundle to all members of its destination multicast endpoint is the responsibility of all of the multicast-aware nodes of the network. These nodes are additionally configured to be nodes of a single multicast spanning tree overlaid onto the dtnet. A single multicast tree serves to forward bundles to all multicast groups: each node of the tree manages petitions indicating which of its "relatives" (parent and children) are currently interested in bundles destined for each multicast endpoint, either natively (due to membership in the indicated group) or on behalf of more distant relatives.

Unicast

We begin unicast route computation by attempting to compute a dynamic route to the bundle's final destination node. The details of this algorithm are described in the section on Contact Graph Routing, below.

If no dynamic route can be computed, but the final destination node is a "neighboring" node that is directly reachable, then we assume that taking this direct route is the best strategy unless transmission to that neighbor is flagged as "blocked" for network operations purposes.

Otherwise we must look for a static route. If the bundle's destination node number is in one of the ranges of node numbers assigned to exit nodes, then we forward the bundle to the exit node for the smallest such range. (If the exit node is a neighbor and transmission to that neighbor is not blocked, we simply queue the bundle for transmission to that neighbor; otherwise we similarly look up the static route for the exit node until eventually we resolve to some egress plan.)

If we can determine neither a dynamic route nor a static route for this bundle, but the reason for this failure was transmission blockage that might be resolved in the future, then the bundle is placed in a "limbo" list for future re-forwarding when transmission to some node is "unblocked."

Otherwise, the bundle cannot be forwarded. If custody transfer is requested for the bundle, we send a custody refusal to the bundle's current custodian; in any case, we discard the bundle.

Multicast

Multicast route computation is much simpler.

- The topology of the single network-wide multicast distribution tree is established in advance by invoking tree management library functions that declare the children and parents of each node. These functions are currently invoked only from the imcadmin utility program. (Manual configuration of the multicast tree seems manageable for very small and generally static networks, such as the space flight operations networks we'll be seeing over the next few years, but eventually an automated tree management protocol will be required.) Each relative of each node in the tree must also be a neighbor in the underlying dtnet: multicast routing loops are avoided at each node by forwarding each bundle only to relatives other than the one from which the bundle was received, and currently the only mechanism in ION for determining the node from which a bundle was received is to match the sender's convergence-layer endpoint ID to a plan in the unicast forwarding database - i.e., to a neighbor.

■ When an endpoint for the "imc" scheme is added on an ION node - that is, when the node joins that multicast endpoint - BP administrative records noting the node's new interest in the application topic corresponding to the endpoint's group number are passed to all of the node's immediate relatives in the multicast tree. On receipt of such a record, each relative notes the sending relative's interest and forwards the record to all of its immediate relatives other than the one from which the record was received, and so on. (Deletion of endpoints results in similar propagation of cancelling administrative records.)

■ A bundle whose destination endpoint cites a multicast group, whether locally sourced or received from another node:

  ■ Is delivered immediately, if the local node is a member of the indicated endpoint.

  ■ Is queued for direct transmission to every immediate relative in the multicast tree other than the one from which the bundle was received (if any).

3.4.7.5 Delivery Assurance

End-to-end delivery of data can fail in many ways, at different layers of the stack. When delivery fails, we can either accept the communication failure or retransmit the data structure that was transmitted at the stack layer at which the failure was detected. ION is designed to enable retransmission at multiple layers of the stack, depending on the preference of the end user application.

At the lowest stack layer that is visible to ION, the convergence-layer protocol, failure to deliver one or more segments due to segment loss or corruption will trigger segment retransmission if a "reliable" convergence-layer protocol is in use: LTP "red-part" transmission or TCP (including Bundle Relay Service, which is based on TCP)[1]
.

Segment loss may be detected and signaled via NAK by the receiving entity, or it may only be detected at the sending entity by expiration of a timer prior to reception of an ACK. Timer interval computation is well understood in a TCP environment, but it can be a difficult problem in an environment of scheduled contacts as served by LTP. The round-trip time for an acknowledgment dialogue may be simply twice the one-way light time (OWLT) between sender and receiver at one moment, but it may be hours or days longer at the next moment due to cessation of scheduled contact until a future contact opportunity. To account for this timer interval variability in

---

[1] In ION, reliable convergence-layer protocols (where available) are by default used for every bundle. The application can instead mandate selection of "best-effort" service at the convergence layer by setting the BP_BEST_EFFORT flag in the "extended class of service flags" parameter, but this feature is an ION extension that is not supported by other BP implementations at the time of this writing.

retransmission, the ltpclock task infers the initiation and cessation of LTP transmission, to and from the local node, from changes in the current xmit and recv data rates in the corresponding Neighbor objects. This controls the dequeuing of LTP segments for transmission by underlying link service adapter(s) and it also controls suspension and resumption of timers, removing the effects of contact interruption from the retransmission regime.

Note that the current OWLT in Neighbor objects is also used in the computation of the nominal expiration times of timers and that ltpclock is additionally the agent for LTP segment retransmission based on timer expiration.

It is, of course, possible for the nominally reliable convergence-layer protocol to fail altogether: a TCP connection might be abruptly terminated, or an LTP transmission might be canceled due to excessive retransmission activity (again possibly due to an unexpected loss of connectivity). In this event, BP itself detects the CL protocol failure and re-forwards all bundles whose acquisition by the receiving entity is presumed to have been aborted by the failure. This re-forwarding is initiated in different ways for different CL protocols, as implemented in the CL input and output adapter tasks. If immediate re-forwarding is impossible because transmission to all potentially viable neighbors is blocked, the affected bundles are placed in the limbo list for future re-forwarding when transmission to some node is unblocked.

In addition to the implicit forwarding failure detected when a CL protocol fails, the forwarding of a bundle may be explicitly refused by the receiving entity, provided the bundle is flagged for custody transfer service. A receiving node's refusal to take custody of a bundle may have any of a variety of causes: typically the receiving node either:

1. has insufficient resources to store and forward the bundle,

2. has no route to the destination, or

3. will have no contact with the next hop on the route before the bundle's TTL has expired.

In any case, a "custody refusal signal" (packaged in a bundle) is sent back to the sending node, which must re-forward the bundle in hopes of finding a more suitable route.

Alternatively, failure to receive a custody acceptance signal within some convergence-layer-specified or application-specified time interval may also be taken as an implicit indication of forwarding failure. Here again, when BP detects such a failure it attempts to re-forward the affected bundle, placing the bundle in the limbo list if re-forwarding is currently impossible.

In the worst case, the combined efforts of all the retransmission mechanisms in ION are not enough to ensure delivery of a given bundle, even when custody transfer is requested. In that event, the bundle's "time to live" will eventually expire while the bundle is still in custody at some node: the bpclock task will send a bundle status report to the bundle's report-to endpoint, noting the TTL expiration, and destroy the bundle. The report-to endpoint, upon receiving this report, may be able to initiate application-layer retransmission of the original application data unit in some way. This final retransmission mechanism is wholly application-specific, however.

3.4.7.6 Rate Control

In the Internet, the rate of transmission at a node can be dynamically negotiated in response to changes in level of activity on the link, to minimize congestion. On deep space links, signal propagation delays (distances) may be too great to enable effective dynamic negotiation of transmission rates. Fortunately, deep space links are operationally reserved for use by designated pairs of communicating entities over pre-planned periods of time at pre-planned rates. Provided there is no congestion inherent in the contact plan, congestion in the network can be avoided merely by adhering to the planned contact periods and data rates. Rate control in ION serves this purpose.

While the system is running, transmission and reception of bundles is constrained by the current capacity in the throttle of each convergence-layer manager. Completed bundle transmission activity reduces the current capacity of the applicable throttle by the capacity consumption computed for that bundle. This reduction may cause the throttle's current capacity to become negative. Once the current capacity of the applicable throttle goes negative, activity is blocked until non-negative capacity has been restored by bpclock.

Once per second, the bpclock task increases the current capacity of each throttle by one second's worth of traffic at the nominal data rate for transmission to that node, thus enabling some possibly blocked bundle transmission and reception to proceed. bpclock revises all throttles' nominal data rates once per second in accord with the current data rates in the corresponding Neighbor objects, as adjusted by rfxclock per the contact plan.

Note that this means that, for any neighboring node for which there are planned contacts, ION's rate control system will enable data flow only while contacts are active.

3.4.7.7 Flow Control

A further constraint on rates of data transmission in an ION-based network is LTP flow control. LTP is designed to enable multiple block transmission sessions to be in various stages of completion concurrently, to maximize link utilization: there is no requirement to wait for one session to complete before starting the next one. However, if unchecked this design principle could in theory result in the allocation of all memory in the system to incomplete LTP transmission sessions. To prevent complete storage resource exhaustion, we set a firm upper limit on the total number of outbound blocks that can be concurrently in transit at any given time. These limits are established by ltpadmin at node initialization time.

The maximum number of transmission sessions that may be concurrently managed by LTP therefore constitutes a transmission "window" - the basis for a delay-tolerant, non-conversational flow control service over interplanetary links. Once the maximum number of sessions are in flight, no new block transmission session can be initiated - regardless of how much outduct transmission capacity is provided by rate control - until some existing session completes or is canceled.

Note that this consideration emphasizes the importance of configuring the aggregation threshholds and session count limits of spans during LTP initialization to be consistent with the maximum data rates scheduled for contacts over those spans.

3.4.7.8 Storage Management

Congestion in a dtnet is the imbalance between data enqueuing and dequeuing rates that results in exhaustion of queuing (storage) resources at a node, preventing continued operation of the protocols at that node.

In ION, the affected queuing resources are allocated from notionally non-volatile storage space in the SDR heap and/or file system. The design of ION is required to prevent resource exhaustion by simply refusing to enqueue additional data that would cause it.

However, a BP router's refusal to enqueue received data for forwarding could result in costly retransmission, data loss, and/or the "upstream" propagation of resource exhaustion to other nodes. Therefore the ION design additionally attempts to prevent potential resource exhaustion by forecasting levels of queuing resource occupancy and reporting on any congestion that is predicted. Network operators, upon reviewing these forecasts, may revise contact plans to avert the anticipated resource exhaustion.

The non-volatile heap storage used by ION serves several purposes: it contains queues of bundles awaiting forwarding, transmission, and delivery; it contains LTP transmission and reception sessions, including the blocks of data that are being transmitted and received; it contains queues of LTP segments awaiting radiation; it may contain CFDP transactions in various stages of completion; and it contains protocol operational state information, such as configuration parameters, static routes, the contact graph, etc.

Effective utilization of non-volatile storage is a complex problem. Static preallocation of storage resources is in general less efficient (and also more laborintensive to configure) than storage resource pooling and automatic, adaptive allocation: trying to predict a reasonable maximum size for every data storage structure and then rigidly enforcing that limit typically results in underutilization of storage resources and underperformance of the system as a whole. However, static preallocation is mandatory for safety-critical resources, where certainty of resource availability is more important than efficient resource utilization.

The tension between these two approaches is analogous to the tension between circuit switching and packet switching in a network: circuit switching results in underutilization of link resources and underperformance of the network as a whole (some peaks of activity can never be accommodated, even while some resources lie idle much of the time), but dedicated circuits are still required for some kinds of safety-critical communication.

So the ION data management design combines the two approaches as following.

1. A fixed percentage of the total SDR data store heap size (by default, 40%)

is statically allocated to the storage of protocol operational state information, which is critical to the operation of ION.

2. Another fixed percentage of the total SDR data store heap size (by default, 20%) is statically allocated to "margin", a reserve that helps to insulate node management from errors in resource allocation estimates.

The remainder of the heap, plus all pre-allocated file system space, is allocated to protocol traffic[2].

The maximum projected occupancy of the node is the result of computing a congestion forecast for the node, by adding to the current occupancy all anticipated net increases and decreases from now until some future time, termed the "horizon" for the forecast.

The forecast horizon is indefinite - that is, "forever" - unless explicitly declared by network management via the ionadmin utility program. The difference between the horizon and the current time is termed the "interval" of the forecast.

Net occupancy increases and decreases are of four types:

1. Bundles that are originated locally by some application on the node, which are enqueued for forwarding to some other node.

2. Bundles that are received from some other node, which are enqueued either for forwarding to some other node or for local delivery to an application.

3. Bundles that are transmitted to some other node, which are dequeued from some forwarding queue.

4. Bundles that are delivered locally to an application, which are dequeued from some delivery queue.

The type-1 anticipated net increase (total data origination) is computed by multiplying the node's projected rate of local data production, as declared via an ionadmin command, by the interval of the forecast. Similarly, the type-4 anticipated net decrease (total data delivery) is computed by multiplying the node's projected rate of local data consumption, as declared via an ionadmin command, by the interval of the forecast. Net changes of types 2 and 3 are computed by multiplying inbound and outbound data rates, respectively, by the durations of all periods of planned communication contact that begin and/or end within the interval of the forecast.

Congestion forecasting is performed by the ionwarn utility program. ionwarn may be run independently at any time; in addition, the ionadmin utility program automatically runs ionwarn immediately before exiting if it executed any change in the contact plan, the forecast horizon, or the node's projected rates of local data production or consumption. Moreover, the rfxclock daemon program also runs ionwarn automatically whenever any of the scheduled reconfiguration events it dispatches result in contact state changes that might alter the congestion forecast.

---

[2]Note that, in all occupancy figures, ION data management accounts not only for the sizes of the payloads of all queued bundles but also for the sizes of their headers.

If the final result of the forecast computation - the maximum projected occupancy of the node over the forecast interval - is less than the total protocol traffic allocation, then no congestion is forecast. Otherwise, a congestion forecast status message is logged noting the time at which maximum projected occupancy is expected to equal the total protocol traffic allocation.

Congestion control in ION, then, has two components:

First, ION's congestion detection is anticipatory (via congestion forecasting) rather than reactive as in the Internet.

Anticipatory congestion detection is important because the second component - congestion mitigation - must also be anticipatory: it is the adjustment of communication contact plans by network management, via the propagation of revised schedules for future contacts.

(Congestion mitigation in an ION-based network is likely to remain mostly manual for many years to come, because communication contact planning involves much more than orbital dynamics: science operations plans, thermal and power constraints, etc. It will, however, rely on the automated rate control features of ION, discussed above, which ensure that actual network operations conform to established contact plans.)

Rate control in ION is augmented by admission control. ION tracks the sum of the sizes of all zero-copy objects currently residing in the heap and file system at any moment. Whenever any protocol implementation attempts to create or extend a ZCO in such a way that total heap or file occupancy would exceed an upper limit asserted for the node, that attempt is either blocked until ZCO space becomes available or else rejected altogether.

3.4.8 BP/LTP Detail - How It Works

Although the operation of BP/LTP in ION is complex in some ways, virtually the entire system can be represented in a single diagram. The interactions among all of the concurrent tasks that make up the node - plus a Remote AMS task or CFDP UT-layer task, acting as the application at the top of the stack (see Figure 3.8). (The notation is as used earlier but with semaphores added. Semaphores are shown as small circles, with arrows pointing into them signifying that the semaphores are being given and arrows pointing out of them signifying that the semaphores are being taken.)

Further details of the BP/LTP data structures and flow of control and data appear in the following sections (see Figures 3.9, 3.10, 3.11, 3.12, 3.13, 3.14 and 3.15). (For specific details of the operation of the BP and LTP protocols as implemented by the ION tasks, such as the nature of report-initiated retransmission in LTP, please see the protocol specifications. The BP specification is documented in Internet RFC 5050 [244], while the LTP specification is documented in Internet RFC 5326 [224].)

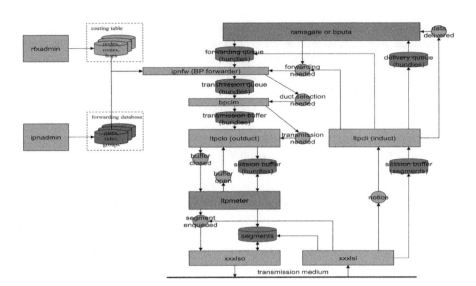

Figure 3.8: ION node functional overview.

3.4.8.1 Databases

Figures 3.9 and 3.10 show the BP database and LTP database, respectively.

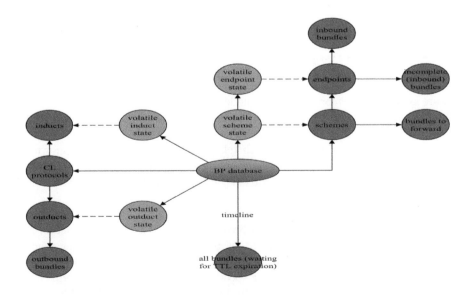

Figure 3.9: Bundle protocol database.

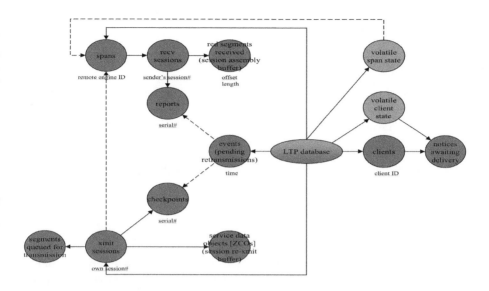

Figure 3.10: Licklider transmission protocol database.

3.4.8.2 Control and Data Flow

Figure 3.11 shows the BP forwarder and the main steps of the forwarding procedure.

1. Waits for forwarding needed semaphore.

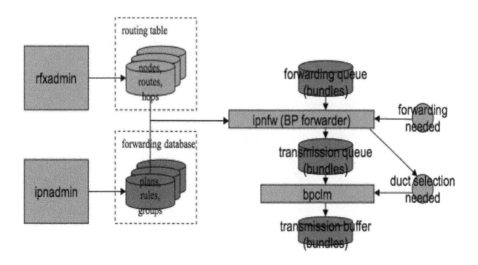

Figure 3.11: Bundle protocol forwarder.

2. Gets bundle from queue.

3. Consults routing table and forwarding table to determine all plausible proximate destination - routing.

   ■ A plausible proximate destination is the destination node of the first entry in a contact sequence (a list of concatenated contact periods) ending in a contact period whose destination node is the bundle's destination node and whose start time is less than bundle's expiration time.

4. Appends bundle to transmission queue (based on priority) for best plausible proximate destination.

5. Gives duct selection needed semaphore for that transmission queue.

6. Convergence-layer daemon waits for duct selection needed semaphore.

7. Gets bundle from queue. Imposes flow control, fragments as needed. Consults mission-provided code (if provided), selects outduct to use for transmission of bundle to this proximate destination.

8. Inserts bundle into transmission buffer for selected outduct.

9. Gives transmission needed semaphore for this buffer.

Figure 3.12 shows the BP convergence layer output.

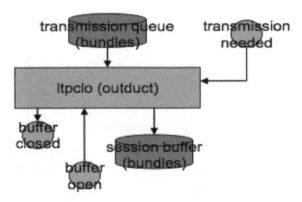

Figure 3.12: BP convergence layer output.

1. Waits for buffer open semaphore (indicating that the link's session buffer has room for the bundle).

2. Waits for transmission needed semaphore.

3. Gets bundle from queue, subject to priority.

4. Appends bundle to link's session buffer - aggregation. Buffer size is notionally limited by aggregation size limit, a persistent attribute of the Span object: implicitly, the rate at which we want reports to be transmitted by the destination engine.

5. Gives buffer closed semaphore when buffer occupancy reaches the aggregation size limit.

Figure 3.13 shows the LTP transmission metering.

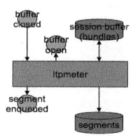

Figure 3.13: LTP transmission metering.

1. Initializes session buffer, gives buffer open semaphore.

2. Waits for buffer closed semaphore (indicating that the buffer session is ready for transmission).

3. Segments the entire buffer into segments of managed MTU size fragmentation.

4. Appends all segments to segments queue for intermediate transmission.

5. Gives segment enqueued semaphore.

Figure 3.14 shows the LTP link service output.

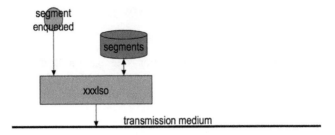

Figure 3.14: LTP link service output.

1. Waits for segments enqueued semaphore (indicating that there is now something to transmit).

2. Gets segment from the queue.

3. Sets retransmission timer if necessary.

4. Transmits the segment using link service protocol.

   Figure 3.15 shows the LTP link service input.

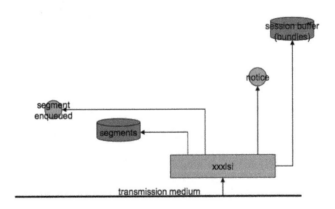

Figure 3.15: LTP link service input.

1. Receives segment using link service protocol.

2. If data, generates report segment and appends it to queue - reliability. Also in-

serts fata into reception session buffer "red part" and, if that buffer is complete, gives notice semaphore to trigger bundle extraction and dispatching by ltpcli.

3. If a report, appends acknowledgment to segments queue.

4. If a report of missing data, recreates lost segments and appends them to queue.

5. Gives segment enqueued semaphore.

3.5 μ DTN : Sensor Networks

μ Sensor Networks DTN is a bundle protocol implementation for 8 and 16 bits microcontrollers. It is integrated into the Contiki OS [2] and enables transmission of bundles in IEEE 802.15.4 radio frames. It was designed to meet the hardware and energy constraints of wireless sensor nodes which use the Compressed Bundle Header Encoding (CBHE) [58]. CBHE enables smaller bundles facilitating their processing and reducing communication overhead.

A variant of μ Sensor Networks DTN is called *miniDTN* [236]; it runs in FreeR-TOS [4] using the same architecture. μ Sensor Networks DTN is compatible with IBR-DTN on Linux. A security layer, named μDTNSec [243] was designed to operate with μ Sensor Networks DTN. μDTNSec consists of asymmetric encryption and signatures with elliptic curve cryptography [124] and hardware-backed symmetric encryption.

3.5.1 μ Sensor Networks DTN Architecture

Figure 3.16 shows the μSensor Networks DTN architecture [283], which is composed of seven main components.

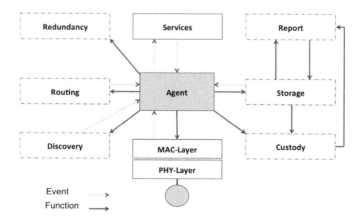

Figure 3.16: μDTN architecture [283].

1. **Agent:** core functionality that cannot be modified. It provides a communication interface for other components. It is implemented as a single daemon protothread process that enables event-based bundle protocol with underlying layers of the stack to send and receive radio frames. Also, it can receive events from other components and interpret them. As a result the appropriate functions are executed.

2. **Discovery:** enables the node to discover other nodes via IEEE 802.15.4. Two discovery mechanisms are implemented:

   (a) Reactive node discovery: once a node has a bundle to send, it starts to send beacon messages to all neighbors. Those neighbors that receive the beacon will respond and start to exchange bundles.

   (b) IP neighbor discovery: similar to the node discovery mechanism in IP networks.

3. **Routing:** makes the decision of which bundles will be sent to which nodes. The bundles are enqueued and sent to the agent as events. The daemon then interprets the events and forwards the bundles to the neighbors. For each bundle successfully transmitted the routing component receives a notification from the agent to keep track of the current status of the enqueued bundles.

4. **Storage:** enables storage of bundles in a persistent memory. The storage component uses an internal ID to identify each bundle to improve the access to the bundle.

5. **Redundancy:** performs a check to see if a given bundle has been already received.

6. **Custody:** implements the custody procedure. In addition, the custody component has capability of defining whether or not a node can or cannot be a custodian. Each DTN node contains a list with metadata of all bundles for which it has assumed custody.

7. **Report:** handles bundle status reports. Status reports are bundles of a special type, providing information about the status of a bundle.

The communication between the Contiki node and the Agent happens through events. First the Contiki node service registers as an endpoint using the daemon. After the registration, the Contiki node is able to receive bundles as well as to send bundles.

3.6 Other Implementations

3.6.1 Bytewalla

The Bytewalla platform [103] [208] is an implementation in Java of DTN bundle protocol developed at KTH University in Sweden. The main idea of this platform is to provide connectivity to rural areas using Android phones. The Android phones are used as data mules to carry data, for example, from the village to the city or vice-versa. Once the Android phone is in the village, it receives the bundles from a server via 802.11 access point, and when it returns to the city, the bundles are uploaded via gateway APs.

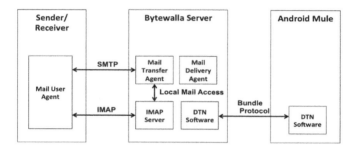

Figure 3.17: Bytewalla high level village architecture [103].

Figure 3.17 shows the Bytewalla village architecture that contains three components.

1. **Sender/Receiver** that implements a mail user agent.

2. **Bytewalla Server** that contains the DTN implementation, a mail transfer agent, a mail delivery agent and an Internet Message Access Protocol (IMAP) server.

3. **Android Mule** that implements DTN.

The DTN software module located in both the Bytewalla Server and the Android Mule components is composed of three software components: DTNService, DTN-Manager and DTNApps.

1. **DTNService** is a backend application implemented as an Android service [1]. It uses Android based-TCP/IP network communication. It is based on DTN2 and it contains nine sub-modules: bundle daemon, contact manager, TCP convergence layer, discovery node, persistent storage, registration, bundle router, fragmenter and APILib.

2. **DTNManager** is a front end application that enables the user to configure and manage the DTNService module.

3. **DTNApps** is an application on top of the DTNService module. It includes DTNSend and DTNReceive functions.

3.6.2   DTNLite

The DTNLite platform is a custody-based reliable transfer mechanism [206] designed for sensor networks. It was developed taking into account the following design principles:

■ *Reliable custody transfer*: implementation of one hop transfer of a bundle from one overlay DTNLite hop to another.

■ *Persistent storage management*: implementation of flash storage system to support lower level operations that enable to record the store bundles in a consistent way.

■ *Duplicate management*: implementation mechanism to detect duplicated bundles at the base station.

■ *Application awareness*: implementation of mechanisms for long term storage of data using in-network aging or data compression.

The DTNLite architecture is based on data collection through a centralized base-station and it implements a light version of the DTN architecture. Figure 3.18 shows the DTNLite architecture designed to TinyOS that includes three main components.

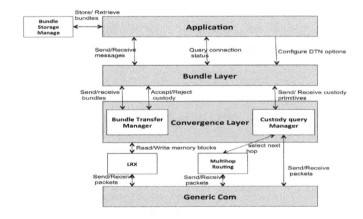

Figure 3.18: DTNLite architecture [206].

1. **Bundle Storage Manager** component which is responsible for providing persistent storage for bundles.

2. **Bundle Agent** component responsible for implementing custody transfer operation, providing an interface for sending and receiving bundles.

3. **Convergence Layer** component works with the bundle agent to provide primitives for transferring a bundle to another hop, for sending and receiving custody requests and custody acknowledgments. It also provides node discovery. The convergence layer component is composed of two other components:

   ■ **LRX** that enables reliable multi-hop transfer using windowing scheme and selective NACKs.

   ■ **Multihop** that enables the node to send and to receive custody queries. Basically, it provides routing schemes.

3.6.3 *ContikiDTN*

ContikiDTN is a platform that implements DTN architecture for sensors running Contiki OS. It uses bundle protocol over TCP convergence layer in IEEE 802.15.4 networks. Figure 3.19 shows an overview of the ContikiDTN architecture. This architecture is composed of:

μIP is a component that implements a TCP/IP stack designed for tiny microcontroller systems. For sending and receiving data, Contiki uses functions provided by the μIP component.

Protosockets is a programming interface for μIP. When a TCP connection is started in ContikiDTN a protosocket function is also started to handle the processing of the TCP connection. It provides a reliable send and receive TCP byte stream.

Bundle Daemon component provides the overlay routing layer. This component is responsible for sending and receiving bundles. It also makes decisions about routing and forwarding bundles. Each DTN daemon assumes a unique EID which is statically configured with routing information. This daemon component is composed of bundle layer and the convergence layer. The convergence layer is a general interface between the bundle daemon and the transport layer of the underlying network. It enables the node to receive bundles from the bundle layer and transmit them.

Bundle Store component provides persistent storage for bundles through a primary memory resource available in Contiki.

Application represents the DTN applications which make use of the services provided by the bundle daemon. This application component does not have direct access to the bundles in the memory.

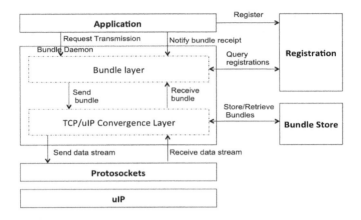

Figure 3.19: ContikiDTN architecture [186].

Registrations component allows the application to register its intention to receive
bundles. Before receiving a bundle an application must register at the bundle
daemon. The daemon checks the destination of incoming bundles for the registered applications.

ContikiDTN is based on DTN2 and it implements the TCP convergence layer,
which uses a handshake mechanism before bundle transmission. The handshake enables the exchange of contact information between a pair of nodes.

3.6.4 6LoWDTN

6LoWDTN is a DTN transport layer for WSNs built on-top of the Low-power Wireless Personal Area Networks (LoWPAN) that has been adapted to support multiple
stacks such as IPv6 over LoWPAN (6LoWPAN [168][199]); it is based on the Contiki operation system. The underlying IPv6/6LoWPAN layer is also employed for
node addressing. 6LoWDTN introduces additional overlay routing, forwarding and
synchronization on top of the underlying layers. Figure 3.20 shows 6LoWDTN as a
bundle layer providing transport functionalities to applications, based on UDP communication.

The 6LoWDTN overlay routing algorithm is responsible for making the decision
of whether or not to forward a bundle. The decision is based on the evaluation of
a metric named Estimated Delivery Delay (EDD), moderated by a score result provided by a utility function. The EDD is an estimate of the total delay required to
transport a bundle from its origin to its destination. The utility function is used to
take into account in the routing decision process a variety of heterogeneous decision
metrics (eg. local buffer occupation, residual battery power, etc.). In order to keep
the metrics up to date, nodes can exchange link local broadcast UDP beacons among

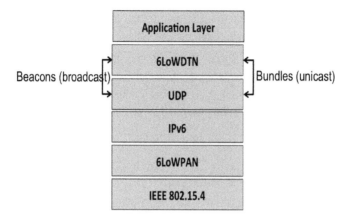

Figure 3.20: 6LoWDTN network stack [50].

themselves. The process of bundle forwarding is performed through UDP unicast packets where the sink nodes are the bundles' final destinations.

Figure 3.21 shows a reference implementation of 6LoWDTN architecture within the Contiki operating system. This architecture is composed of three building blocks.

1. **Communication** is the main building block, performing all the network operations that are based on the Contiki stack. It contains two processes: beacon handler and bundle handler. The beacon handler is responsible for broadcasting beacons on the network, receiving and processing beacons. The bundle handler deals with bundle forwarding.

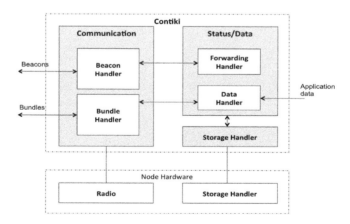

Figure 3.21: 6LoWDTN reference implementation architecture [50].

2. **Storage Handler** building block enables the node to store, access, and handle data in persistent storage based on Contiki COFFEE File System (CFS) [276].

3. **Status/Data** block contains two modules. First is the data handler, which serves as an interface for application data and is responsible for managing updated information such as power and storage occupation. Second is the forwarding handler, which enables the EDD calculation and the next hop selection for bundle forwarding.

3.6.5   CoAP over BP

The Constrained Application Protocol (CoAP) over BP package addresses the implementation of a BP binding for CoAP as a means to enable Delay Tolerant IoT. CoAP [250] provides an application layer protocol that enables resource-constrained devices to interact asynchronously. It is designed for machine-to-machine (M2M) communications and is compliant with the Representational State Transfer (REST) architecture style. CoAP defines a simple messaging layer, with a compact format, that runs over UDP (or DTLS when security is enabled). Its low header overhead and low complexity simplify the processing of CoAP messages for constrained nodes. On top of this message layer, CoAP uses request/response interactions between clients and servers.

A version of CoAP integrated with BP, named BoAP, is described in [36]. BoAP is based on IBR-DTN and it is implemented in Java. Figure 3.22 shows the BoAP client-server architecture that contains three components.

1. **Client node** is based on CoAP client using BoAP API to manage resources. The communication with the Relay node is performed through IBR-DTN API and BP daemon using TCP socket.

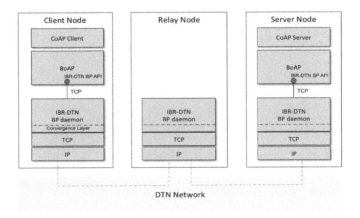

Figure 3.22: BoAP client-server architecture [36].

2. **Relay node** works as a middleware between the client and the server. It contains an IBR-DTN daemon that implements the BP, storing bundles and exchanging them with other IBR-DTN BP daemons in the DTN network.

3. **Server node** is based on CoAP server using BoAP to manage resources. The communication with the Relay node is performed through IBR-DTN API and BP daemon using TCP socket that enables client-server communication.

Basically, CoAP over BP substitutes BP for UDP as an underlying transport for CoAP, in order to enable a delay-tolerant Internet of Things. Each CoAP message is encapsulated as a bundle to preserve all the CoAP features that have been designed for a UDP binding.

3.7 Summary

This chapter presented some existing DTN platforms that have been used in testbeds. DTN2 is the reference implementation, a product of the Delay Tolerant Networking Research Group (DTNRG) of the Internet Research Task Force (IRTF). ION is a platform that has been developed at NASA's Jet Propulsion Laboratory for IPN and/or deep space communication and it is interoperable with DTN2. The other platforms include lightweight, portable versions of BP that can be used in portable devices like smartphones and wireless sensors: IBR-DTN (which was developed with embedded systems in mind) and DTNLite. These platforms have been tested in a large range of application [284] that includes: connectivity in developing areas, remote and disconnected areas, environment and wildlife monitoring, undersea communications and military applications.

Chapter 4

Case Study: Interplanetary Networks

Aloizio P. Silva

University of Bristol

Scott Burleigh

JPL-NASA/Caltech

CONTENTS

4.1 Introduction

Currently communication systems allow human beings to talk to almost anyone, in any corner of the Earth, almost instantly because of the Internet and other advances in electronic technologies. These systems have also enabled communication beyond Earth (e.g., Mars), thanks in large part to a NASA telecommunication system named Deep Space Network, or DSN.

Scientists are now looking for a way to improve communication beyond Earth, that will better utilize the DSN. Ideally the next phase of the Internet would take humankind to the far reaches of the solar system and lay the foundation for a communications system for manned and autonomous missions to Mars and planets beyond.

Communication in space moves slowly compared to communication on Earth. There are several reasons for this:

■ Distance: On Earth, people are only a fraction of a light second apart, making Earth communication nearly instantaneous over the Internet. The most common unit of measurement for distances within the solar system is the astronomical unit (AU). One AU equals the average distance from the sun to Earth, about 150,000,000 km. Another option to indicate distances within the solar system is terms of light time, which is the distance light travels in a unit of time at the rate of 300,000 km per second. As an object moves farther out into space, there is a delay of minutes or hours in communicating with it because light has to travel millions of miles, instead of thousands of miles, between transmitter and receiver.

■ Line of sight: Electromagnetic transmission, including light, generally travels in a straight line. The rays or waves may be diffracted, refracted, reflected, or absorbed by atmosphere and obstructions with material and they generally cannot travel over the horizon or behind obstacles. Anything that blocks the space between the signal transmitter and receiver can interrupt communication.

■ Mass: The cost of launching a satellite increases with its mass. The high-powered antennas that would improve communication with deep space probes may be too heavy to send on a cost-constrained space mission.

For these reasons, extending the Internet beyond the Earth to form an interplanetary network is difficult. In particular, the Internet architecture does not perform well in extreme environments such as deep space (this will be explained in detail in the next sections). That was the key insight of Vint Cerf and his colleagues at the InterPlanetary Internet (IPN) team.

Naively, we might think of the Interplanetary Internet as a series of Internets linked by gateways and using the Internet Protocol (IP) suite as its basis. By using the IP protocol, we could leverage existing technology so as to speed up development of the Interplanetary Internet. The problem is that IP is not a good fit for interplanetary communications.

This chapter describes "IPN" architecture, which is one way of configuring and using DTN. In particular, we discuss one implementation of DTN named Interplanetary Overlay Network (ION) which provides the main functions to allow networked communications in deep space. The principal components of ION are discussed as well as their configurations.

4.2 Deep Space Communications Main Features

As we move from space discovery to exploration, and perhaps even human settlements in space, engineers need to rethink how missions can better communicate. To this end, extending networked communications to other planets might be the best approach. Due to the astronomical distances and unique operational environment, deep space communication presents several particular characteristics as compared with terrestrial and near space communications. The main features, as described below, should be considered when extending Internet communication to other planets or extreme environments. These features directly impact the way researchers and engineers design protocols and applications for space communications. Furthermore, they will confound earthbound protocols and applications in ways most network specialists rarely conceive of.

1. **Astronomical distance** The distances between planets vary depending on where each planet is in its orbit around the Sun, simply because their distances from the Sun - and hence their orbital periods - are very different. Table 4.1 shows the eight planets and the average distance between them. The AU[1] column is the distance in astronomical units. The distance between an entity pair (origin and destination) located on different planets in deep space is typically many millions of kilometers away from the Earth, while on Earth each entity pair is separated by no more than a fraction of a light second even if geostationary communication satellites are involved. When astronomical distances are involved the distance that the signals need to travel is measured in light minutes or hours. This means that highly interactive protocol operations cannot work.

2. **Long propagation delays**

    Because of the interplanetary distances, the signal needs to travel for a long period of time until it arrives at its destination - that is, signal propagation delay is high (see Table 4.2 [310]). The propagation delay from Earth to Mars, for example, is approximately 4 minutes when Earth and Mars are at their closest approach. The one-way light time can exceed 20 minutes when Earth and Mars are in opposition. The propagation delay to the outer planets is significantly higher. Propagation delay to Jupiter varies between approximately 30 and 45 minutes; to Saturn it is between 70 and 90 minutes. Propagation delays in these ranges have a significant effect on the applicability of the standard Internet suite for use over interplanetary distances.

3. **Low Signal-to-Noise Ratios**

    The term "signal-to-noise ratio" compares a level of signal power versus a level of noise power. Higher numbers generally mean a "better" signal, since there is more useful information (the signal) than there is unwanted data (the noise). In a communication system, the transmitter sends data to the receiver,

---

[1] 1 AU is the distance from the Sun to Earth, which is 149,600,000 km.

Table 4.1: Planet distance table

FROM	TO	AU	KM	MILES
Mercury	Venus	0.34	5029000	31248757
Mercury	Earth	0.61	91691000	56974146
Mercury	Mars	1.14	170030000	105651744
Mercury	Jupiter	4.82	720420000	447648234
Mercury	Saturn	9.14	1366690000	849221795
Mercury	Uranus	18.82	2815640000	1749638696
Mercury	Neptune	29.70	4443090000	2760936126
Venus	Earth	0.28	41400000	25724767
Venus	Mars	0.8	119740000	74402987
Venus	Jupiter	4.48	670130000	416399477
Venus	Saturn	8.80	1316400000	817973037
Venus	Uranus	18.49	2765350000	1718388490
Venus	Neptune	29.37	4392800000	2729685920
Earth	Mars	0.52	78340000	48678219
Earth	Jupiter	4.2	628730000	390674710
Earth	Saturn	8.52	1275000000	792248270
Earth	Uranus	18.21	2723950000	1692662530
Mars	Jupiter	3.68	550390000	342012346
Mars	Saturn	7.99	1196660000	743604524
Mars	Uranus	17.69	2645610000	1643982054
Mars	Neptune	28.56	4273060000	2655279484
Jupiter	Saturn	4.32	646270000	401592178
Jupiter	Uranus	14.01	2095220000	1301969708
Jupiter	Neptune	24.89	3722670000	2313267138
Saturn	Uranus	9.7	1448950000	900377530
Saturn	Neptune	20.57	3076400000	1911674960
Uranus	Neptune	10.88	1627450000	1011297430

Table 4.2: Propagation delay between the Earth and other planets

Planets	Minimum Delay (min)	Maximum Delay (min)
Mercury	5.617	12.378
Venus	2.20	14.50
Mars	3.311	22.294
Jupiter	32.983	53.778
Saturn	86.661	92.172
Uranus	143.994	175.283
Neptune	239.161	260.783
Pluto	238.772	418.617

where each data corresponds to some number of information bits. The receiver should decide, based on the received signal and noise, what the transmitted data were in order to rebuild the original information stream. All systems are subject to noise that perturbs the transmitted signal before it reaches the receiver; this noise can cause the receiver to make errors in its decisions about the received data. A parameter that is a measure of how likely the receiver is to making errors is the ratio of the received signal power to the noise power at the receiver (see Equation 4.1 where P is the average power), or signal-to-noise ratio (SNR). The higher the SNR, the less likely it is that the receiver will make a mistake in decoding a received data. For a given data rate, coding, and modulation scheme, there is a mapping from received SNR to the error characteristics of the decoded information stream. When errors are relatively uncommon and widely spaced, it is appropriate to refer to the bit error rate (BER) of the system as the rate at which the receiver makes mistakes in the decoded bitstream. While fiber-optic systems can achieve bit error rates as low as 10^{-12} to 10^{-15}, deep space missions typically operate with uncoded bit error rates on the order of 10^{-1}. They use a concatenated code composed of a rate 7, constraint-length $1/2$ inner convolutional code and a $223, 255$ Reed-Solomon outer code to bring the error rates down to the order of 10^{-9} or better.

$$SNR = \frac{P_{signal}}{P_{noise}} \qquad (4.1)$$

4. **Intermittent connectivity** The term "Intermittent connectivity" refers to the ability to establish and maintain a continuous communication path between a pair of local-remote endpoints. The entities or endpoints in deep space are moving along their own orbits. According to Newton's laws of motion the orbital motions are defined as ellipses and at least six parameters are needed to predict the location of an entity along its orbital route at a given time.

   (a) **Orbital inclination**: measures the inclination of an object around a celestial entity and it is one of the elements that must be specified in order to define the orientation of an elliptical orbit. It is expressed as an angle between two planes as shown in Figure 4.1.

   (b) **Longitude of ascending node**: specifies the orbit of an entity in space and measures the distance from the Vernal equinox, that is when the Sun crosses the equator from south to north, to the point where a satellite crosses the equator going from south to north, measured from the center of Earth, in degrees. Figure 4.2 shows this orbital element.

   (c) **Eccentricity**: determines the amount by which an orbit around another body deviates from a perfect circle. A value of 0 means a circular orbit, values between 0 and 1 form an elliptical orbit (as shown in Figures 4.3 and 4.4), 1 is a parabolic escape orbit, and greater than 1 is a hyperbola (as shown in Figure 4.5). The eccentricity value is given by Equation 4.2

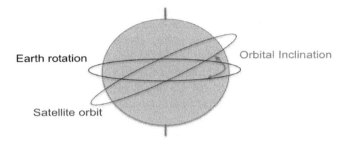

Figure 4.1: Orbital inclination.

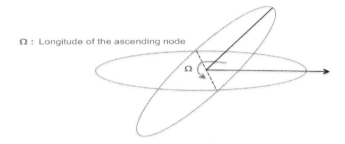

Figure 4.2: Longitude of ascending node.

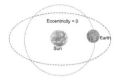

Figure 4.3: Circular orbit eccentricity when e = 0.0.

Figure 4.4: Transition from circular orbit to elliptical orbit.

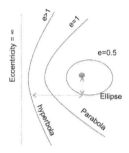

Figure 4.5: Eccentricity greater than 1.0.

where "b" is the length of the semi minor axis and "a" is the length of the semi major axis.

$$e = \sqrt{1 - \frac{b^2}{a^2}} \qquad (4.2)$$

(d) **Semi major axis**: is the measure of one half of the ellipse's long axis as shown in Figure 4.6.

(e) **Periapsis' argument**: is the angle between a ray extending through the ascending node and another ray extending through the orbit's perihelion. It is also known as the argument of perihelion or the argument of perifocus. It is measured in the plane of the orbit in the orbit's direction of angular motion. In Figure 4.7 the ω represents the argument of the periapsis.

(f) **Periapsis' time**: is the instant when the object is found at the perihelion of its orbit. It is usually expressed as a Julian Date, which is a running count of mean solar days started in ancient times. 1 January 2014 at 12h GMT is JD 24566590.

Due to the relative motion of communication entities and the rotation of the

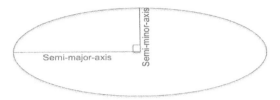

Figure 4.6: Semi major axis.

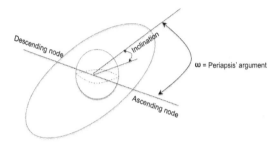

Figure 4.7: Periapsis' Argument.

planets on their own axes, the communication channel between entities is sub-ject to disruption. For instance, the planet's motion can result in orbital obscu-ration, in which communication systems lose line-of-sight due to the positions of the planets, generating a source of intermittent connectivity. According to Newton's laws and the six parameters discussed before it is possible to predict when a communication channel opportunity will be available between entities in deep space. Unlike terrestrial communication networks which are denser and whose entities have a high mobility degree resulting in more communi-cation link opportunities, the deep space networks are sparse and are prone to fewer communication link opportunities.

5. **Asymmetric data channel**

The term "asymmetric data channel" indicates that a system may have different data rates for outbound traffic and inbound traffic. Asymmetrical data flow can, in some instances, make more efficient use of the available infrastructure than symmetrical data flow, in which the speed or quantity of data is the same in both directions, averaged over time. In deep space communication the links are not only often asymmetrical but also fairly limited. The outbound traffic from Earth typically has lower bandwidth requirement than inbound traffic because the outbound link only transfers low data rate telecommand while the inbound link transmits telemetry information and high data rate exploration data.

6. **Scarce resource**

   Resources in deep space environment are generally scarce due to the physical environment and also engineering tradeoffs like mass, volume and power. When deep space environment is considered the processing capability of communication system is very limited due to the restrictions (mass, volume and power) mentioned before.

7. **Extreme environment**

   An extreme environment is characterized by harsh and challenging conditions. Humans, computing devices and other species need to adapt in order to survive and/or operate adequately in it. The challenging conditions could be from the ecosystem, climate, landscape or location. The moons and planets in the Solar System are extreme environments compared to Earth's environment. Mercury, for instance almost lacks an atmosphere. The lack of atmosphere means that it is unable to retain the heat and as a result, it exhibits extreme temperature fluctuations. These kinds of physical conditions can directly impact the way communication networks are designed by constraining the communication hardware that can be used.

4.3 Deep Space Network

The Deep Space Network (DSN) is a network of antennas used by the National Aeronautics and Space Administration (NASA) of the United States to track data and control navigation of interplanetary spacecrafts. It was designed to allow for continuous radio communication with spacecraft and has its origin in the Deep Space Instrumentation Facility (DSIF) constructed around 1963. NASA established the concept of DSIF as a separately managed and operated communications system that would accommodate all deep space missions, thereby avoiding the need for each flight project to acquire and operate its own specialized space communications network. In this context DSN has become the core space communication system deployed by NASA. It consists of antenna arrays allowing the spacecraft teams to control unmanned space probes in Earth's orbit or beyond as well as to exchange data with NASA's missions in space. It is formed by a set of three communication complexes with 16 huge and advanced antennas. These three antenna complexes, located in Goldstone - California - USA, Madrid - Spain, and Canberra - Australia, make up the DSN as shown in Figure 4.8. All three facilities are administrated by the Jet Propulsion Laboratory in Pasadena, with USA based antennas located at the Goldstone Deep Space Communications Complex (GDSCC) in the Mojave Desert. The three complexes are placed 120 degrees longitude apart to enable light-of-sight to be sustained between spacecraft and the DSN continuously. Each complex is equipped with one 34 meter diameter high efficiency antenna, one 34 meter beam waveguide antenna, one 26 meter antenna, one 70 meter antenna and one 11 meter antenna.

Using the DSN, JPL is able to access key information it needs to control the

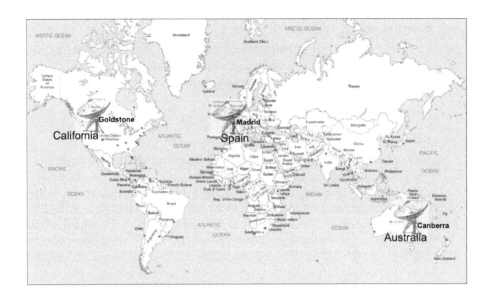

Figure 4.8: Deep space network facilities.

spacecraft and instruments that have been sent to space, making it possible to study planets and the solar system. In addition, other kinds of information about asteroids, meteoroids and comets are available.

The idea of expanding DSN in the context of IPN is not completely new since this expansion brings a lot of advantages. Mainly, it improves network communication and interoperability among space agencies. The interoperability among space agency enables cross-support of spacecraft such that agencies can gain the benefit of shared resources and infrastructure. In order to enable cross-support of network communication, the following must be established [128]:

■ Interoperable physical communications layers including spectrum allocation, modulation and channel coding and communication data link layer standards.

■ Interoperable network architectures and agreement on a set of common datagrams (e.g. IP, Space Packet, DTN Bundle) that will be cross-supported.

■ Agreement on a standard set of cross-support services at the data link, datagram, and possibly application layer (e.g file transfers and store-and-forward delivery services).

■ Agreement on the behaviors that can be expected by a user accessing an infrastructure relay node that is providing cross-support. Examples of these behaviors are route determination, traffic prioritization, and store and forward.

■ Agreement on the operational configuration and management of shared infrastructure providing cross-support, including router/network addressing, naming and name resolution, route path determination heuristics, link establishment and scheduling, end-to-end status, and data accountability.

Interoperability among spacecraft enables researchers to deal with the interconnection of multiple worlds populated by rovers or robots, space vehicles, or even humans exchanging data among themselves over a cross-supportable internetworking architecture called Interplanetary Internet or Interplanetary Network (IPN).

4.4 Interplanetary Network

One of the problems with space communication was the limited use of standards. Typically, new communication software was written for each new mission. In the Internet, standards such as the TCP/IP protocol suite packet switching and store-and-forward methods enable devices from multiple manufacturers to interact with each other. The idea of IPN is to develop a suite of protocols that will allow us to have the kind of network flexibility in space that we have on Earth.

In 2003 the US Defense Advanced Research Projects Agency (DARPA) became the main sponsor of work on the technical architecture of an Interplanetary Internet. The principal idea was to extend the ongoing work on space communication standardization with new standards for automated Internet-compatible network operations. The term "Interplanetary Internet" is meant to connote the integration of space and terrestrial communication infrastructures to support information flow throughout the Solar System. To this end, the Interplanetary Internet is structured as a "network of internets" where:

- the traditional internet protocols can be used in planetary surface environments characterized by low delay and relatively low noise;

- an interplanetary backbone of long-haul wireless links can be used to interconnect the local internets;

- a new overlay network protocol named "bundle protocol" operates over both the planetary networks and the backbone, enabling end-to-end data flow.

Unlike the terrestrial backbone, which is characterized by continuous connectivity and negligible delay, the interplanetary backbone links are characterized by intermittent connectivity, high propagation delays and noisy data channels. The intermittent connectivity is due to the fact that the hubs on the backbone are rotating, moving with respect to each other, sometimes occulted, and not always configured for data transmission and/or reception. The planets move in fixed orbits and sometimes bodies like the Sun cause line of sight obstruction that lasts for a long period of time. In addition, landed vehicles on remote planetary surfaces will move out of the sight of Earth as the body rotates and they may have to communicate through local relay satellites that only provide data transmission contacts for a short period of time. In this case the parties cannot always communicate with others, the communication is both delayed and potentially disrupted.

The high propagation delays are directly related to the astronomical distances. The distance signals need to travel is measured in light minutes or hours; interactive protocols do not work. On Earth we are separated from one another by no more than a fraction of a light second. The delay or latency between request for and receipt of information (i.e. a telephone call) is usually not noticeable. In contrast, the distance between the planets is very large. For instance, when Earth and Mars are closest, it takes about 4 minutes for a radio signal to travel at the speed of light to propagate from one to the other. If one device "A" on Mars sends a message to another device "B" on Earth, a minimum of 8 minutes will pass before "A" hears a response. When Earth and Mars are farthest apart, the round trip takes about 40 minutes.

The traditional Internet has been wildly successful. However, it only works well whe the assumptions on which it was based are valid.

The core protocols of the Internet (termed "TCP/IP") are the Internet Protocol, which establishes the paths by which packets of data are sent from location to location in the network, and the Transport Control Protocol, which guarantees that the data gets to its destination, providing ways for errors to be detected and packets to be resent when necessary. The key assumptions underlying the design of these protocols are:

1. End-to-end connectivity always exists between devices attempting to communicate with each other. When the connection breaks it is necessary to establish the entire communication session again.

2. Round-trip times (RTT) between any two nodes of the network are brief, usually less than 1 second.

3. Bit error rates on transmission channels are generally very low.

4. Data rates on each link are generally symmetrical: messages from A to B are transmitted at about the same rate as messages from B to A.

For space operations, the IP suite certainly works in environments that are close to what it was designed for: richly connected, short-delay, bidirectional, always-on and chatty data communications. Unfortunately, many space communications environments display almost the inverse set of characteristics: sparsely connected, medium to long delay, often unidirectional and intermittently available.

The new delay-tolerant networking protocols ceratinly do not obsolete Internet protocol suite, which will continue to be used very heavily in the environments for which they were designed (in which almost all humans live). But for communication beyond the surface of Earth we need something more.

4.5 IPN Architecture

The basic IPN concept consists of a network of Internets connected through an interplanetary backbone (see Figure 4.9). A backbone is generally defined as a set of

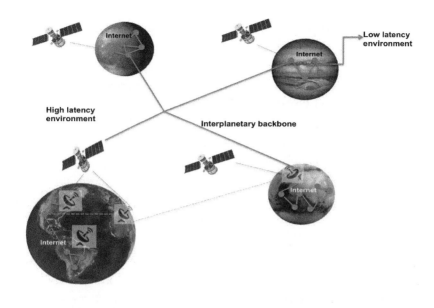

Figure 4.9: Basic IPN scenario.

high-capacity, high availability links between network traffic hubs that exchange Internet traffic between the countries, between continents, and across oceans. In the case of the terrestrial backbone links are between hubs like Rio de Janeiro and London. Interplanetary backbone links are between hubs like Earth and Mars.

The three main goals of the IPN project can be summarized below.

1. Deploy standard internets in low-latency remote environments, for example, on other planets and on remote spacecraft.

2. Connect these distributed internets via an interplanetary backbone that handles the high latency deep space environment.

3. Create gateways and relays to interface between low and high latency environments.

A general infrastructure identifying the architecture elements for space Internet is presented in [158]. The architecture itself was defined in terms of several types of communication subnetworks, called architectural elements. These elements were identified by NASA to support three of NASA's enterprises:

1. **Earth Science Enterprise (ESE)**: the purpose of NASA's Earth science program is to develop a scientific understanding of Earth's system and its response to natural or human-induced changes, and to improve prediction of climate, weather, and natural hazards. A principal block of NASA's Earth Science Division is a coordinated series of satellite and airborne missions for long-term

global observations of the land surface, biosphere, solid Earth, atmosphere, and oceans.

2. **Human Exploration and Development in Space (HEDS)**: the purpose of the HEDS program is to open the space frontier by exploring, using, and enabling the development of space, and to expand the human experience into the far reaches of space. HEDS's main goals can be summarized as following.

   ■ Prepare to conduct human missions of exploration to planetary and other bodies in the solar system;

   ■ Use the environment of space to expand scientific knowledge;

   ■ Provide safe and affordable human access to space, establish a human presence in space, and share the human experience of being in space;

   ■ Enable the commercial development of space and share HEDS knowledge, technologies, and assets that promise to enhance the quality of life on Earth.

3. **Space Science Enterprise (SSE)**: the SSE addresses advances in cosmology, planetary research, and solar terrestrial science which form a baseline for the enterprise. Furthermore, SSE is looking for answers to some of the fundamental questions that science can ask: how the universe began and is changing, what are the past and future of humanity, and whether we are alone. The four goals of the SSE can be summarized as following.

   ■ Establish a virtual presence throughout the solar system, and probe deeper into the exploration of the universe and life on Earth and beyond;

   ■ Pursue space science programs that enable and are enabled by future human exploration beyond low-Earth orbit;

   ■ Develop and utilize revolutionary technologies for missions;

   ■ Contribute measurably to achieving the science, mathematics, and technology education goals of Nation, and share the findings of the missions and discoveries.

The architecture elements are listed below. They are the main building blocks of the three NASA's enterprises based on the Internet and its technology. In this direction this general infrastructure provides an horizontal structure capable of holding flexibility and interoperability standards.

■ Backbone network: the backbone consists of ground network, space network, NASA's Intranet and virtual private networks, the Internet, and any foreign communication and commercial system that may be employed.

■ Access Networks: the local area network onboard a spacecraft or other vehicle and the wireless/optical interfaces between the backbone and the mission vehicles.

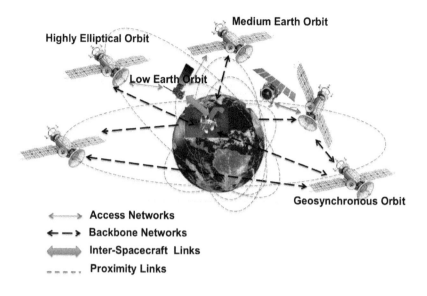

Highly Elliptical Orbit

Medium Earth Orbit

Low Earth Orbit

Geosynchronous Orbit

```
←——→  Access Networks
←— —→  Backbone Networks
◁══▷   Inter-Spacecraft Links
- - - -  Proximity Links
```

Figure 4.10: IPN Architectural Elements.

- Inter-spacecraft network: the network of spacecraft flying in a constellation or cluster.

- Proximity Network: an ad hoc network consisting of space vehicles (rovers, airplanes, aerobots), landers and sensor nodes.

Figure 4.10 displays a simple scenario that includes these architectural elements. The figure shows an Earth vicinity communication infrastructure for observation and exploration missions. This architecture is expected to provide high availability and high speed communication between Earth and Mars base station and other Mars vehicles.

A common architecture for interplanetary networking was proposed in 2003 [24]. This proposed architecture was based on the architectural elements presented before. The main idea is to build the space Internet on top of Internet technologies enabling any space mission to have high quality of service and reduced cost. The Interplanetary Internet architecture is presented in Figure 4.11. It includes the interplanetary backbone network, the interplanetary external networks and the planetary networks.

The interplanetary backbone network provides a common infrastructure for communication among the Earth, moons, satellite, outer-space planets and relay stations at gravitationally Lagrangian points[2] of planets. It includes the data link between elements with long haul capabilities.

---

[2]A Lagrange point is a location in space where the combined gravitational forces of two large bodies, such as Earth and the sun or Earth and the moon, equal the centrifugal force felt by a much smaller third body [141].

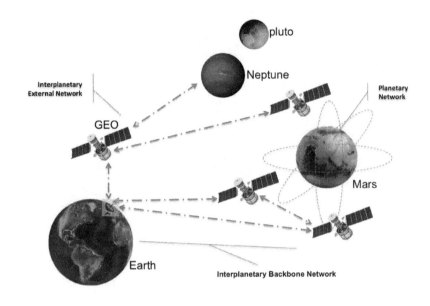

Figure 4.11: IPN Architecture by [24].

The interplanetary external network (the same as that Inter-spacecraft network) refers for example to a group of spacecraft flying between planets or a set of sensor nodes, etc.

The planetary network can be deployed in outer planets allowing interconnection and cooperation among satellites and elements on a planet surface. It consists of a planetary satellite network and a planetary surface network as shown in Figure 4.12. The planetary satellite network includes the links between orbiting satellites and links between satellites and surface elements. The satellites are positioned in layers of orbital altitude enabling intermediate storage and relay service between Earth and the planet, relay service between the in situ mission elements, and location management of planetary surface networks. The planetary surface network comprises the communication links between high power surface elements (i.e. rovers and landers) generally organized in clusters that are able to connect with satellites as well as the ones that are not able to directly communicate with satellites.

4.5.1 Solar System Internetwork Architecture

This section presents a high level architecture of the Solar System Internetwork (SSI) architecture, for comparison with the previously presented architectures. The SSI architecture has been described in detail in the CCSDS Space Internetworking Services-Delay-Tolerant Networking (SIS-DTN) working group [128]. Basically, the SSI is a single network system designed to enable communication in the exploration of space.

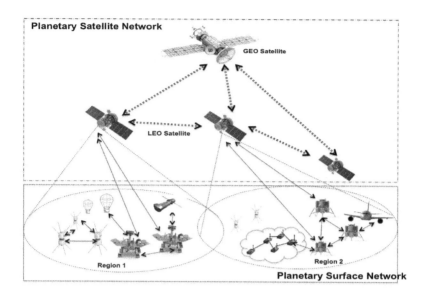

Figure 4.12: The planetary network architecture.

In November 2008 the Space Internetworking Strategy Group (SISG) released a report that includes a recommendation that the space flight community adopt DTN to address the problem of interoperability and communication scaling, especially in mission environment where exist multiple spacecraft operating in an orchestration model [130]. More specifically, SIS provides the participants in space mission the capability of exchanging information in a standard manner, including:

■ crewed and robotics space vehicles;

■ ground antenna stations;

■ centralized ground-based mission operation centers on Earth;

■ science investigators at laboratories on Earth.

Figure 4.13 shows the SSI network system elements. More specifically, this figure indicates which protocols are able to run on top of which others. The DTN applications are implemented to use CFDP and other DTN application layer services over a BP network. The BP network runs over LTP in space and also over TCP/IP. Note that the SSI architecture comprises two major blocks: Internet and DTN. The core Internet protocols include:

■ TCP: this protocol ensures reliability on the end-to-end transmission and provides congestion control;

■ IP: this protocol allows to forward datagrams from source nodes to destination nodes using routing protocols;

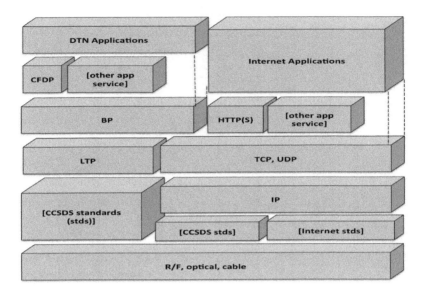

Figure 4.13: SSI protocol stack.

■ HTTP: the underlying protocol used by the World Wide Web to define how messages are formatted and transmitted;

■ FTP: is a standard network protocol used to transfer files between Internet nodes;

■ IPSec: the Internet security protocol for secure exchange of packets.

The reference architecture for space communications is defined as a framework that must be used by the space agencies when presenting space communications systems and space communications scenarios. This architecture defines four views (view Figure 4.14) as follows.

1. **Physical View** is used to visualize the physical configuration of space communications systems and scenarios and its physical characteristics. This view is used to show the physical elements used in space communication, both their physical characteristics and also their topology and connectivity. Orbiting elements (such as Earth orbiter, Lunar orbiter, planet orbiter), Landed elements (such as Lunar lander/rover, planet lander/rover) and Ground elements (such as Ground station, control center, science facility) are some examples of physical elements.

2. **Service View** is used to describe services and their functional characteristics, often associated with physical elements. In particular, it includes the services provided and used by space communication systems, the functional characteristics of services, the performance characteristics of services, methods and/or

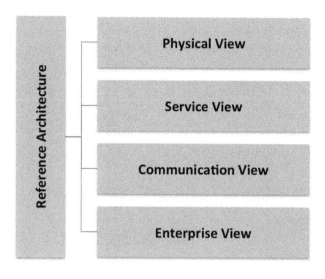

Figure 4.14: The four views defined by the reference architecture.

standards for using services, and methods and/or standards for managing services. Relay services, routing services, store and forward services (frames, packets, files, ...), positioning and timing services (orbit determination, clock synchronization, ...), and management services are examples of services.

3. **Communication View** is used to describe the protocols, protocol parameter values, and modulation/coding methods used in communication between physical elements. Examples of communication protocols include modulation methods (Phase Modulation (PM) , Quadrature Phase Shift Keying (QPSK), Binary Phase Shift Keying (BPSK), ...), coding methods (Bose-Chaudhuri-Hocquenghem (BCH), convolutional, Reed-Solomon (RS), Turbo, ..), data link protocols (TeleCommand (TC), TeleMetry (TM), Advanced Orbiting Systems (AOS), Proximity-1, ...), network protocols (space packet, IP, ...), transport protocols (TCP, UDP, Space Communications Protocol Specifications (SCPS)-TP, ...) and application protocols (CCSDS File Delivery Protocol (CFDP), Asynchronous Message Service (AMS), ...).

4. **Enterprise View** is used to describe the organizational structure of space communication systems and scenarios and their administrative characteristics. It addresses the organizations involved in space communication systems and scenarios, the physical and administrative interfaces between organizations, and the documents exchanged between the organizations. Space agencies, commercial service providers and science institutes are some examples.

The simple communication model used by space flight missions is characterized

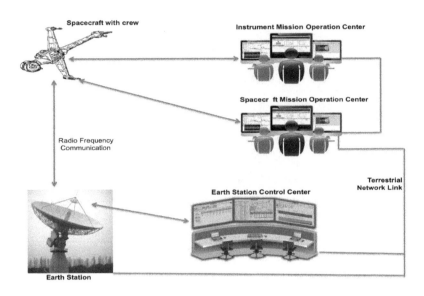

Figure 4.15: Simple mission communication model.

as depicted in Figure 4.15 [128]. The idea is to evolve to a more complex operation scenario that includes a more intricate communication model.

This simple communication model (Figure 4.15) represents a mission and its space communication services, managed by a single space agency. The mission is responsible for the provisioning and utilization of the space link. It also operates a single spacecraft which communicates with a single Mission Operation Center (MOC). The spacecraft contains a human crew or/and investigative instruments. The downlink data from the spacecraft to the MOC take the form of telemetry packets that are encapsulated in telemetry frames. The uplink data from the MOC to the spacecraft take the form of mission specific commands (packets) that are encapsulated in telecommand frames. The spacecraft is able to communicate with one or more Earth stations, which forward telemetry and telecommand frames between Earth station and MOC during the intercontact time. In this case, all the Earth station resources are devoted to the current mission. The data are immediately delivered to their destination. The initiation and termination of contacts between the spacecraft and the Earth station, as well as the selection of data to be sent during the intercontact time, are managed through spacecraft command and staff operations on the ground. Data retransmission is initiated by mission operator commands.

A more complex operation scenario is shown in Figure 4.16 where interoperability is one of the main features. In this communication model missions are able to operate with different space agencies and MOCs. Space communications Earth station services may be provided by multiple space agencies. Missions can comprise several spacecrafts which may collaborate on mission objectives. Data can be for-

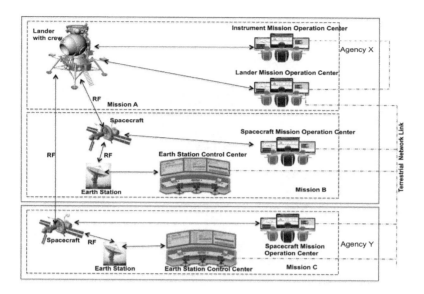

Figure 4.16: A more intricate mission communication model.

warded via different spacecraft independent of the mission. The data exchanged may be more complex, for instance, streaming of video.

The SSI architecture specification defines three stages (1. Mission functionality; 2. Internetworking functionality; 3. Advanced functionality) to support the organizations in the deployment of the SSI as shown in Figure 4.17. Note that these stages address some functionalities that must be implemented by the organization participating in the SSI. The organization can operate in any of these stages, e.g., they do not need to operate at stage 1 before transitioning to stage 2. However they must implement the functionalities required in the previous stage in order to operate in the next one.

- **Stage 1 - Mission Functionality**: the SSI protocols are implemented in the MOC to automate mission data communication for each space flight mission. At this stage the Earth station does not implement the SSI protocol. The automation of the basic communication process includes:

  (a) starting and ending a transmission;

  (b) data selection for transmission;

  (c) segmentation and reassembly of data for transmission;

  (d) data forwarding by one entity pre-selected by management.

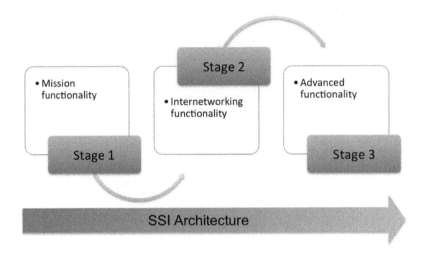

Figure 4.17: The SSI architecture transition stages.

Stage 1's Principles

(a) The coordination and provision of the mission data communication is internal to the mission.

(b) Interagency SSI cross-support may be provided, but the agreements governing it are ad hoc and privately negotiated.

■ **Stage 2 - Internetwork Functionality**: The Earth station implements the SSI protocols to enable network-layer cross-support. As a result multiple space flight missions and agencies are able to interoperate.

Stage 2's Principles

(a) A provider node (i.e., a SSI node whose network protocol entity is configured to forward network protocol data units (PDUs) received from other entities) may support a mission that is managed by a different authority.

(b) Provider nodes in the solar system internet may be operated by multiple authorities.

(c) In this stage the coordination of mission data communications is not automated.

(d) The coordination function must guarantee successful negotiations among member authorities.

■ **Stage 3 - Advanced Functionality**: the coordination of mission data communication is automated enabling the emergence of a unified solar system wide communication network.

Stage 3's Principles

(a) The coordination of mission data communications is an automated process.

4.6 Summary

This chapter presented a case study that describes IPN as a kind of DTN to support deep space exploration. The end-to-end architecture development for the Interplanetary Internet supported by the Consultative Committee for Space Data Systems (CCSDS) was presented [74]. In this architectural model, the routing handle identifies an "IPN Region," as an area of the Interplanetary Internet in which the administrative name is resolvable, and in which a route can be formed from anywhere within the region to the address returned when the administrative name is resolved.

Exploration of space will eventually lead to the need for communication among planets, satellites, asteroids, robotic spacecraft and crewed vehicles. To this end the development of a stable interplanetary backbone network is fundamentally important. The previous chapters show that simply extending the Internet suite to operate end-to-end over interplanetary distances is not feasible. Rather, the necessary vision is of a "network of disconnected internets" where internets are interconnected by a system of gateways that cooperate to form the stable backbone across interplanetary space. Each internet's protocols are terminated at its local gateway, which then uses a specialized long-haul transport protocol to communicate with peer gateways. The end-to-end bundle protocol can operate above the transport layer to carry necessary information from one internet to another.

Many elements of the current terrestrial Internet suite of protocols as have been shown throughout this chapter are expected to be useful in low-delay space environments, such as local operations on and around other planets or within free flying space vehicles. However, the speed-of-light delays, intermittent and unidirectional connectivity, and error-rates characteristic of deep-space communication make their use infeasible across deep-space distances. All of these effects combine to make the design of an interplanetary backbone communication system a considerable challenge that has been addressed by the CCSDS community.

Chapter 5

Routing in Delay and Disruption Tolerant Networks

Aloizio P. Silva

University of Bristol

Scott Burleigh

JPL-NASA/Caltech

Katia Obraczka

University of California Santa Cruz

CONTENTS

5.1 Introduction

Routing, which is one the core functionalities provided by computer networks, is the process of finding a path between the data source and its destination(s). Routing usually builds "minimum cost" paths between a pair of nodes where "cost" can be a function of distance (e.g., in number oh network hops), latency, bandwidth, remaining battery level at intermediate nodes, and a combination thereof. Data forwarding, another core network functionality, uses paths determined by routing to forward data from sources to destinations. Routing and forwarding are typically performed by specialized network appliances called *routers*.

This chapter discusses routing in DTNs and its challenges. Since DTN scenarios and appications can vary widely in terms of their characteristics and requirements, no single solution that fits all DTN scenarios has yet been identified. This chapter also reviews existing DTN routing approaches and protocols. We start with a brief overview of routing in the Internet.

5.2 Internet Routing

In the TCP/IP Internet, routers participating in the routing process build *routing tables* which, for a given destination address, determine the router's outgoing interface. When a router receives a data packet, it looks up the packet's destination address in its routing table to determine how it will forward the packet to its next hop. This process is called *store-and-forward* and is typically carried out at the network layer.

To address scalability and administrative decentralization, Internet routing is based on a two-level hierarchy, where routers under the same administative authority are organized as an independent *Autonomous System, or AS* as illustrated in Figure 5.1, which shows three ASes, namely AS1, AS2, and AS3. Each AS establishes its own internal routing policies and selects its internal routing protocol independently from other ASes. The AS's *Internal Routers, or IRs* (shown in blue in Figure 5.1) are responsible for routing within the AS, also known as intra-AS routing which is carried out using interior gateway protocols (IGPs). Some routers within an AS are designated as *Border Routers, or BRs* (shown in red in Figure 5.1) and are responsible for inter-AS routing which are carried out using exterior gateway protocols (EGPs). As illustrated in Figure 5.1, communication between AS1 and AS2 goes over the green link connecting BR1-AS1 and BR1-AS2.

In the Internet, it is assumed that there is always an end-to-end path between the data source and its destination. However, the arbitrarily long delays and frequent connectivity disruptions that set DTNs apart from traditional networks imply that

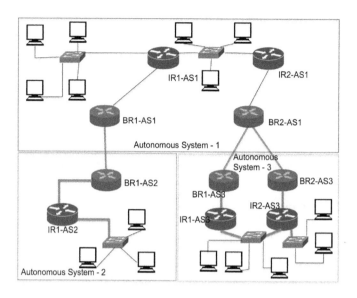

Figure 5.1: TCP/IP Internet Routing Architecture
IR: Internal Router   BR: Border Router   AS: Autonomous System

there is no guarantee an end-to-end path between a given pair of nodes exists at a given point in time. Instead, nodes may connect and disconnect from the network over time due to a variety of factors such as mobility, wireless channel impairments, nodes being turned off to save energy or running out of power, etc. Consequently, in DTNs, the set of links connecting DTN nodes, also known as "contacts", varies over time. This fundamental difference between DTNs and conventional networks like the Internet results is a major paradigm shift in the design of core networking functions such as routing and forwarding.

5.3 Routing in DTN

Since DTNs operate in environments where episodic connectivity and long delays are the typical rather than the exception, DTN routing must be able to yield adequate performance under such "extreme" conditions. To this end, a new routing paradigm for DTNs called *store-carry-and-forward* has been proposed as an extension to the Internet's *store-and-forward* approach. In *store-carry-and-forward*, when a DTN node receives data which needs to be forwarded to its ultimate destination, the node decides whether to forward the message to one of its directly connected neighbors (assuming the node is not partitioned from the network), or store the message and "carry" it until it has an "opportunity" to forward it to another node or deliver it to the ultimate destination.

Note that in DTNs, the concept of *links* used from traditional networks (wired or wireless) is replaced with the notion of *contacts*, which are established when a DTN node "encounters" another DTN node and a communication link is established between the two. Consequently, as DTN topology is constantly changing, the set of links (or contacts) connecting DTN nodes varies over time. Contacts in DTN can be classified as: *scheduled*, *probabilistic*, and *random*.

Scheduled contacts happen in DTNs where node encounters follow a schedule that is known a-priori. For instance, in deep space communication scenarios, where the orbits of planets and space vehicles is well defined, their encounters are deterministic events.

Probabilistic encounters follow some statistical distribution that determines the probability that a node will encounter another node. These probability distributions can be derived from past encounter history.

Random contacts characterize DTNs where there is no a-priori knowledge of when nodes encounter each other and node encounters do not necessarily follow any pattern.

Since end-to-end connectivity in DTNs is not guaranteed, routing decisions are made on a hop-by-hop basis based on a variety of factors that can be classified as destination-dependent and destination-independent [145] [262]. Examples of destination-dependent considerations include the frequency at which a node encounters the message's destination, when the node last encountered the destination, the node's "social networks" which may detemirne how likely it is that the node will encounter the destination, etc. Destination-independent factors include the node's level of mobility (i.e., how much or how far away the node moves), the node's capabilities (e.g., battery life, buffer space, etc.), and the node's trustworthyness). Note that, since contact duration are finite and may be arbitrarily short, a node may need to choose which messages to forward based on some priority.

Existing DTN routing approaches focus on trying to maximize end-to-end data delivery under episodic end-to-end connectivity. As such, DTN routing can be classified in two major categories, namely:

Replication-Based Routing protocols generate a number of message copies, or replicas that are forwarded at each encounter or contact opportunity. Each relay node then maintains the message's replicas in its buffer until a new contact opportunity arises or the message expires. Even though routing protocols that fit in this category do not require network topology information to operate, their performance can benefit from destination-dependent or independent knowledge.

Forwarding-Based Routing approaches do not replicate the original message and thus are more efficient in terms of memory- and network resources. However, compared to replication-based approaches, access to destination-dependent and/or independent knowledge of the network is more critical.

5.3.1 Replication–Based Routing

As previously discussed, DTN routing protocols that fit in this category are based on message replication techniques. The main goal is to reduce delivery delay and increase delivery ratio. They can be further divided into two groups as follows.

1. Unlimited Replication: routing protocols that belong to this group authorize each network node to create copies of messages and send them through the network to reach their destination. In the worst case a message can be replicated $n - 1$ times before reaching the destination, where n is the number of network nodes.

2. Quota-Based Replication: routing protocols in this group use a *quota* to limit the number of message copies that can be created. Typically, *quota* is a pre-specified parameter of the protocol.

5.3.1.1 Unlimited Replication

■ **Epidemic:** in this routing protocol each node replicates the message to every other node it encounters [278]. Thus all nodes eventually receive all messages and the destination node is guaranteed to receive the message, provided it comes in contact with at least one other node that has received a copy of the bundle, prior to expiration of the bundle's time to live. Epidemic routing works as following.

1. When two nodes encounter they exchange lists of the IDs of the bundles stored in their buffers (termed "summary vectors").
2. Checking the summary vector they exchange the messages they do not have.
3. At the end, both nodes have the same messages in their buffers.

The epidemic routing approach is simple. However it consumes a huge amount of resources due to the great number of message copies inserted into the network. As a result of consuming a large amount of buffer space, large amounts of bandwidth and power are required. Another issue with epidemic is that it continues to disseminate message copies through the network even after the destination has received it. To mitigate this problem the concept of *death certificates* has been proposed. The idea of a *death certificate* is to propagate a notification that informs the network nodes to delete the message that has been delivered. Generally, the notification size is smaller than the original message, resulting in less resource consumption.

In the literature, there are several variations of epidemic protocol that try to be more efficient than the original version [63][51][258][95][181][270][134].

■ **PRoPHET:** Probabilistic Routing Protocol using History of Encounters and

Transitivity (PRoPHET) [181]. The PRoPHET protocol exploits the non-randomness in node mobility in DTN by maintaining a set of probabilities for delivery and replicating messages during encounters only when the newly discovered neighbor node has a better chance of delivering the bundle than the node that currently holds a copy. Delivery probability is computed for each prospective relay node. The relay node 'N' stores delivery probabilities for each known destination 'D' given by $p(N,D)$. If the relay node has not stored a predictability value for a destination $p(N,D)$ is assumed to be zero. The delivery probabilities used by each relay node are recalculated at each encounter according to the following rules.

1. When the node 'N' encounters another node 'B', the probability for B is increased according to $p(N,B)_{new} = p(N,B)_{old} + (1 - p(N,B)_{old}) * X_{encouters}$ where 'X' encounters is an initialization constant.

2. The probabilities for all destinations 'D' other than 'B' are *aged* by $p(N,D)_{new} = p(N,D)_{old} * Y^k$ where 'Y' is the aging constant and 'K' is the number of time units that has elapsed since the last aging.

3. Probabilities are exchanged between 'N' and 'B' and the transitive property of probability is used to update the probability of destinations 'D' for which B has a $p(B,D)$ value on the assumption that 'N' is likely to meet 'B' again by $p(B,D)_{new} = p(B,D)_{old} + (1 - p(B,D)_{old}) * p(N,B) * p(B,D) * \beta$ where β is a scaling constant.

■ **RAPID:** Resource Allocation Protocol for Intentional DTN (RAPID) routing. RAPID routing protocol considers DTN routing as a resource allocation problem. Routing metrics (i.e. average delay, missed deadlines, maximum delay) are translated into per packet utilities determining how packets should be replicated in the system. The main idea of RAPID is based on the concept of a utility function. The utility function assigns a utility value according to the routing metric to be optimized. Thus the messages that have the highest utility value are replicated first.

The RAPID routing process works as follows. The utility function assigns a utility value U_i to every message 'i'. The value U_i is defined as the expected contribution of message 'i' to the routing metric. Then RAPID replicates the messages that result in the highest increase in utility. For example, assume the metric to optimize is average delay. The utility function defined for average delay is $U_i = -D(i)$, the additive inverse of the average delay. In this case, the protocol replicates the message that results in the greatest decrease in delay. The main steps of RAPID routing protocol can be summarized as follows.

1. Initialization: Information are exchanged to help estimate message utilities.

2. Direct Delivery: Messages destined for immediate neighbors are directly transmitted.

3. Replication: Messages are replicated based on marginal utility (the change in utility over the message's size).

4. Termination: The protocol ends when contacts break or all messages have been replicated.

■ **MaxProp:** also known as Maximum Priority [51]. In MaxProp routing, when an encounter between two nodes occurs, all messages held during the encounter will be replicated and transferred. Each node sets an encounter's probability to all the other network nodes and also exchanges this information with its neighbors. The probability value is used to calculate a destination path cost. To obtain the estimated path likelihood, each node maintains a vector of size $n-1$ (where 'n' is the number of nodes in the network) consisting of the likelihood the node has of encountering each of the other nodes in the network. Each of the $n-1$ elements in the vector is initially set to $\frac{1}{|n|-1}$, meaning the node is equally likely to meet any other node next. When the node meets another node 'j', the jth element of its vector is incremented by 1, and then the entire vector is normalized such that the sum of all entries add to 1. Observe that this phase is local and does not require transmitting routing information between nodes.

When two nodes meet, they first exchange their estimated likelihood vectors. Ideally, every node will have an up-to-date vector from every other node. With these vectors the node can then compute a shortest path via a depth-first search where path weights indicate the probability that the link does not occur. These path weights are summed to determine the total path cost, and this cost is computed over all possible paths to the destinations for all messages currently being held. The path with the least total weight is chosen as the cost for that particular destination. The messages are then ordered by destination costs and are transmitted and dropped in that order. Messages ranked with highest priority are the first to be transferred during the encounter. Messages ranked with lowest priority are the first to be removed. The encountered nodes are notified via broadcast acknowledgment by MaxProp protocol to clear out the existing copies of the messages already delivered.

■ **Bubble Rap:** taking into account that human interaction is dynamic in terms of hubs and communities, Bubble Rap combines the concept of community structure with node centrality to make routing decisions [144]. Centrality can be defined as a measurement of the structural importance to identify the key node to bridge a message in the network. There are two principal principles considered by Bubble Rap routing protocol:

1. The roles and popularities that people have in society are replicated in networks. In this case, the Bubble Rap strategy first forwards messages to nodes which are more popular than the current node.

2. People are always forming communities in their social lives. The same behavior it is observed in the network layers. In this case, the Bubble Rap

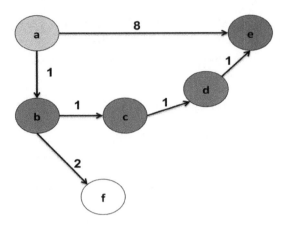

Figure 5.2: Direct Contact routing example

looks for the members of destination nodes' communities and uses them as relay nodes.

The Bubble Rap routing process can described as follows. If a source node has a message to send to another node, this source node first bubbles the message up the hierarchical ranking tree using the global ranking until it reaches a node which is in the same community (Bubble) as the destination node. Then the local ranking system is used instead of the global ranking, and the message continues to bubble up through the local ranking tree until the destination is reached or the message expires. Each node does not need to know the ranking of all other nodes in the system, but it should be able to compare ranking with the node encountered and to push the message using a greedy approach. In order to reduce cost, whenever a message is delivered to the community, the original source node can delete the message from its buffer to prevent further dissemination. This assumes that the community member can deliver the message.

5.3.1.2 *Quota-Based Replication*

■ **Direct Contact:** this is a very simple routing protocol that does not consume a lot of network resources. The sender node only deliver the message directly to the destination without any intermediation. In this case the message transfer process only uses one hop [260]. Figure 5.2 shows one example of direct contact routing. In this example, assume that node 'a' (blue node) needs to delivery a message to node 'e' (green node). In this case, node 'a' can only delivery message to 'b' and 'e'. Note that while faster to deliver a message to 'e' via 'b', it cannot deliver to 'c' and 'd' (red nodes).

■ **Two-Hop Relay:** in this approach the message will be delivered at its destina-

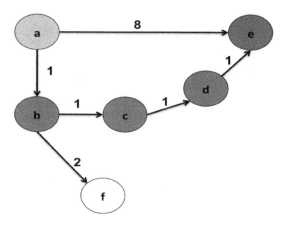

Figure 5.3: Two-Hop Relay routing example

tion after at most two hops; after it is forwarded to a relay node, it will not be forwarded from the relay node to another node unless that node is the destination node. Basically, the source node forwards copies of the message to every node it encounters, but each relay node holds the message until it can forward it to the destination node (or the message's time to live expires). Figure 5.3 shows an example of two-hop relay routing. In this example, node 'a' (blue node) has a message to be delivered to node 'f' (green node). Node 'a' sends copies to nodes 'a', 'c' and 'e'. When node 'b' (red node) encounters node 'c' (red node) it will not send it the message since node 'c' is not the destination. Likewise, when node 'c' encounters node 'd' (red node) it will not send the message since node 'd' is not the destination. The message will be delivered to node 'f' (green node) when node 'c' encounters it. Note that nodes 'a', 'b', 'c' and 'd' have received each one a copy of the message. And also node 'a' is able to reach all other nodes with two-hops. However, bundles from node 'a' cannot reach nodes 'h' and 'i' since those nodes are three hops from node 'a'.

■ **Tree-Based Flooding:** in the tree-based flooding routing protocol the source node shares the responsibility of generating message copies with other nodes [259]. When a message is sent to a relay node, a constraint about the number of copies that the relay node can generate is defined. The set of relay nodes forms a tree which has as its root the source node. There are different strategies for limiting the number of message copies. For instance, each node may be allowed to make unlimited copies, but the message only can travel 'n' hops from the source node [278]. As a result, this limits the tree's depth and establishes no limit to the tree's breadth. The tree-based approach enables delivery of messages to destinations that are located several hops away, unlike the Direct Contact and Two-Hop Relay approaches. Figure 5.4 shows an example of Tree-Based Flooding. In this example, node 'a' (blue node) wishes

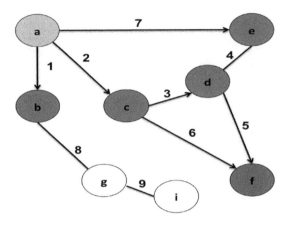

Figure 5.4: Tree-Based Flooding routing example

to send a message to node 'f' (green node) and it is allowed to make seven message copies. First, node 'a' sends a copy to node 'b' and notes that node 'b' can make one copy. Second, node 'a' sends a copy to node 'c' and notes that node 'c' can make two copies. Then node 'a' hold two copies. Third, node 'c' encounters node 'd' and one copy is delivered to node 'd'. Node 'c' still keeps one copy. Fourth, node 'd' encounters node 'e' but it cannot send a copy to node 'e' since it has no copy to forward. Fifth, node 'd' encounters node 'f' and delivers the message. Sixth, node 'c' encounters node 'f' and tries to deliver its last copy. Seventh, node 'a' encounters node 'e' since it does not know that the message has been delivered. It sends the copy to node 'e'. At the end, node 'a' still has one copy available for the next encounter.

■ **Spray and Wait:** this routing protocol defines a maximum number of copies each message can have in the network. Initially, each time the source node generates a new message, the routing protocol assigned a number 't' to the new message where 't' indicates the maximum message copies allowed in the network. The Spray and Wait (SaW) protocol is [263] divided into two phases, as follows:

1. Spray Phase: in this phase the source *sprays* one copy to the first 'N' relay nodes that it encounters. When these relay nodes receive the message copy they enter the *wait* phase.

2. Wait Phase: in this phase the relay node stores the message copy until it encounter the final destination (direct delivery).

The previous description is the basic version of SaW routing protocol where only the source node can spray the message copies. An operational overview of the basic SaW routing is shown in Figure 5.5. In this example node 'a' has

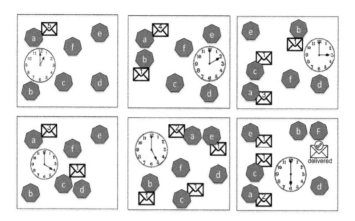

Figure 5.5: Basic SaW routing example

a message to send to node 'f' where the number of allowed message copies is '5'. The following steps take place in the example showed at Figure 5.5.

1. 1PM: Node 'a' generates a message with destination node 'f' and number of copies '5'

2. 2PM: Node 'a' encounters node 'b' and sends one copy to it (node 'a' has 4 copies remaining and relay node 'b' has 1 copy)

3. 3PM: Node 'a'' encounters node 'c' and sends one copy to it (node 'a' has 3 copies remaining and relay node 'c' has 1 copy)

4. 4PM: Node 'c' encounters node 'd'. However it cannot send the copy to it only to the destination (node 'c' has one copy remaining and relay node 'd' does not receive any copy)

5. 5PM: Node 'a' encounters node 'e' and sends one copy to it (node 'a' has 2 copies remaining and relay node 'e' has 1 copy)

6. 6PM: Node 'b' encounters node 'f' and sends the copy to it. The delivery is complete since node 'f' is the destination.

A variant of version of SaW, named Binary SaW, builds a binary tree which enables message copies to be equally sprayed by relay nodes rather than only allowing the source node to take the responsibility for the spraying. The source of a message initially starts with 'L' copies. Any node that has $n > 1$ message copies, upon encountering another node B with no copies, sends to B $\lfloor \frac{n}{2} \rfloor$ and keeps $\lceil \frac{n}{2} \rceil$ for itself; when it is left with only one copy, it switches to the direct transmission mechanism. Figure 5.6 shows an example of binary SaW described as follows.

1. 1PM: Node 'a' generates a message with destination node 'f' and number of copies '6'

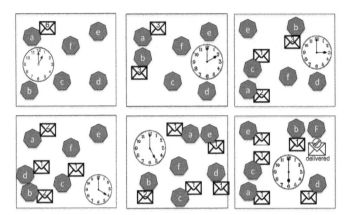

Figure 5.6: Binary SaW routing example

2. 2PM: Node 'a' encounters node 'b' and sends 3 ($\lfloor \frac{N}{2} \rfloor$) copies to it (node 'a' has 3 copies remaining and relay node 'b' has 3 copies)

3. 3PM: Node 'a'' encounters node 'c' and sends one copy to it (node 'a' has 2 copies remaining and relay node 'c' has 1 copy)

4. 4PM: Node 'b' encounters node 'd' and sends one copy to it (node 'b' has 2 copies remaining and relay node 'd' has one copy)

5. 5PM: Node 'a' encounters node 'e' and sends one copy to it (node 'a' has 1 copies remaining and relay node 'e' has 1 copy)

6. 6PM: Node 'b' encounters node 'f' and sends the copy to it. The delivery is complete since the node 'f' is the destination.

Another version of SaW called Spray and Focus (SaF) [261] replaces the "Wait" phase of SaW with a "Focus" phase. In SoF, a relay node switches to the Focus phase when it has only one replica of a message left. Unlike the Wait phase of SaW where the relay has to deliver its last copy of a message to the message's ultimate destination, in the Focus phase of SaF, a message can be forwarded to another relay on its way to the destination.

5.3.2   Forwarding-Based Routing Protocols

This category includes routing protocols that try to use network topology information to forward a single copy of the message through the "best" path. In this category no node will generate replicates of the messages. The following routing protocols belong to this category.

■ **Location-Based Routing:** this protocol assigns coordinates to each network node and then uses a physical distance function to estimate the cost of delivering messages among nodes. Basically, a node forwards a message to the next hop if that hop is physically closer to the coordinate space than the current custodian. The coordinates can be GPS coordinates. To select the best path to send the message a node needs three items of information: its own coordinates, the destination's coordinates, and the next hop's coordinates. In addition to requiring information about physical location (a non-trivial constraint), location-based routing is not easy to implement for two main reasons:

1. Node mobility: nodes may move, in which case their coordinates will change. As a result the network topology changes, complicating the routing process between origin-destination.

2. No guarantee of communication: two nodes can be close to each other but they maybe not be able to communicate due to some signal obstruction. In this case physical location does not always correspond to the network topology.

■ **Source Routing:** in source routing the path through the network is set by the source node, which dictates to the network the desired path. It is assumed that the source of the message knows about the network topology and can specify the best path for the message. The source routing process comprises two steps, route discovery and route maintenance. In route discovery, the source node sends packets toward the destination. Each intermediate node appends its address in the packet. When the packet arrives at the destination a complete route is recorded and the destination sends a message containing path information back to the source. In route maintenance the source node is notified if the discovered route has broken due to topology changes. As a result the source node can use another existing route or execute route discovery again.

■ **Per-Hop Routing:** in per-hop routing each intermediate node takes the responsibility of choosing the next hop to which the message will be forwarded for a specific destination.

■ **Per-Contact Routing:** in per-contact routing each intermediate node checks the current active contacts and selects the most appropriate hop for forwarding the message.

■ **Hierarchical Routing:** the hierarchical routing approach divides the DTN network into clusters taking into account each link's property and communication characteristics [142]. The routing process is based on the premise that the probability of communication inside of each cluster is higher than the probability of inter-cluster communication. Each cluster is managed by a Cluster Head (a particular node that belongs to the cluster and has the highest stability and quality). The cluster head store three sets of information:

1. Identifications of all nodes that belongs to that cluster
2. Information about other cluster heads

3. Index of the received messages

Basically, the routing decision is assigned to the cluster head. Each message generated inside of each cluster has a sequence number and a TTL (the message will be discarded when the TTL expires). When a node needs to send a message it first sends a request to the cluster head. Then the cluster head makes the decision on how to route that message, whether intra-cluster or inter-cluster.

■ **Nectar:** the routing approach adopted by Nectar is based on the concept of a neighborhood index table [98]. Each node has a neighborhood index table that stores the information about the frequency of the node's encounters with other nodes in the network. During each encounter, the nodes first transmit messages whose destination is the node with which contact has been established, then exchange information about the neighborhood (Neighborhood Index), and then forward other messages. The exchange of the Neighborhood Index provides more accurate knowledge of network topology. When a node needs to send a message it checks the table to identify the node with the highest encounter frequency and sends the message to it.

■ **DTLSR:** DTLSR is an abbreviation for Delay-Tolerant Link State Routing, an extension of link-state routing [100] that is implemented in DTN2. DTLSR considers that each node in the system is assigned to an administrative area, and a link state protocol instance operates only within a single area. As the network state changes, link state announcements are flooded throughout the network. Each node maintains a graph representing its current view of the state of the network, and uses a shortest path computation (i.e. Dijkstra algorithm) to find routes for the messages.

■ **Gradient Routing:** this protocol assigns a weight to each network node, indicating the node's ability to deliver messages to a specific destination [216]. When a node 'A' carrying a message encounters another node 'B' that is better able to deliver the message to the destination (has a higher gradient), node 'A' forwards the message to node 'B'. Otherwise, node 'A' keeps the message. The idea is to improve the utility function values (the delivery capability) toward the message's destination. This approach requires a level of knowledge of the network since each node must store delivery capability for all potential destinations and proper information must be propagated throughout the network to enable each node computes its delivery capability for all destinations. As the utility functions values can take a long time to propagate, the Gradient routing process can be slow to find a potential relay node.

■ **Contact Graph Routing:** this approach uses a *contact plan* provided by network management describing the current connectivity and future connectivity schedule. The precondition for CGR is that future contacts can be anticipated with very high confidence because they will be established by network management according to an operational schedule, as in space flight operations.

Contact Graph Routing (CGR) [33] makes forwarding decisions based on an earliest arrival time metric where messages/bundles are routed over the time-varying connectivity graph. CGR was designed to operate in space networks, but it can be used in terrestrial applications where the contacts are known in advance. The core strategy of CGR is to take advantage of the fact that, since communications are planned in detail, the communication routes between any pair of nodes that have been previously announced to all nodes can be inferred rather than discovered. CGR additionally uses probabilistic contacts (inferred from historical contact information) and neighbor discovery to address routing over non-scheduled links/encounters. The CGR protocol is being standardized by CCSDS. The next section describes CGR in greater detail.

5.4   Contact Graph Routing (CGR)

CGR was initially documented in 2009 and was updated in 2010 as an IETF Internet Draft [57]. CGR is a dynamic routing system that computes routes through a time-varying topology of scheduled communication contacts in a DTN network. It is designed to support operations in a space network based on DTN, but it also could be used in terrestrial applications where operation according to a predefined schedule is preferable to opportunistic communication, as in a low-power sensor network.

The basic strategy of CGR is to take advantage of the fact that, since communication operations are planned in detail, the communication routes between any pair of "bundle agents" in a population of nodes that have all been informed of one another's plans can be inferred from those plans rather than discovered via dialogue (which is impractical over long-one-way-light-time space links).

CGR has two fundamental components (contact plan and contact graph) which are described as follows.

Contact Plan: in the Internet, operations are driven by information that is promptly discovered on time, with low latency due to short distances and continuous end-to-end connectivity. In DTN operations cannot be accurately based on topological information reported in real time because the reporting latency may be very high: information may become inaccurate by the time it is received. As an alternative, protocol operations may be driven by information that is pre-placed at the network nodes and tagged with the date and times at which it becomes effective. This information may include "contact plans" that provide a schedule of planned changes in network topology. CGR depends on the propagation of contact plan messages to perform routing. The contact plan messages are of two types: contact messages and range messages.

The contact message includes the following information:

■ The starting Coordinated Universal Time (UTC) of the contact interval identified by the message.

■ The stop time of this interval, again in UTC.

■ The transmitting node number. (For the purposes of CGR, nodes are identified by numbers rather than text strings.)

■ The receiving node number.

■ The planned rate of transmission between the two nodes in bytes per second.

The range message includes the following information:

■ The starting UTC time of the interval identified by the message.

■ The stop time of this interval in UTC.

■ Node number A

■ Node number B

■ The estimated distance between nodes A and B over this interval, in light seconds.

Note that range messages may be used to declare that the "distance" in light seconds between nodes A and B is different in the $B \to A$ direction from the distance in the $A \leftarrow B$ direction. While direct radio communication between A and B will not be subject to such asymmetry, it's possible for connectivity established using other convergence-layer technologies to take different physical paths in different directions, with different signal propagation delays.

Contact Graph: Contact graphs are actually conceptual. One contact graph is virtually projected from the contact plan for each destination node to which the local node is required to forward bundles. The vertices of each such graph are the contacts that may be used to forward bundles to the destination node, while the edges of the graph are periods of time during which bundles may reside at intermediate nodes while awaiting the next.

5.4.1 Routing Tables

Each node uses Range and Contact messages in the contact plan to build a "routing table" data structure.

The routing table is constructed locally by each node in the network and it is a list of entry node lists, one route list for every other node D in the network that is cited in any Contact or Range in the contact plan. Entry node lists are computed as they are needed, and the maximum number of entry node lists resident at a given time is the number of nodes that are cited in any Contacts or Ranges in the contact plan. Each entry in the entry node list for node D is a list of the neighbors of local node X; included with each entry of the entry node list is a list of one or more routes to D through the indicated neighbor, termed a route list.

Each route in the route list for node D identifies a path to destination node D, from the local node, that begins with transmission to one of the local node's neighbors in the network– the initial receiving node for the route, termed the route's entry node. For any given route, the contact from the local node to the entry node constitutes the initial transmission segment of the end-to-end path to the destination node. Additionally noted in each route object are all of the other contacts that constitute the remaining segments of the route's end-to-end path.

Each route object also notes the forwarding cost for a bundle that is forwarded along this route. In the current version of ION, CGR is configured to deliver bundles as early as possible, so best-case final delivery time is used as the cost of a route. Other metrics might be substituted for final delivery time in other CGR implementations. Note, however, that if different metrics are used at different nodes along a bundle's end-to-end path it becomes impossible to prevent routing loops that can result in non-delivery of the data. Finally, each route object also notes the route's termination time, the time after which the route will become moot due to the termination of the earliest-ending contact in the route.

5.4.2 Key Concepts

Expiration Time Every bundle transmitted via DTN has a time-to-live (TTL), the length of time after which the bundle is subject to destruction if it has not yet been delivered to its destination. The expiration time of a bundle is computed as its creation time plus its TTL. When computing the next-hop destination for a bundle that the local bundle agent is required to forward, there is no point in selecting a route that can't get the bundle to its final destination prior to the bundle's expiration time.

OWLT Margin One-way light time (OWLT) – that is, distance – is obviously a factor in delivering a bundle to a node prior to a given time. OWLT can actually change during the time a bundle is en route, but route computation becomes intractably complex if we can't assume an OWLT "safety margin" – a maximum delta by which OWLT between any pair of nodes can change during the time a bundle is in transit between them.

We assume that the maximum rate of change in distance between any two nodes in the network is about $150,000$ miles per hour, which is about 40 miles per second. (This was the speed of the Helios spacecraft, the fastest man-made object launched to date.) At this speed, the distance between any two nodes that are initially separated by a distance of N light seconds will increase by a maximum of 80 miles per second of transit (in the event that they are moving in opposite directions). This will result in data arrival no later than roughly $(N+2Q)$ seconds after transmission – where the "OWLT margin" value Q is $(40 * N)$ divided by $186,000$ – rather than just N seconds after transmission as would be the case if the two nodes were stationary relative to each other. When computing the expected time of arrival of a transmitted bundle we simply use $N+2Q$, the most pessimistic case, as the anticipated total in-transit time.

Capacity The capacity of a contact is the product of its data transmission rate (in bytes per second) and its duration (stop time minus start time, in seconds).

Estimated Capacity Consumption (ECC) The size of a bundle is the sum of its payload size and its header size[1], but bundle size is not the only lien on the capacity of a contact. The total estimated volume consumption (EVC) for a bundle is the sum of the sizes of the bundle's payload and header and the estimated convergence-layer overhead. For a bundle whose header is of size M and whose payload is of size N, the estimated convergence-layer overhead is defined as 3% of $(M + N)$, or 100 bytes, whichever is larger.

Residual Capacity The residual capacity of a given contact between the local node and one of its neighbors, as computed for a given bundle, is the sum of the capacities of that contact and all prior scheduled contacts between the local node and that neighbor, less the sum of the ECCs of all bundles with priority equal to or higher than the priority of the subject bundle that are currently queued on the outduct for transmission to that neighbor.

Excluded Neighbors A neighboring node C that refuses custody of a bundle destined for some remote node D is termed an excluded neighbor for (that is, with respect to computing routes to) D. So long as C remains an excluded neighbor for D, no bundles destined for D will be forwarded to C – except that occasionally (once per lapse of the RTT between the local node and C) a custodial bundle destined for D will be forwarded to C as a "probe bundle". C ceases to be an excluded neighbor for D as soon as it accepts custody of a bundle destined for D.

Critical Bundles A Critical bundle is one that absolutely has got to reach its destination and, moreover, has got to reach that destination as soon as is physically possible[2].

For an ordinary non-Critical bundle, the CGR dynamic route computation algorithm uses the routing table to select a single neighboring node to forward the bundle through. It is possible, though, that due to some unforeseen delay the selected neighbor may prove to be a sub-optimal forwarder: the bundle might arrive later than it would have if another neighbor had been selected, or it might not even arrive at all.

For Critical bundles, the CGR dynamic route computation algorithm causes the bundle to be inserted into the outbound transmission queues for transmission to all neighboring nodes that can plausibly forward the bundle to its final destination. The bundle is therefore guaranteed to travel over the most successful route, as well as over all other plausible routes. Note that this may result in multiple copies of a Critical bundle arriving at the final destination.

---

[1]The minimum size of an ION bundle header is 26 bytes. Adding extension blocks (such as those that effect the Bundle Security Protocol) will increase this figure.

[2]In ION, all bundles are by default non-critical. The application can indicate that data should be sent in a Critical bundle by setting the BP_MINIMUM_LATENCY flag in the "extended class of service" parameter, but this feature is an ION extension that is not supported by other BP implementations at the time of this writing.

5.4.3 *Dynamic Route Selection Algorithm*

The computation of a new route in CGR happens as follows. Assume that a node has a bundle to send to a destination 'D' and a well-defined contact plan exists.

First, if no contact to node 'D' exists in the contact plan, CGR is not able to find a route for the bundle.

Second, if the contact plan has been recently updated in any way since routes were computed for any node, all routes for each destination node are discarded and route recomputation is required.

An empty list of proximate nodes (network neighbors) to send the bundle to is created.

A list of excluded nodes (i.e. nodes through which a route will not be computed for this bundle). The list of excluded nodes is initially populated with:

- the node from which the bundle was directly received (so that we avoid cycling the bundle between that node and the local node) – unless the Dynamic Route Selection Algorithm is being re-applied due to custody refusal as discussed later;

- all excluded neighbors for the bundle's final destination node.

If all routes need to be recomputed for node 'D' due to contact plan updates then following steps must be performed.

- An abstract contact graph for node 'D' is constructed. It is a directed acyclic graph whose root is a contact from the local node to itself and whose other vertices are all other contacts that can be part of some end-to-end path to node 'D'. A terminal vertex is also included in the graph, representing the contact from node 'D' to itself.

- Several executions of the Dijkstra shortest-path algorithm are performed over the constructed graph. For each execution, the lowest cost route that starts at the root and ends at the terminal vertex is found, the computed route is added to the node's list of routes, and the entry node for the computed route is removed from consideration in subsequent Dijkstra searches. As soon as no more routes can be found, the process ends. During each execution the following actions take place:

  ⋆ The lowest cost route is the one for which the bundle will arrive at the destination 'D' at the earliest possible time.

  ⋆ The earliest possible arrival time for the bundle on a given contact is computed as the sum of the bundle's earliest possible transmission time plus the range in light seconds from the contact's sending node to its receiving node, plus the applicable one-way light time margin (an interval that accounts for possible movement of the receiving node in the course of propagation of the bundle).

Now, if the list of routes to node 'D' is empty then CGR cannot be used to find a route for the current bundle. Otherwise the computed routes for transmission of bundles to node 'D' are examined as follows.

■ All contacts whose termination time is in the past are deleted, as are all routes that include the deleted contacts. When a route to 'D' through entry node 'B' is deleted, the Dijkstra search is executed again to find the best remaining route through 'B' (if any).

■ Any route for which the sum of delivery time and aggregate radiation time is after the bundle's TTL expiration time is ignored. Aggregate radiation time is the sum of the product of payload size and contact transmission rate over all contacts in the route.

■ For each remaining route, the earliest transmission time (or "forfeit time") for the bundle is computed as the initial contact's transmission rate multiplied by the sum of the sizes of all bundles already queued for that route's entry node whose priority is greater than or equal to that of this bundle. The bundle's projected bundle delivery time over each route is then computed based on the route's earliest transmission time, and the route providing the earliest projected delivery time for the bundle is selected; the bundle is queued for transmission to the entry node of this route.

If, at the end of this procedure, the proximate nodes list is empty, then we have been unable to use CGR to find a route for this bundle; CGR route selection is abandoned. Otherwise:

■ If the bundle is flagged as a critical bundle, then a cloned copy of this bundle is enqueued for transmission to every node in the proximate nodes list.

■ Otherwise, the bundle is enqueued for transmission on the outduct to the most preferred neighbor in the proximate nodes list:

  ⋆ If one of the nodes in this list is associated with a best-case delivery time that is earlier than that of all other nodes in the list, then it is the most preferred neighbor.

  ⋆ Otherwise, if one of the nodes with the earliest best-case delivery time is associated with a smaller hop count than every other node with the same best-case delivery time, then it is the most preferred neighbor.

  ⋆ Otherwise, the node with the smallest node number among all nodes with the earliest best-case delivery time and smallest hop count is arbitrarily chosen as the most preferred neighbor.

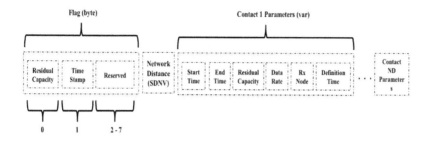

Figure 5.7: CRG extension block format

5.4.4 Contact Graph Routing Extension Block

The main focus of CGR extension block [46] is "source path routing", i.e., annotating the bundle with the best computed route as determined at the source node. The information stored in the CGR extension block can be classified into three types as follows.

1. Implementation of a source path routing system where the nominal message delivery path is encoded with the bundle.

2. Synchronization of CGR across the network: this refers to the fact that the source path approach requires a downstream node be able to understand the nominal path in the context of its local contact graph. This requires updated path information (link availability, data rate, etc) which is determined from the contact graph of the node that populated the extension block. Thus the downstream node may choose to use the information in the extension block to update its local information in case it is more recent.

3. Inference of congestion metrics across overload paths. Since the nominal paths equate to the anticipated path of a bundle through the DTN, each node can use the path information to determine the predicted link capacity in its local contact graph, enabling a certain level of congestion management.

Figure 5.7 shows the CGR extension block format conform RFC 5050. The block type code is $0xED$. The block control flag must be set to 1 (bit 0 means that the block must be replicated in every fragment).

Flags identify the amount of contact parameter information available for residual capacity and timestamp. Both can be 1 (provided) or 0 (not provided).

Network Distance (ND) specifies the number of contacts remaining in the block.

Contact specifies the information about the *nth* contact. The contact parameters are listed below:

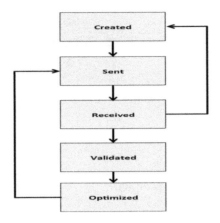

Figure 5.8: CRG extension block lifecycle

- Start Time: specifies the time at which the contact will begin.
- End Time: specifies the time at which the contact will end.
- Residual Capacity: specifies the estimated link capacity. Optional parameter.
- Data Rate: the data rate in *bps* for the current contact.
- Definition Time: indicates the time at which the information in this extension block was asserted. Optional parameter.

Figure 5.8 illustrates the CGR extension block lifecycle, comprising four states (created, sent, received, validated and optimized).

Created In this state the BP creates a CGR extension block. The block describes the path from the local node to the bundle's destination.

Sent The BP adds the CGR extension block to the outgoing bundle according to the processing flag. Only one extension block is allowed in each bundle at any time.

Received After receiving a bundle with a CGR extension block the BP agent validates this block's content through a block validation procedure before transitioning to the "Validated" state. The validation procedure checks the contacts in the block against the local contact plan. If the block fails in its validation, it is discarded and a new block is created by the local node.

Validated Once the block is validated, this state indicates that the path have been validated as well.

Optimized Once the path in the block has been validated, an optimization procedure determines whether or not a better path exists to the next hop.

5.4.5   CGR Route Exception Handling

Conveyance of a bundle from source to destination through a DTN can fail in a number of ways, many of which are best addressed by means of the Delivery Assurance mechanisms described earlier. Failures in Contact Graph Routing, specifically, occur when the expectations on which routing decisions are based prove to be false. These failures of information fall into two general categories: contact failure and custody refusal.

1. Contact failure: A scheduled contact can, for example, start later or finish earlier than planned. In this case, the bundle transmission may not occur as planned. For instance, the bundle's TTL can expire before the contact's start time, or the contact's end time can be reached before the bundle has been transmitted. When a bundle's forfeit time is reached before the bundle has been transmitted to the proximate node, the bundle is removed from the outbound transmission queue and a new route is computed for the bundle.

2. Custody refusal: A node that has received a bundle may not be able to forward it; for example, it may not be able to compute a forwarding route, or it may have too little available buffer space to retain the bundle until it can be forwarded. In such a case, the bundle may be discarded; if the discarded bundle was marked for custody transfer, a custody refusal signal must be forwarded back to the current custodian. When a custody refusal signal is received by the current custodian for a bundle, a new route - excluding the node that sent the custody refusal - is computed for that bundle and the bundle is re-forwarded.

   When the affected bundle is non-Critical, the node that receives the custody refusal re-applies the Dynamic Route Computation Algorithm to the bundle so that an alternate route can be computed – except that in this event the node from which the bundle was originally directly received is omitted from the initial list of Excluded Nodes. This enables a bundle that has reached a dead end in the routing tree to be sent back to a point at which an altogether different branch may be selected.

For a Critical bundle no mitigation of either sort of failure is required or indeed possible: the bundle has already been queued for transmission on all plausible routes, so no mechanism that entails re-application of CGR's Dynamic Route Computation Algorithm could improve its prospects for successful delivery to the final destination. However, in some environments it may be advisable to re-apply the Dynamic Route Computation Algorithm to all critical bundles that are still in local custody whenever a new contact is added to the contact graph: the new contact may open an additional forwarding opportunity for one or more of those bundles.

5.4.6   CGR Remarks

The CGR routing procedures respond dynamically to the changes in network topology that the nodes are able to know about, i.e., those changes that are subject to mission operations control and are known in advance rather than discovered in real time. This dynamic responsiveness in route computation should be significantly more effective and less expensive than static routing, increasing total data return while at the same time reducing mission operations cost and risk.

Note that the non-Critical forwarding load across multiple parallel paths should be balanced automatically:

■ Initially all traffic will be forwarded to the node(s) on what is computed to be the best path from source to destination.

■ At some point, however, a node on that preferred path may have so much outbound traffic queued up that no contacts scheduled within bundles' lifetimes have any residual capacity. This can cause forwarding to fail, resulting in custody refusal.

■ Custody refusal causes the refusing node to be temporarily added to the current custodian's excluded neighbors list for the affected final destination node. If the refusing node is the only one on the path to the destination, then the custodian may end up sending the bundle back to its upstream neighbor. Moreover, that custodian node too may begin refusing custody of bundles subsequently sent to it, since it can no longer compute a forwarding path.

■ The upstream propagation of custody refusals directs bundles over alternate paths that would otherwise be considered suboptimal, balancing the queuing load across the parallel paths.

■ Eventually, transmission and/or bundle expiration at the oversubscribed node relieves queue pressure at that node and enables acceptance of custody of a "probe" bundle from the upstream node. This eventually returns the routing fabric to its original configuration.

Although the route computation procedures are relatively complex they are not computationally difficult. The impact on computation resources at the vehicles should be modest.

5.5   Summary

DTN routing follows the *store-carry-and-forward* paradigm, an extension of traditional *store-and-forward*. In *store-carry-and-forward*, DTN nodes store messages locally in the absence of contact opportunities and may forward data from node to node when contacts are established. Many DTN routing protocols have been proposed, each of which targeting different DTN applications. As such, a comparative analysis of DTN routing protocols is sometimes difficult, since they are designed to operate in different DTN scenarios. In general, repliaction-based routing protocols may yield higher end-to-end delivery when compared to forwarding-based strategies in the absence of reliable topology knowledge. However, they consume far more network resources and may cause increased network contention and congestion.

Although contact is a precondition for DTN routing, contact alone does not guarantee that the bundle will be forwarded to the next node: the next node must have enough buffer space to accommodate the incoming bundle until the next contact arises. To this end the consideration of congestion control in the context of DTN routing is fundamentally important.

Inherently, it is hard to know the topological location of any node in a delay-tolerant network (other than a current neighbor) due to unknown mobility patterns and dynamic changes in the network topology. Routing protocols that are based on network information typically fail except under specific circumstances. Machine learning techniques might help to mitigate this issue if they enable a routing protocol that is able to infer the current DTN operational scenario and switch among different routing strategies, adapting to different parameters, depending on the scenario.

At this time, most proposed DTN routing protocols have not been deployed in a real DTN. The design and simulation of these protocols are nonetheless an important step for the definition and deployment of an operational DTN routing protocol in a real scenario. To this end, building an experimental platform that enables validation and analysis of these routing protocols is critical for determining a general and suitable DTN routing protocol that is able to operate across a heterogeneous interplanetary/terrestrial delay-tolerant network.

Chapter 6

DTN Coding

Marius Feldmann

TU Dresden

Felix Walter

TU Dresden

Tomaso de Cola

German Aerospace Center

Gianluigi Liva

German Aerospace Center

CONTENTS

6.1   Introduction

Error control coding lies at the core of nearly any modern digital communication system. In fact, reliable communication over a link affected by noise cannot be established without a suitable mechanism for the recovery of transmission errors. When dealing with error correction, two main approaches shall be distinguished, i.e.

- Automatic retransmission protocols;

- Forward error correction (also known as channel coding).

While the first approach requires the presence of a feedback channel to signal to the transmitter which data units arrived corrupted at the receiver, the latter addresses the error correction problem by adding a certain amount of redundancy (proactively) to the transmitted information which is then used at the receiver to detect and (possibly) correct errors. Retransmission protocols have the inherent advantage of entailing a low complexity. However, their applicability is limited, especially in satellite and space communications, due to the large propagation delays which render buffering at the transmitter side (for future possible retransmission requests) extremely costly.

In the following two subsections two related categories of approaches applicable to forward error correction will be introduced and discussed. Section 6.2 focuses on *Network Coding*. This term refers to techniques for recombining sets of packets transmitted within a network to new packets to e.g. increase the overall end-to-end delivery probability. In section 6.3 several approaches to coding at the physical layer as well as packet-level coding for space links implementing the CCSDS protocol stack will be introduced. Besides providing fundamentals about the underlying theory and principles, in both sections, a dedicated focus is directed to the approaches' application in the DTN domain.

6.2 Network Coding

Network coding takes a fundamentally different perspective on the packet transmission compared to classical approaches. While in conventional packet-based networks the transfer of data between nodes has been seen as serialization of a packet on the transmitter side and its deserialization on the receiver side, network coding reinterprets this step: Packets are interpreted as functions and, thus, mathematical operations can be applied on a set of them to generate new functions. As a central characteristic of network coding in contrast to source-based approaches, packets can be combined (i.e. coded) on any node on the path. Using a set of received coded packets, the original data can be reconstructed if a sufficient amount of packets is available. In the following sections we would like to shed some further light on the overall approach and discuss its advantages in the delay-tolerant networking (DTN) domain.

In a first step (section 6.2.1), network coding will be introduced and the underlying idea will be clarified. Furthermore, basic terminology will be discussed and selected advantages of network coding in delay-tolerant networks will be pointed out. In a second step (section 6.2.2), an overview of the state of the art of network coding in the DTN context is given. This section will specifically provide a general classification of approaches used in this domain. The chapter concludes with a brief focus on implementation aspects and practical considerations of network coding in DTNs (section 6.2.3.2), including a concrete case study.

6.2.1 Fundamentals on Network Coding

In this section, a brief overview of network coding is given. Though scenarios out of the DTN context are presented, the section discusses network coding in general. The information is intended for everybody without background knowledge about network coding and associated topics. However, due to limited space some of the background information can be only *mentioned without full explanation*. In these cases extensive references are provided for further studying. Furthermore [119] is recommended as a good introduction to network coding.

6.2.1.1 Network Coding by Example

Often cited literature discussing network coding introduces the overall idea and the advantage of this technique based on the so-called Butterfly network (e.g. Figure 1.1 in *Network Coding: An Introduction*, Ho and Lun [140]). This example network has been already used in 2000 by Ahlswede et al. in their paper published in the

IEEE Transactions on Information Theory journal [21]. It discusses the reduction of bandwidth usage rendered possible by encoding message sets using linear codes (in this case: *XOR* operation on two messages). Thus, we seem to break a sort of *"Hello, World!"* principle of the network coding community as we do not refer to the basic Butterfly network for explaining the overall approach, but to an extended variant sketched in Figure 6.1. The depicted example is a time-varying modification plus extension by two source nodes of a sample network that was as well discussed (beside the Butterfly network) by Ahlswede et al. (see Figure 8 of [21]).

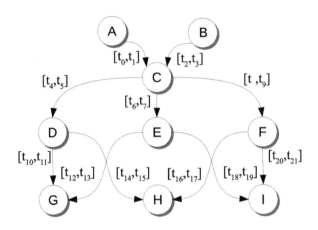

Figure 6.1: Example topology with two information sources and three sinks

Before we focus on the steps taken by network coding, core characteristics of the chosen scenario should be pointed out: The network consists of nine nodes with two nodes forming an information source (node A and B) and three information sinks (nodes G, H and I). The edges are time-varying and only exist during the specified time periods $[t_i, t_j]$ defined by a start time t_i and an end time t_j with $t_i < t_j$ for each $i < j$. In this scenario, node A and node B transmit one message each. These two messages m_1 and m_2 with identical size *msglen* are handed over to node C. They are supposed to be transferred to each of the information sinks. It is assumed that each link has a constant bit rate b. As a limitation, the link capacity for each link existence period $[t_i, t_j]$ is $C = (t_j - t_i) * b$ with $C = msglen$. Hence, each link can only transfer exactly one message.

We have selected the sketched scenario because it reflects topologies of a very interesting application area for network coding in a good manner. For example, it could be assumed that the nodes represent ground stations and LEO satellites as follows: Node C could be a ground station which receives sensor data from two earth observation satellites A and B that have to be transferred via communication satellites (nodes D, E, and F) to three remote ground stations (G, H, and I). Thus, the sketched network may reflect a simple LEO-based delay-tolerant communication network.

An interesting characteristic of this network is that there does not seem to be any way to transfer both messages m_1 and m_2 to the sink nodes. Whichever combination is tried, one message is missing in one of the sinks. The explanation is simple: Node C has only three links to further nodes D, E, and F. Thus, one of the two messages cannot be transferred more than once. As a consequence, this message is available only once in the nodes D, E, or F. Each of these nodes has a maximum of two links to the information sinks. Thus, one of the information sinks does not receive the message that was initially transferred only once by node C.

At a first glance, this issue seems not solvable. However, network coding provides a solution to this problem by generating three encoded messages from the two original ones. Receiving an arbitrary two of these three encoded messages is enough to reconstruct the two original messages. Expressed more formally, the encoding step can be seen as a function nc taking a set of messages plus a parameter C defined below as input and yielding a set of encoded messages. How this is done will be explained in detail in the next sections. Based on this general idea, we can generate three encoded messages y_1, y_2, and y_3 with $nc(m_1, m_2, C) = (y_1, y_2, y_3)$. Each of these three messages has the identical length $msglen$. In the discussed scenario, this operation is executed by node C and, thus, on an intermediate node on the path through the network. The resulting packets that contain the encoded messages are transferred to the neighbors of node C, one packet per neighbor. From there on, two packets are sent to each of the three information sinks.

If at least as many linearly independent[1] messages are received by an information sink as there have been messages initially used during the encoding step, these messages can be reconstructed using a function $ndc(y_i, y_j, C) = (m_1, m_2)$ with $1 \leq i, j \leq 3$ and $i \neq j$. Due to this requirement being fulfilled in the given scenario, each of the three information sinks can reconstruct the original two messages m_1 and m_2.

The core question is which functions nc and dec render the mentioned scheme possible. The answer is simple: For encoding, arbitrary linear combinations of messages can be created, enabling decoding by solving a (sub-)set of these equations. In the following, we will focus on this class of network coding, called *linear network coding (LNC)*. Before we address this approach in detail, we would like to briefly discuss the underlying mathematical domain: *Finite Fields*.

6.2.1.2 Basic Network Coding Operations in Finite Fields

Finite fields are mathematical structures containing a finite number of elements. Both addition and multiplication are defined on these structures. The results of these operations are again elements of the selected finite field. In the context of network coding, the use of finite fields guarantees that encoded messages have the same size as the original messages. This section provides an overview of finite fields. If more details about this domain are required, we recommend e.g. [177].

A finite field is referred to by $\mathbb{F}_q$ with q being a power of a prime number: $\mathbb{F}_q =$

---

[1]For an explanation of *linear independence* see the last part of the following section.

$\mathbb{F}_{p^n}$ with p prime and $n \geq 1$. Finite fields with $n = 1$ are called prime fields. Thus, the most simple finite field is $\mathbb{F}_2$. In this field with its elements 0 and 1, addition $(+)$ is defined as logical *XOR*. Multiplication $(\times)$ is defined as logical *AND*. By this, it is guaranteed that the results of these operations are again elements of $\mathbb{F}_2$.

For network coding, non-prime fields are widespread. The elements of these fields can be represented as polynomials over $\mathbb{F}_p$ with a degree less than n. Assuming e.g. the finite field $\mathbb{F}_{16}$, thus, $\mathbb{F}_{2^4}$. This field contains elements such as $x^3 + x + 1$ and $x^3 + x^2$, which correspond to 1011 and 1100 in binary representation. In order to transfer the polynomial representation to binary, each power of x is expressed by one bit. Addition in these fields is defined by the logical *XOR* operation of the coefficients with same power. Thus, adding $x^3 + x + 1$ and $x^3 + x^2$ results in $x^2 + x + 1$.

If multiplication in these types of fields would have been defined in a naïve way, an obvious problem would arise: The results would not be elements of $\mathbb{F}_q$ anymore as coefficients of higher powers become part of the result. To avoid this contradiction against the definition of finite fields, multiplication is instead defined by calculating the product of the two polynomials modulo a so-called *irreducible polynomial P* of degree n. The term *irreducible* comes from the fact that it cannot be factored into two polynomials $p_1, p_2 \in \mathbb{F}_q$. For each $\mathbb{F}_q$ at least one irreducible polynomial exists. In the case of $\mathbb{F}_{2^4}$, e.g. the polynomial $x^4 + x^3 + 1$ is irreducible. Assuming this irreducible polynomial, $x + 1$ is the result of multiplying $x^3 + x + 1$ and $x^3 + x^2$ in $\mathbb{F}_{16}$. With a brief focus on the practical side of multiplication in finite fields, it has to be noted that multiplication is much more computationally expensive than addition. Thus, it is the subject of performance improvement efforts. A very simple example for that is the use of predefined tables reducing the effort of multiplication to a simple table lookup. However, this requires to store a value for every individual operand combination in $\mathbb{F}_q$.

After this short overview of finite fields, we have everything together to direct the focus back to network coding. In the following, we assume the application of a finite field $\mathbb{F}_{2^n}$. As it has been mentioned already, finite fields are used in network coding to allow working with plain and encoded messages of identical size. Expressing these characteristics more formally with the symbols introduced above, a message with *msglen* bits can be divided into $\frac{msglen}{n}$ symbols with each symbol $s \in \mathbb{F}_{2^n}$. Obviously, this is only valid if *msglen* is a multiple of n. If this is not the case, padding can be used to extend the messages to fulfill the criterion. To rephrase, each message has to be dividable into a set of symbols which are elements of the underlying finite field.

Following the discussion of the overall construction principle for code words, we can shed light on the above-mentioned functions *nc* and *ndc*: In linear network coding, the coding function just performs a multiplication of the matrix C and a matrix representation of the messages selected to be encoded together. In the network coding domain, the messages which are encoded together are called a *generation*. Their count is called *generation size*. The generation size is an important property of network coding algorithms. The generation size is equivalent to the minimal amount of messages that have to be received to finally render the decoding step possible.

The messages selected for encoding are represented, as discussed before, using symbols which are elements of the selected finite field. The decoding function performs e.g. the Gauss-Jordan elimination algorithm to extract the original messages from received equations consisting of the matrix C and the encoded messages.

To provide a practical example for this step, we assume that the above-mentioned messages are $m_1 = 01000001$ (the 8-bit-ASCII character "A") and $m_2 = 01000010$ (the 8-bit-ASCII character "B"). Representing these messages using elements of $\mathbb{F}_4$, we get $m_1 = 1001$ and $m_2 = 100x$ as it is shown in Figure 6.2. The selected finite field is $\mathbb{F}_{2^2}$ in order to keep the overall example clear and comprehensible.

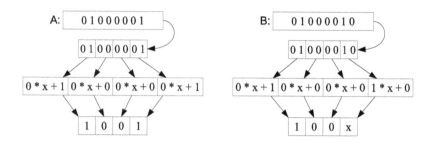

Figure 6.2: Representation of ASCII characters A and B as elements of $\mathbb{F}_{2^2}$

In this example, the multiplication matrix C consists of three encoding vectors $(x+1,0)$, $(x+1,x+1)$, and $(x,1)$. The overall encoding step is realized by multiplying m_1 and m_2 with C, as shown in equation 6.1. Only a single irreducible polynomial exists in the selected finite field, $x^2 + x + 1$. This is used to perform the multiplication operations.

$$
\begin{pmatrix} x+1 & 0 \\ x+1 & x+1 \\ x & 1 \end{pmatrix} \times \begin{pmatrix} 1 & 0 & 0 & 1 \\ 1 & 0 & 0 & x \end{pmatrix} = \begin{pmatrix} x+1 & 0 & 0 & x+1 \\ 0 & 0 & 0 & x \\ x+1 & 0 & 0 & 0 \end{pmatrix} \tag{6.1}
$$

Thus, in binary representation the generated messages are $y_1 = 11000011$, $y_2 = 00000010$, and $y_3 = 11000000$. Two of these three encoded messages are transferred to each of the three sink nodes in the scenario as depicted in Figure 6.1 together with the applied encoding vector of the encoding matrix C. It should be obvious that in any two combinations of the received packets information from both m_1 and m_2 is contained. Interestingly, the elements in the encoding matrix C can be selected randomly from the applied finite field as it has been shown in [139]. The vectors used to create a specific encoded message are transferred together with the encoded messages to the receiving nodes.

As soon as two packets are received in any of the nodes G, H, or I, the decoding operation can take place. If it is assumed that node H has received messages y_2 and y_3, the system of linear equations to be solved is:

$$
\begin{aligned}
(x+1) \times m_1 + (x+1) \times m_2 &= y_2 \\
x \times m_1 + 1 \times m_2 &= y_3
\end{aligned}
\tag{6.2}
$$

Using the Gauss-Jordan algorithm, this can be transformed to a row-echelon form and solved by:

$$
\left(\begin{array}{cc|cccc}
x+1 & x+1 & 0 & 0 & 0 & x \\
x & 1 & x+1 & 0 & 0 & 0
\end{array}\right) \rightarrow \left(\begin{array}{cc|cccc}
1 & 0 & 1 & 0 & 0 & 1 \\
0 & 1 & 1 & 0 & 0 & x
\end{array}\right)
\tag{6.3}
$$

From this representation, the original messages m_1 and m_2 can be extracted directly. The same operation is executed for any two packets received by G, H, and I. Thus, all three nodes have both messages originating from nodes A and B available.

To sum up, the *trick* applied by linear network coding is to perform a matrix multiplication between a selected $m \times n$ coefficient matrix and m messages, resulting in a matrix which contains information from the former messages in each of its n rows. The sinks have to receive a set of encoded messages plus the used coefficient matrix in order to solve the set of equations for the messages represented as unknowns.

It is obvious that at least m packets have to be available in order to decode the original messages. These packets have to be *linearly independent*. This is the case due to linearly independent coefficient vectors: In the above-mentioned example, $(x+1,0)$, $(x+1,x+1)$, and $(x,1)$ are linearly independent. In the network coding domain, a packet that is linearly independent from already received packets is called *innovative*.

6.2.1.3 Advantages and Drawbacks of Network Coding

There are two core benefits of applying network coding in contrast to plain packet forwarding or source-based coding approaches, resulting in some application contexts: An increase in *throughput* and an increase in *reliability*. Further advantages discussed in the state of the art are aspects such as an increase in security or a reduction in complexity. We focus on the two firstly mentioned striking advantages in the following as we consider them the two most important benefits in selected scenarios.

Throughput

Increasing a network's throughput by applying network coding was the first reason for justifying research in this domain. From the discussion of the example provided above in Figure 6.1, this advantage gets very obvious. In the given network scenario, packets cannot be routed in a classical way to the destination nodes. In contrast to this, transferring the information to the sink nodes was rendered possible by network coding. By extending the scenario in a simple manner, the same could be achieved by classical routing: If e.g. one contact from E to H would be added allowing for transmission between t_{10} and t_{11}, all three sink nodes could have received messages m_1 and m_2. However, it is obvious that in this vein the overall used bandwidth would have increased: With an optimal network coding solution, the added link would not have been used while achieving the same result. Thus, in this scenario, network coding would use 11 contacts while classical routing would use 12. This conforms to a reduced bandwidth consumption by 8 % if network coding is applied.

The relevant characteristic of the shown example network is that it is a multi-sink network in which all sinks receive the same information. In these multicast networks, network coding is capable to achieve the theoretical upper bound of throughput between two nodes. In order to discuss this point further, some background information on the *max-flow* characteristic of networks is necessary. Details about this topic can be found in [21].

Assuming our current understanding of the physical world, it is evident that in all computer networks, the amount of data that can be transferred between any two nodes is limited. Thus, every edge has a maximum capacity of *n units* assigned to it. For the amount of units that are transferred between two nodes, the term *flow* is used. It is obvious that the following trivial statement holds true: The maximum flow between two directly connected nodes cannot exceed the capacity of the connecting edge.

The core question here is concerned about the amount of the maximum flow (max-flow for short), i.e. the upper bound on the amount of data transferable between two nodes connected by a path of at least one further intermediate node. The answer to this question is provided by the *max-flow min-cut theorem* [117].

This theorem states that the maximum flow between any two nodes in a network with given edge capacities is equal to the so-called *minimum cut* of the network. A cut of a network is a partition of it into two disjoint sets of nodes by removing all edges connecting the two node sets. A minimum cut in a flow network is a cut that is minimal in the sense of the capacity that the removed edges have in sum. Thus, the minimum cut in the network depicted in Figure 6.1 between node A and node E is 2, if we assume that the capacity of each edge is 1. Consequently, it is sufficient to remove specific edges with a total capacity of 2 to separate the source from the sink node. In the case of unicast communication between two nodes in a flow network, a path with maximum flow can be computed using the *Ford-Fulkerson algorithm* [117]. In the case of multicast communication, the situation is different: Using regular routing, the upper bound defined by the max-flow cannot be reached in all cases. In cases where it can be reached, the underlying problem (packing Steiner trees) is *NP-hard* and thus computationally expensive (see e.g. [295]). In contrast, the application of network coding is a straightforward approach in order to reach this upper bound in any case, as soon as the amount L of receiving nodes is greater than 1. Ahlswede et al. state in [21], p. 3: "For $L \geq 2$, network coding is in general necessary in an optimal multicast scheme." This core feature makes network coding very interesting for multi-hop multicast networks.

Reliability

Increasing end-to-end reliability is desired in most networks with lossy or opportunistic links. In order to sketch the core benefits of applying network coding to such a network, the simple scenario depicted in Figure 6.3 is assumed. It shows four nodes, each connected to its direct neighbors via four links existing during the time-spans $[t_x, t_y]$. Each of these links has a probability value p_c assigned to it. This value represents the probability with which a packet may be transmitted successfully via the link. In the scenario, the probability values for all links from the same source to the same destination node are assumed to be identical.

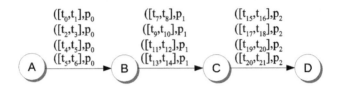

Figure 6.3: Simple network example with contact probabilities

In the given example, two packets should be transmitted from node A to node D. The size of the packets exactly fits the capacity of each of the depicted links. Thus, e.g. the first two contacts between A and B, the first two contacts between B and C, and the first two contacts between C and D would be a valid path. The overall end-to-end transmission probability for a single packet on this path is $p_0 * p_1 * p_2$. If we assume the values $p_0 = 0.9$, $p_1 = 1$ and $p_2 = 0.4$, the overall delivery probability for receiving both packets is $0.36^2 = 0.1296$.

In order to achieve a higher end-to-end delivery probability via this channel, two approaches could be considered: Either a source-based or a network coding-based approach. In the following, a very simple source-based *message replication* approach is assumed to clearly demonstrate the advantages of network coding. If the two messages are replicated one time each, the four resulting messages can be scheduled via the four contacts available between each pair of nodes. Following this approach, nine cases exist in which the two messages are received by D: Either all (1 case) or at least three messages (4 cases) are received, or two different messages are received (4 cases). Thus, the overall probability for the reception of m_1 and m_2 is $0.36^4 + 0.36^3 * 0.64 * 4 + 0.36^2 * 0.64^2 * 4 = 0.35$.

In contrast to this, network coding has a core advantage: For receiving the two original messages, it is valid to receive an arbitrary combination of network coded messages. Assuming that four network coded packets are generated from the original two messages, there exist eleven cases in which two messages are received. Thus, the overall delivery probability for m_1 and m_2 is: $0.36^4 + 0.36^3 * 0.64 * 4 + 0.36^2 * 0.64^2 * 6 = 0.45$. In consequence, the probability for successfully transmitting both messages has been increased by 0.1 without increasing the used bandwidth.

However, the same effect may be achieved by source-based encoding. An additional core advantage of the network coding approach can be pointed out based on the scenario sketched in 6.3: In the case of source-based message replication, all four contacts between nodes A and B as well as between B and C are used because both messages have to be replicated to avoid losing the single copy of either m_1 or m_2. Due to the high delivery probability of the contacts between A and B (0.9), creating three messages by network coding should be enough to transmit at least two messages to node B. Furthermore, no replication has to be done on the link between nodes B and C. On node C another packet can be generated by re-encoding the received messages. By this approach, the amount of packets can be optimized within the network based on the delivery probability of the next contact. As a consequence, the bandwidth consumption in the overall network may be heavily reduced, though the delivery probability is increased in comparison to regular message replication.

Drawbacks of Network Coding

Beside the mentioned benefits, network coding has some essential drawbacks. First of all, encoding and decoding implies a computational overhead. Specifically, solving equations on the receiver side using the Gauss-Jordan elimination algorithm does not offer essential possibilities for optimization.

A further problem when employing network coding is the need for a specific amount of encoded packets to render decoding possible. As it has been stated above, n innovative (linearly independent) packets have to be available at any sink node with n being the size of the generation. If the amount of received innovative packets is less than n, the received messages are useless.

Furthermore, network coding may result in a delay while a node is gathering messages to fill the generation buffer. However, this aspect can be typically ignored in delay-*tolerant* networks.

6.2.1.4 Application of Network Coding

Before the state of the art of network coding in the DTN domain is discussed in section 6.2.2, an overview of the application of network coding is provided below. In the context of applying network coding, four points of view are discussed, summarized in Figure 6.4.

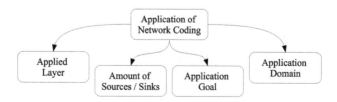

Figure 6.4: Points of view regarding the application of network coding

Firstly, network coding is not bound to a specific layer of the network stack. For several years already, network coding is being applied in different layers, starting from the physical layer (e.g. [252]) up to the application layer (e.g. [314]). In the analysis presented in section 6.2.2 we focus on its application on the *bundle layer* in DTNs as it has been defined in RFC 4838 [76].

Secondly, it can be differentiated if network coding is applied to a topology with a single or multiple information sources and with a single or multiple information sinks. Based on the type of networks, the taken coding approaches may differ fundamentally. In case of multiple information sources, network coding may be applied between different flows, thus, called *inter-flow network coding*: Messages from different sources are used as input parameters for the network coding function. In contrast to this, in the case of single-source networks the coding is always done on single flows, thus, so-called *intra-flow network coding* is applied.

Thirdly, the application goal of using network coding can be leveraged for differentiation. Many presented approaches focus on the above-mentioned advantages (increased throughput and increased delivery probability). Furthermore, security is one of the aspects pointed out in several state-of-the-art approaches.

The last differentiation criterion is the concrete application domain. Though network coding has been initially seen as a general mechanism for improving packet transmission in multicast networks, specific exploitation for innovative software applications has emerged. Besides *peer-to-peer* networks [300], a very promising application domain is the field of distributed storage solutions such as the one discussed in [79].

After this very brief overview of the overall approach taken by linear network coding, the following sections will purely focus on its application in delay-tolerant networks.

6.2.2 Survey on Network Coding in the DTN Context

For getting a detailed overview of the work published on network coding in the DTN context, this section aims to survey the state of the art in a concise fashion. As discussed previously, the DTN architecture and, thus, any protocol aware of the delay-tolerant overlay network and its characteristics operates on the application layer. We explicitly focus our analysis on network coding approaches and protocols which extend the DTN stack. Hence, approaches on lower layers such as "physical" network coding (i.e. message recombination by applying radio signal interference), and those which do not take into account the delay-tolerant characteristics of the network are considered out of scope.

To be able to see the different approaches from a more practical perspective, we focus on a characterization by the targeted DTN use case and the topological knowledge required by the specific approaches. This also reflects the separation of DTNs into deterministic and opportunistic networks in a straightforward manner – algorithms requiring more knowledge may only be employed in networks where it is available. On the contrary, algorithms requiring no specific topological knowledge can be applied in practically any topology. However, it can be assumed that their performance will not be optimal for the given scenario.

In section 6.2.2.1, the central criteria which differentiate network coding approaches in the DTN field are introduced in short. Secondly, in sections 6.2.2.2 and 6.2.2.3 a classification of state-of-the-art approaches based on these criteria is presented. Section 6.2.2.4 provides a summary of identified approaches. Finally, in section 6.2.2.5, open research topics and possibilities to extend the state of the art are discussed.

6.2.2.1 Central Criteria for Differentiation

The delay-tolerant networking architecture is employed in a variety of topologies with the nodes possessing different levels of knowledge concerning their neighbors and the rest of the network. The performance of network coding, however, vastly depends on topology and the used routing strategy. More precisely, it is of great relevance which paths messages take and at which nodes there are opportunities for re-encoding and decoding them. Due to this, numerous approaches to network coding in DTNs have been published, addressing and optimizing for different scenarios and metrics. On the scenario side, this includes highly opportunistic networks such as ad-hoc and sensor networks with varying delay and mobility characteristics as well as vehicular networks and pre-configured, purely deterministic topologies.

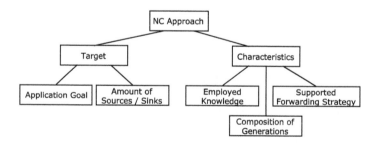

Figure 6.5: Differentiation of coding approaches in the DTN field

As discussed in Section 6.2.1.4, we differentiate four points of view regarding the application of network coding. The application domain has been restricted to the DTN context in this section. The applied layer is restricted as well to the DTN-specific *bundle layer*. Hence, we only discuss the *application goal* and the *number of sources and sinks* as criteria for differentiation. These can also be seen as targets of coding approaches as shown in Figure 6.5. Additionally, we characterize approaches by their inherent characteristics such as the employed topological knowledge, the composition of generations, and the supported forwarding strategies. In the following sections, criteria which deserve a more specific explanation are discussed in short.

Application Goal

Every coding approach focuses on one or several metrics or optimizes for a given property. These objectives partly overlap with those of DTN routing algorithms which are, for example, discussed in [72]. This shows the close relation between routing and network coding: both need to be employed in concert to achieve the best possible performance and no approach suits all possible topologies perfectly.

In section 6.2.1.4 we identified increases in throughput and reliability as the two central advantages that network coding may offer. These two advantages conform to the objectives of most of the existing coding approaches. The goal of increased throughput can be achieved in different ways:

1. **Reduction of transmission overhead**. With network coding, in scenarios with multiple different paths fewer copies of individual messages have to be transmitted as they can be combined with one another before transmission. This can effectively reduce message transmission overhead as shown for example in [280] and [291].

2. **Decreased buffer utilization**. Network coding allows arriving messages to be coded together in the local node's buffer. This can improve buffer utilization and even allows for accepting messages when the buffer is already exhausted, under the assumption that at least one message of the same generation is already contained in the buffer.

3. **Decreased delay**. In conventional replication-based non-coding routing schemes, multiple identical messages may arrive over different paths. In contrast, when employing a random coding scheme, the probability that a second message arriving at the destination does not contribute information is significantly lower. Thus, the *block delivery delay* can be decreased by using network coding in the case that a destination needs to receive multiple messages and the information can take different paths.

An increase in throughput can furthermore result in energy efficiency by reducing the total time required for exchanging a given amount of data in the network. Additionally, by offering possibilities to lower replication rates, network coding allows to reduce the count of messages transmitted per contact, especially in broadcast scenarios. This can increase energy efficiency as well, as shown by Fragouli et al. in [118]: with simple decentralized algorithms, network coding can bring benefits of factors up to $log\,n$ (with n being the number of nodes) in terms of energy efficiency. In unicast communication over unreliable paths, fewer messages need to be transmitted for achieving the same delivery probability which in turn also leads to reduced energy consumption.

On the other hand, a higher delivery probability can be achieved with network coding. In networks with opportunistic contacts and multiple available paths from source to destination, the information contained within messages can take more different paths while maintaining or even reducing the amount of exchanged data, resulting in higher probabilities for successful message delivery.

Employed Knowledge of Network Topology

Jain et al. have shown in [10] that having more topological knowledge available allows for better routing decisions. Intuitively, this also holds for network coding as future coding opportunities can be planned proactively if knowledge about them is available.

Most publications on network coding in the DTN context focus on sparse ad-hoc networks such as vehicular networks. Thus, the resulting algorithms cannot make use of comprehensive topological knowledge. However, in other types of DTNs (e.g. interplanetary or satellite networks), such information is commonly available.

In section 6.2.2.2, a classification by used topological knowledge is done which shows that only few algorithms are targeted at deterministic use cases, i.e. such having scheduled contacts. One reason why such networks have not been evaluated thoroughly for network coding might be that there often are only few favorable paths from a source to a destination and, thus, employing network coding in the first place does not make much sense. Though, topologies such as Ring Road networks (see 6.2.3.2) offer many more transmission opportunities in combination with a great amount of knowledge regarding future changes in topology. The application of network coding in these networks may bring additional benefits regarding overhead reduction and an increase in resilience against unstable channels.

Use and Composition of Generations

To make effective use of network coding it has to be decided which messages are allowed to be combined. On one hand, when combining too few messages, the benefits of coding become negligible. On the other hand, when combining more messages the required storage, message transmission, and decoding overhead increase through the increasing size of encoding matrices. Additionally, in unicast setups, nodes may have to decode messages not destined for them which generates further overhead. Thus, there is a trade-off between generation size, overhead, and performance of decoding.

The most straightforward approach to determine a *generation* is to only allow the combination of messages belonging to the same flow. In this vein, a *flow* refers to a single transmission from a source to a group of destinations. Most published algorithms follow this approach. Others, however, allow to combine messages belonging to different flows as well. This is also called *inter-flow* network coding and especially makes sense if the flows carry data with the same characteristics, e.g. sensor data that is collected from multiple sources. Commonly, the generations are formed using flows with the same destination. Combining messages to different destinations has been considered as well, although the authors in [316] have shown that this may even result in decreased performance compared to a non-coding scheme.

Supported or Required Forwarding Algorithm

A great amount of research on routing and coding in DTNs has been done for opportunistic networks, e.g. *mobile ad-hoc networks (MANETs)*, leading to algorithms designed to be used with a replication-based forwarding algorithm such as Epidemic Routing [279] or Spray and Wait [263]. Selected works evaluate probability-based algorithms (e.g. PRoPHET) or are flexible in regard to the applied routing approach, while others develop own algorithms for generating the forwarding decisions.

Considering deterministic routing, unicast algorithms which are based on a single path through the network and do not replicate messages are unsuitable to employ network coding. Though this would allow for recombination at intermediate nodes, all sent messages have to be delivered to the respective destinations. To successfully apply network coding in a deterministic unicast setup, algorithms need to be extended and have to be aware of coding opportunities. An approach where this is realized is presented in section 6.2.3.2.

Further Criteria Identified in Literature

Beside the management of coding generations, Zhang et al. [313] identify further criteria in their discussion of the design space for DTN routing algorithms. Most of these need to be approached from a different perspective when employing network coding. These further identified criteria are the following:

1. **Control signaling**. This includes which information is exchanged by the nodes. The authors distinguish *no signaling* (no information is exchanged), *normal signaling* (only presence information is exchanged) and *full signaling* (presence information and data about stored messages are exchanged). Such a mechanism can be used to determine coded messages and their composition in the buffers of other nodes to restrict transmission to only innovative messages. Though a coded message has different contents compared to the ones already contained in the local buffer, it may not contribute to the total information in case it is linearly dependent. Thus, its exchange would result in unnecessary overhead.

2. **Replication control**. This describes approaches to limit the number of copies transmitted of a single message. Such mechanisms are important to prevent buffer and link capacity exhaustion through over-replication of messages. In the case of network coding it has to be considered that coded messages may consist of several source messages.

3. **Transmission scheduling and buffer management**. Nodes have to decide which messages to transmit, to accept, and to drop when encountering other nodes. There may be different policies using varying levels of knowledge. These range from random selection to the transmission and retention of messages with the least number of copies in the network. When applying network coding, messages may be combined directly upon reception which lets nodes even accept messages without dropping others if their buffer is exhausted.

4. **Recovery scheme**. To allow freeing resources (i.e. dropping buffered copies of messages) when a given message has been delivered, a recovery scheme may be used. The most straightforward approach to that is broadcasting an acknowledgment for each message, however, more extensive schemes have been developed. If network coding is applied, it is difficult to drop a single message

because it might have been combined with others. Commonly, generations are dropped as a whole on successful delivery. It should be noted that this aspect is not relevant for broadcast transmissions as here all nodes are interested in the contents of the messages.

It should be noted that there are interdependencies between the design criteria. For example, a transmission scheduling and buffer management algorithm may depend on the information gained from control signaling and may again produce information which then can be exchanged in beacon messages. Thus, it is beneficial to consider combinations of multiple criteria.

6.2.2.2   Classification of Approaches by Knowledge

This section aims to classify published coding algorithms by the leveraged topological knowledge. Three groups have been identified: Algorithms which do not use topological knowledge in the coding decisions, algorithms which only take the local neighborhood into account, and algorithms which accumulate information about the whole network. The last group can be further divided by the type of data that is collected. Commonly, only a single metric (such as a probability value) is stored per node, whereas in other cases more extensive information and calculations based on it (e.g. by applying Dijkstra's algorithm on a specifically crafted graph) are employed.

No Knowledge Used in Coding Decisions

By far the hugest group of algorithms is the one which does not use any topological knowledge for the coding decision. This results from most publications focusing on intermittently connected mobile ad-hoc networks where such knowledge is usually not available.

A subset of the approaches in this group uses some form of information about further nodes, however, in that case only for the forwarding decision. For example, if Epidemic Routing [279] is employed as forwarding algorithm, nodes check at the beginning of a connection which messages the neighbor already has and exchange only those which are not already contained in its buffer. A problem which arises here is that when applying network coding, messages may be exchanged which, though they are not contained in the neighbor's buffer, do not increase information in it. Such messages are *not innovative*, i.e. they are linear combinations of other coded messages. Approaches that take care of this are discussed in more detail in the next section.

Lin et al. [179] present a protocol based on Epidemic Routing which on each transmission randomly combines all packets in a node's buffer into a new packet that is then transmitted. Received packets are only re-encoded if no buffer space is left. For this approach, a node does not need to know which messages its neighbors already have. They show that employing network coding brings benefits especially when the buffer sizes are limited: the delays of replication-based approaches are largely influenced by the size of the buffer whereas for their network coding-based approach the delay stays approximately constant.

Sheu et al. [253] developed an algorithm based on Epidemic Routing as well (called *Network Coding based on Limited Buffer (NC-LB)*) that only uses local information to decide which packets to transmit. The rank of the local encoding matrix is computed and determines the count of linear combinations which are transmitted. If the receiving node's buffer is full, coding is applied to store the arriving packets as combinations with ones already contained in the buffer. By simulation, the authors show that their scheme performs better in terms of delivery delay than Epidemic Routing and is especially useful if the buffers are constrained.

Another commonly used forwarding strategy is Spray&Wait [263]. This algorithm limits replicas of messages by setting a maximum replication count which is updated on each transmission. The authors in [178] employ a Spray&Wait-based network coding approach they call *Efficient Network Coding Based Protocol (E-NCP)*. Every packet is assigned to an index and a *spray counter*, i.e. a maximum count of replicas. On forwarding, packets are recombined by random linear coding and the spray counters get decreased accordingly. The authors show with a custom discrete event-based simulator that their approach has a much lower overhead than one based on Epidemic Routing and only slightly increases delay. Compared to a similar replication-based approach, the coding approach features a much lower delivery delay as pointed out in the same publication.

Other publications contain algorithms that are independent of the underlying routing or forwarding strategy. As such, they are more flexible. However, they cannot make precise assumptions or control how packets are forwarded. In [280], Vazintari et al. introduce an approach which does not depend on a specific routing algorithm. Non-innovative packets are dropped at the receiver, thus, the sending nodes do not need to know which messages the receivers already have. An evaluation is performed in combination with both Epidemic Routing and PRoPHET in the *Opportunistic Network Environment (ONE)* simulator. Especially the added network coding-based memory management approach reduces overhead drastically while preserving transmission delay characteristics.

Widmer and Le Boudec [291] introduce an own replication strategy that does not need knowledge about connected neighbors. They employ a probability value to determine the count of packets which need to be broadcasted. Using a custom time-based simulator, they show that their coding-based scheme offers benefits in regards to delivery probability and overhead compared to "probabilistic" routing protocols. Other findings of their work are that network coding-based approaches benefit more from node mobility than probabilistic ones and they are a viable solution for networks with high drop rates and long inactivity intervals of nodes.

Only Knowledge of Local Neighborhood Used

As discussed before, not every coded message transmitted by other nodes is *innovative*, i.e. can increase locally available information. Thus, only checking for message identity at the start of a contact will not fully prevent unnecessary transmissions. The encoding vectors have to be compared as well to prevent the transmission of linearly dependent messages. However, this information exchange incurs a non-negligible overhead for larger generations. Several publications on network coding approaches in the DTN context aim to provide a solution for this issue.

Hennessy et al. [137] develop a network coding-based router implementation for the DTN2[2] software. Their approach to reduce unnecessary (non-innovative) packet transmissions while limiting overhead by exchanged encoding vectors is to exchange only a linear projection of the encoding matrix. Though their algorithm may provide *false-negative* results when determining the "innovativeness" of a node, they show that it still brings benefits in regard to delivery latency as compared to other network coding schemes.

Arikan et al. [34] evaluated the exchange of the rank of the encoding matrices against the exchange of the whole matrices on connection establishment in DTNs with group-meetings. In these networks, the characteristic that often more than two nodes are within radio range of each other is used to optimize packet delivery. All nodes within a specific range can receive a single transmission simultaneously. Their results show that besides not drastically impairing delivery delay performance, transmitting only the rank helps to reduce initial transmission overhead significantly. However, the number of transmissions is also increased by the rank-based scheme and, thus, both schemes represent different trade-offs.

Ahmed and Kanhere propose *HubCode* [23], an approach for people-centric DTNs where a portion of the nodes (called *hubs*) has a much greater connectivity compared to others. The authors leverage this characteristic by identifying these *hubs* and forwarding all traffic via them. Hubs re-encode packets when exchanging them with one another. In the paper, two versions of the algorithm are identified: In the first version, hubs do not decode messages, but, to detect innovativeness, have to exchange their encoding matrices at the start of a connection. In the second version, only source message identifiers are exchanged. However, to accomplish that, the hubs need to decode messages. Thus, there is a trade-off regarding required processing power and transmission overhead between the two versions. The authors perform a mathematical analysis of the delivery delay under their approach which is then validated by simulations using a custom event-based simulator and a large vehicular DTN scenario.

Lin et al. [178] introduce an algorithm called *E-NCP*. In their scheme, nodes maintain lists of identifiers of "pseudo source packets" which are contained in their buffers. These "spray lists" are exchanged at the start of a connection to decide which messages to transmit. Additionally, a counter is leveraged to achieve binary spray-like forwarding. With a mathematical model and a simulation it is shown that the approach has better delay and overhead characteristics than one based on Epidemic Routing. Specifically, the transmission delay does not depend on buffer size which it heavily does in the case of replication-based approaches.

In [307], Zeng et al. present a segmented coding approach for stream-like data which they call *DSNC* (dynamic segmented network coding). A stream of data is split into *segments* with a configured maximum count of messages. These segments represent the coding generations of the approach; only messages of the same segment

---

[2]available via https://sourceforge.net/projects/dtn/

can be combined. Additionally the authors employ a pipelining technique: Instead of stopping after the transmission of a segment and waiting for the corresponding acknowledgment, messages of the next segment are forwarded already. During that phase, the new segment grows in size until either the maximum count of messages is reached or the acknowledgment for the previous segment is received. This allows the segment size to be adjusted dynamically according to the current network characteristics. In a simulation using a custom discrete-event simulator the authors show a better performance in terms of delivery delay as compared to Spray&Wait-like, non-pipelined forwarding.

Knowledge Beyond Local Neighborhood Used

A small portion of published network coding approaches for DTNs uses more than current knowledge of the local neighborhood. Especially in highly volatile networks often only local knowledge is available or can reasonably be inferred from history as recurring encounters between the same pair of nodes may have a low probability. However, in networks with more restrained or even deterministic node trajectories, having and employing more topological information is beneficial.

PRoPHET is a well-understood and widely-applied probabilistic forwarding protocol for opportunistic DTNs. It estimates future contact probabilities from historical data and infers a *Contact Predictability* value. Such a scheme is also leveraged by some network coding algorithms.

Chuah et al. [83, 82] introduce the *context-aware network coding (CANCO)* scheme: This approach derives *Delivery Predictability* values for known nodes which are also broadcast via beacon messages. The metric is then employed to derive a *friendliness* of a node, limiting the underlying binary spraying scheme to nodes which exceed a specific threshold regarding this metric. In a simulation using *ns-2*[3] and various node mobility schemes it is shown that the approach has better delivery ratio, delay, and data efficiency characteristics as compared to an erasure-coding-based scheme and another simple Spray&Wait-based network coding approach.

Zhang et al. [312] extend a network coding scheme based on Epidemic Routing by location information. The approach only forwards data in the direction of the destination if its location is known. They show that their approach achieves a much lower delay and greater throughput compared to Epidemic Routing with both metrics staying approximately constant with a varying amount of sent data.

A further algorithm which utilizes node mobility information is introduced by Li et al. in [175]. They specifically focus on vehicular networks and use the direction in which nodes travel to determine whether or not to encode a packet. If two directions are applicable for forwarding a packet, it is encoded and sent over both paths.

The algorithm presented by Chung et al. [84] is based on the inter-contact time between nodes. This metric is accumulated for all known nodes and used to perform Dijkstra's algorithm when forwarding packets.

---

[3]available online via https://www.isi.edu/nsnam/ns/

6.2.2.3   Further Research Domains

Analyses of the Impact of Coding

A further group of publications in the NC-DTN field does not analyze a specific approach but focuses on the usefulness of network coding by itself for the domain of delay-tolerant networks. This has been done in regard to various objectives.

The throughput performance of network coding has been analyzed for multicast flows in [265]. The authors develop a framework based on Markov chains and show that network coding offers significant benefits regarding throughput for the considered networks.

An analysis of the delivery probability under Spray&Wait routing has been done by Altamimi and Gulliver in [30]. They demonstrate that network coding significantly improves the delivery probability in intermittently connected networks using an analytical model, validated with simulations using the ONE.

Pu et al. [218] perform a mathematical analysis of "two-hop single-unicast delay-tolerant networks": Their network model includes a source, a destination, and several relay nodes. A relay can only receive messages from the source and only deliver to the destination. Network coding is applied on reception by combining the incoming message with those in the buffer and on transmission by sending a random linear combination of the messages in the buffer. Based on a stochastic model of packet dependencies, the authors estimate the CDF of block delivery delay. Using a simulation, the accuracy of the model is shown.

Qin and Feng published a comprehensive model for the performance of network coding in combination with Epidemic Routing and unicast traffic [219, 220]. Their model is based on fluid mechanics and especially allows to analyze delay characteristics in networks with limited transmission capacity. A validation is done in [220] using the ONE simulator within an opportunistic network scenario with random movements and shows a high accuracy of the model.

A second model based on fluid mechanics has been published by Sassatelli and Médard in [241] for inter-session network coding under Spray&Wait routing. As an extension to that, Shrestha and Sassatelli published a coding control policy for these networks in [255]. Their validations have been done using a custom simulator in Matlab and show the accuracy of the developed model.

Another analytical model, specifically for analyzing delivery delay, was introduced by Lin et al. in [179]. They also focus on Epidemic Routing and show that network coding brings benefits especially in the case of constrained buffers. The analytical model is validated using a custom event-based simulator. Finally, the authors propose an approach to optimize delay for specific, high-priority messages.

An extensive study of the benefits of network coding in unicast networks has been done by Zhang et al.: In [315] they compare network coded random forwarding with "conventional" random packet forwarding under different combinations of constrained bandwidth and buffers. A result of their simulation is that for networks that are only constrained in regard to the available bandwidth, network coding brings a slight benefit and for networks in which bandwidth as well as buffers are constrained the benefits of coding are significantly greater. Furthermore, the impact of various generation management strategies (single source / single destination, multiple source / single destination, and multiple source / multiple destination) is analyzed. It is shown that employing network coding reduces the delivery delay for the single source / single destination case. Especially for the multiple source / multiple destination case, however, performance may greatly degrade as nodes have to receive information they do not need themselves, just to decode all messages. In [316], an algorithm to calculate the minimum group delivery time is introduced and evaluated. Delivery delay is considered by the authors as "the most important performance metric" in their simulation.

Security

Though network coding can improve the resistance of information flows through a network to some attack scenarios like passive eavesdropping [66], it may also provide surface for various other types of attacks. Gkantsidis and Rodriguez already identified in [127] that jamming (or pollution) of the network is one of the main threats to network coding schemes. In that case, the attacker introduces "fake packets" that seem to be recombinations of valid network-coded messages but only contain random – or even sophistically crafted – data, into the network. As more recombinations are generated by nodes further down the path, using the fake data as part of the input, this results in an increasing amount of wrong packets in the network. Legitimate receivers may then try to decode received messages and fail to do so or even decode the wrong result. Though the latter issue can be counteracted with conventional end-to-end message authentication schemes, DoS attacks by invalid message data cannot. A network coding scheme might get disturbed already by introducing a small amount of invalid messages that seem like legitimate ones as they will be re-encoded with other messages at relays.

Various approaches to mitigate such *pollution attacks* have been published. In [318], the authors present a signature scheme for network coded packets, allowing to verify that two packets have the same origin. However, their approach requires a public key to be exchanged. Sassatelli and Médard [240] discuss this problem for the DTN case: they restrict the public key distribution to a set of "secure nodes" which are then able to verify the packets originating from the authentic source. Czap and Vajda [92] design another approach which allows a *weak verification* of messages: it is only verified that two messages originate from the same source without verifying its identity. This avoids the need to exchange a public key, e.g. via a *public key infrastructure (PKI)*, which may be not feasible to establish in opportunistic DTNs as discussed by the authors. A mathematical verification of the scheme shows its applicability.

Another possible type of attack is *selective dropping*: adversarial nodes choose to not forward specific packets to reduce delivery success rate. Chuah and Yang [82] propose a mitigation scheme for this type of attack by dynamically adjusting redundancy at the sender side. They show in simulations with *ns-2* that their scheme successfully increases delivery probability if attackers are present.

Chuah and Yang [82] further identified the problem that attackers may try to introduce wrong information into the *control plane* of the network, e.g. regarding

Table 6.1: Overview of Broadcast Approaches

Publication	Knowledge[4]	Routing	Generations
Approach in [19]	msgs. in buffer	Epidemic-like	all msgs. of a set
Scheme in [80]	none	broadcast (newer msgs. prioritized)	constant size
Approach in [120]	none	constant fwd. factor	yes, 1 in sim.
NC-DTN [196, 197]	msgs. of neighbors, full topological (OLSR)	OLSR	time-based
Approach in [291]	none	probabilistic-like with forwarding factor	sender + packet identifier

link quality. However, such problems also have to be dealt with under conventional routing approaches.

6.2.2.4   Overview of Approaches

This section aims to give a tabular overview of the characteristics of network coding approaches for DTNs from various publications. A top-level separation is done by the count of sinks of single flows, i.e. whether the approach targets broadcast, multicast, or unicast networks.

Overview of Broadcast Approaches

In broadcast delay-tolerant networks, network coding can result in faster and more reliable message dissemination. Table 6.1 lists a bunch of different, recent approaches in this context, their required knowledge of the network, and the used routing and generation management strategies.

Overview of Multicast Approaches

Multicast communication can be seen as a more generalized form of broadcast communication: Messages are forwarded to a set of multiple, but not all of the available nodes. Table 6.2 provides a list of existing publications on multicast approaches to network coding in DTNs.

---

[4]concerning other nodes in the network

Table 6.2: Overview of Multicast Approaches

Publication	Knowledge[5]	Routing	Generations
Scheme in [27]	not required	Epidemic-like	different sources
Network-Coded Multicast [265]	none	random sender	not mentioned
Approach in [308]	not mentioned	Epidemic	all source packets

Overview of Unicast Approaches

Unicast is the most common type of information flows in current networks. Every packet has exactly one source and sink. Unicast message delivery is also the most widely researched type of network-coded communication flows in the DTN context. It is especially appealing due to the difficult decision which messages to encode together if multiple concurrent flows come into play. Table 6.3 gives an overview of state-of-the-art unicast approaches in the field. See Section 6.2.2.1 for a more detailed discussion of the mentioned criteria.

6.2.2.5 Open Research Problems

In summary, the aforementioned publications address a plethora of problems in the field of network coding in DTNs. However, as an outlook to further work, the following topics can be considered not sufficiently addressed in the state of the art:

1. **Network coding with comprehensive topological knowledge.** As shown by Jain et al. [10], increasing knowledge of the time-varying network topology of a DTN allows for better routing decisions. Such knowledge for example is available in interplanetary network scenarios. In purely deterministic scenarios, messages are forwarded over a single path. However, if unexpected link failures and untrusted nodes have to be taken into account, a certain degree of replication is beneficial. Coping with such characteristics by extending deterministic approaches is currently researched in the course of developing an opportunistic variant of CGR [53]. If network coding could be fully leveraged in these scenarios, additional benefits regarding throughput and delivery probability can be expected. This includes approaches planning all encoding, re-encoding, and decoding events in advance. "Source-based network coding" may also be considered.

---

[5]concerning other nodes in the network

Table 6.3: Overview of Unicast Approaches

Publication	Knowledge[6]	Routing	Generations	Recoding	Recovery
HubCodeV1 [23]	type of neighbors	hub-based	same destination	at "hubs"	not mentioned
HubCodeV2 [23]	type of neighbors, message IDs	hub-based	same destination	at "hubs"	not mentioned
Appr. in [27]	not required	Epidemic-like	no	always	G-SACKs
Appr. in [30]	msgs. of neighbors	Spray&Wait[7]	no	always	not mentioned
Appr. 1 in [34]	encoding matrices	token-based	yes	TX: always, RX: if buffer full	out of scope
Appr. 2 in [34]	matrix rank	token-based	yes	TX: always, RX: if buffer full	out of scope
NCBS [83]	none	Spray&Wait-like	per message	RX: if buffer full	TTL, no ACK
CANCO [83]	delivery probability	Spray&Wait-like	per message	RX: if buffer full	TTL, no ACK
vCF [84]	ICTs[8], msg. in buffer	Dijkstra (ICT)	per message	always	ACK, selective dropping
GER [135]	msgs. of neighbors	Epidemic-like	not mentioned	always	del. after enc.
GFC [135]	none	First Contact	not mentioned	always	not mentioned
TBNC [175]	node pos./directions	trajectory-based	per message	if mult. next hops	TTL

[6]concerning other nodes in the network
[7]in simulation, approach in general flexibly defined
[8]inter-contact times

Scheme					
NCER [179]	not required	Epidemic-like	not discussed	TX: always, RX: if buffer full	TTL, ACK
E-NCP [178]	"spray list"	limited spraying	not discussed	always	TTL, ACK, spray count
Impl. in [214, 285]	not used for NC	random msg.	fragments of one bundle	TX: always	not mentioned
Scheme in [218]	neighbor type	"two-hop single-unicast"	not discussed	always	not mentioned
Appr. in [220]	not required	Epidemic-like, random block	same unicast "session"	TX: always	not discussed
CafNC [223]	"interests" of neighbors, heuristics	custom heuristic-based	same "topic" and receiver	dynamic	not discussed
NC-LB [253]	local msgs.	local decision	same time window	TX: always, RX: if buffer full	drop oldest generation
Appr. in [255]	none	Spray&Wait-like	no (inter-session)	yes	out of scope
Appr. in [280]	none	flexible	same time window & same dst.	always	TTL (hops + time)
Appr. in [304]	msgs. of neighbors	Epidemic-like	comb. of same packets from same source	on RX	replace least recent
DSNC [307]	msgs. of neighbors	prioritized by age and TTL	same "segment"	on reception	ACK, TTL
Scheme in [315]	encoding matrices	Epidemic-like	fixed size	TX: always, RX: if buffer full	broadcast ACK

Appr. in [312]	density & locations	Epidemic-like	subset of a file	on TX	ACK
NTC [317]	msgs. of neighbors	Epidemic-like	same destination	selective (redundancy ratio)	not mentioned

2. **Multicast in DTNs.** Though it has been shown by Ahlswede et al. [21] that for achieving the capacity bound in multicast networks it is mandatory to employ network coding, to the current date only few publications discussed multicast coding in DTNs. Extending research in this direction may be viable to enhance packet delivery in multicast DTNs in the future. This point has already been identified by Zhang et al. in 2012 [313].

3. **Short messages.** A further open topic identified in [313] is the transmission of small packets. Most network coding approaches target generations formed from single large flows. If several small messages need to be transmitted, fragmenting them to form generations results in too much overhead to be viable. However, to gain advantages from employing network coding, whole blocks of data should be transmitted at once. This is the result of the receiver(s) needing to decode the complete generation to make use of the contained data. A coding scheme which effectively addresses these problems is an open research topic.

4. **Network coding in combination with infrastructure-assisted routing.** If infrastructure-assisted routing is used (see e.g. the comprehensive survey by Cao et al. [72]), nodes control the time-varying network topology by themselves. A network coding scheme may as well benefit from this characteristic.

6.2.3   Implementation Considerations and Case Study

After an overview of network coding in general and an up-to-date analysis of associated research in the DTN domain has been provided, selected practical considerations are presented in the following. Based on a general discussion of the processing sequence within network coding-enabled hosts, a bundle protocol extension block is introduced, containing all necessary information. The following content will focus on network coding for bundles with the same source and same destination (*intra-flow network coding*).

6.2.3.1   Processing Flow

In this section, network coding in the context of the bundle protocol is considered. Specifically, a proposal for processing steps taken by DTN hosts to encode bundle payloads is described and the necessary information included into a bundle protocol extension block is discussed.

Figure 6.6 provides an overview of the three stages traversed to create network-coded bundle payloads. The stages are:

1. **Generation preparation:** A set of n bundles is selected and included into one generation. Furthermore, the payload is aligned in regard to its size by adding padding bits.

2. **Encoding:** The sequence of bits forming the payload plus padding bits are interpreted as a sequence of symbols of a selected finite field $\mathbb{F}_q$. Using a coeffi-

cient matrix consisting of m coefficient vectors, encoded messages are created using the approach discussed in section 6.2.1.

3. **Extension block creation:** Information necessary for decoding the messages created in the second step is embedded into a bundle protocol extension block. Furthermore, information in the primary bundle block is adapted.

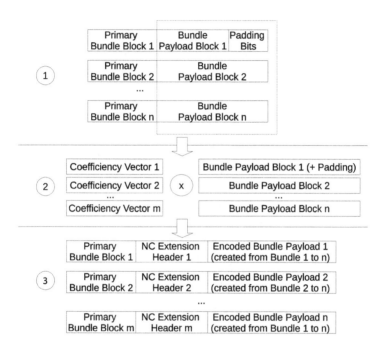

Figure 6.6: Overview of applying network coding to bundles

The processing steps may be applied by a bundle source as well as by any DTN host on the path from source to destination.

In the first step, a host waits for a specific amount of bundles having the same destination, determined by the Endpoint Identifier (EID). This amount, i.e. the generation size, is defined by a parameter. Due to the delay tolerance of the network, this step does not necessarily imply inefficiency. This is specifically the case if a relaying DTN host has several incoming contacts and only few outgoing ones with larger capacity. However, to avoid starvation, a timer should be introduced which independently triggers the transition from the first to the second stage for the accumulated bundles belonging to the same generation.

Another aspect of this first step is the deduction of the optimal size for whole bundles or bundle fragments. Introducing padding bits to achieve same bundle sizes is only a second step after bundles of similar sizes are aggregated together in one generation. Combining several bundles in one generation with heavily varying payload sizes implies a waste of bandwidth if network coding is applied. Thus, the bundles should be of similar size. Three options exist for this: Firstly, bundles of similar size may be created by the source. Secondly, bundles of varying size may be created. However, due to the contact schedule and the bundle creation frequency, a sufficient quantity of bundles of similar size can be aggregated into a single generation by an encoding host. Thirdly, fragmentation with the goal to adjust the bundles' sizes could be applied.

After a generation has been created, encoding can be performed. In this second stage, the payload blocks of the bundles included in one generation are interpreted as sequences of symbols which are elements of the selected finite field, as discussed in section 6.2.1. The generation is multiplied with a randomly generated coefficient matrix. Depending on the implementation of this stage, two variants may be leveraged:

1. $m > n$: Especially if network coding is applied on so-far-not-encoded bundles, it is desired to create more encoded bundles than the size of the generation.

2. $m \leq n$: If re-encoding is done by intermediate nodes, it may be the case that only single – or at least less than the generation size – new encoded bundles are created. By this, the redundancy within the network may be controlled based on the next-hop contact probabilities.

For practical applications of network coding it has to be taken into account that routing/forwarding has quite an impact on both the first and second stage. A mutual dependency between the network coding process and routing/forwarding exists: For determining the bundle payload size, the minimal link capacity of the path to the destination has to be known in order to fragment the bundle in an appropriate way (stage 1). Furthermore, in order to determine the necessary amount of redundancy, probabilities for the selected path to the destination are an important factor one has to take into account (stage 2). Thus, only if the path's characteristics are known, network coding can be optimally applied. However, in that case, further messages are created that have to be handed over to the routing/forwarding component. Unfortunately, this issue has not been solved by any research efforts so far.

As soon as the bundle payload has been network-coded, the modified bundles are created and handed over to the forwarding procedure for selecting the next hop(s). The freshly created bundles have to contain a dedicated extension block including the necessary information for decoding the payload. Based on the discussion in this chapter, the block must contain fields holding the following information:

- **Generation size**. The amount of original bundles has to be known by the decoding DTN host to determine the amount of unknowns.

- **Bundle identifier**. Depending on the version of the bundle protocol, a set of information entities is required which uniquely identifies the original bundle. In the case of bundle protocol version 6, as defined in RFC 5050 [245], the "creation timestamp (...) together with the source endpoint ID and (if the bundle is a fragment) the fragment offset and payload length, serve to identify the bundle." (p. 20 of [245]). Thus, for each bundle included in the generation, this information has to be embedded into the extension block.

- **Selected finite field**. To determine the symbols used in the encoded bundle payload and to solve the equations of n unknowns, the underlying finite field has to be known.

- **Coefficient vector**. It has to be known by the decoding entity which message has been multiplied by which coefficient in order to create the encoded payload included in a received bundle.

- **Original bundle size**. The size of the bundle without the padding bits has to be known. This information may be included in the bundle identifier as it is the case with the payload length information in bundles conforming to the bundle protocol version 6.

For the case study presented in the next section, a prototypical extension of the DTN bundle protocol containing this information has been used.

6.2.3.2   Case Study

After an overview of practical aspects has been given, a simple case study[9] is presented in this section. This case study points out characteristics of network coding in a specific use case: In a LEO satellite network, DTN protocols are applied to render possible world-wide communication by using the satellites as message ferries. By this approach, a low-cost communication network can be established (see e.g. [62]). In the sketched network, only contacts between ground stations and LEO satellites are taken into account. Furthermore, probabilistic links between ground stations and satellites are assumed. Thus, it may happen that a contact that is represented in the contact plan calculated a priori does not appear.

Using a research environment based on the *ONE* DTN simulator [159], the described network has been simulated. Within the simulation, two setups have been created: In the first setup (*5s5g*) five satellites and five ground stations are leveraged, in the second setup (*10s10g*) ten satellites and ten ground stations. The ground stations are randomly placed on the Earth's surface. Trajectory information of real LEO satellites have been gathered as a basis for the simulated satellites' movement. As routing protocol, an own implementation of a CGR-like algorithm has been used. This approach leverages global contact plan knowledge to determine optimal paths based on Dijkstra's algorithm. Global contact plan knowledge implies that every node has access to a database that contains the overall available nodes and their contacts during the simulation time. Furthermore, the overall and remaining capacity of each contact

---

[9]The case study has been partially implemented by Jean Chorin and Olivier De Jonckère under the supervision of Marius Feldmann and Felix Walter. The used network coding library *libtinync* has been implemented by Sebastian Schrader, the extension to The ONE partially by Ricardo Böhm.

is provided. Thus, each node can determine which amount of data may be transferred during each contact between a ground station and a satellite. In the implementation, bundles which are scheduled for a contact that does not appear, are lost.

For the purpose of this case study, bundles are created by one source ground station only. Furthermore, they are transmitted to the same destination ground station. An important characteristic of the bundle creation mechanism is the route validation step: Before a bundle is created at a specific point in time, it is checked whether a path from source to destination exists, offering the necessary contact capacity for transmitting the bundle. If it does not exist, the bundle is not created. This approach avoids network overloads due to freshly created messages.

The following parameters have to be determined for the applied simulations:

- **Generation size (GS)**: Amount of bundles used for the initial coding step.

- **Recoding threshold (RT)**: The percentage value with a range between 0 % and 100 % refers to the generation size and determines the amount of messages that has to be present in order to apply network coding. If e.g. the generation size is 10 and the recoding threshold is 80 %, it results in recoding as soon as 8 messages are available at an intermediary node. A value of 0 % results in recoding as soon as one message is available.

- **Recoded messages (RM)**: Provided as a percentage value with a range from 0 % to 200 %, it defines how many messages are created during the coding step. The value refers to the generation size.

In a first step, the impact of the recoding threshold and of the generation size has been analyzed. For this purpose, these parameters have been varied and different values indicating network characteristics have been gathered. These values are:

- **Packet Delivery Ratio (PDR)**: The relation between initially transmitted and finally received (and, thus, successfully decoded) bundles.

- **Average hops**: The average count of transmissions taken from source to destination through the LEO satellite network.

- **Average end-to-end delay**: The average time in seconds taken by a message to reach the destination node from its source.

- **Messages created at the source**: The total amount of bundles initially created by the sender using the route validation step mentioned above.

- **Messages recoded**: Total amount of bundles that have been created by recoding, either by the source or by intermediary nodes.

- **Total amount of messages**: Sum of the two aforementioned values.

Figure 6.7 shows the results for these six values in the case of the 5s5g scenario with a generation size of 25 and 120 % recoded messages. The contact probability is set to 0.5.

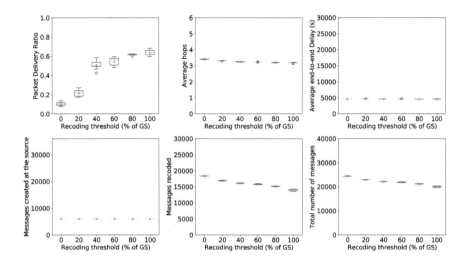

Figure 6.7: Results for scenario with 5 satellites, 5 ground stations, GS = 25, RM = 120 %

The results show a small impact of the recoding threshold on most values except the delivery probability, which is improved significantly when increasing the recoding threshold from 0 % to 40 % and then slightly improves when increasing it further. This finding has been confirmed in any of the further created scenarios with varying generation sizes and amounts of recoded messages.

The situation is very similar regarding the dependency on the generation size. Figure 6.8 shows exemplary results for a setup with a contact probability of 0.5, an RM value of 120 % and an RT value of 40 %.

As it can be seen in Figure 6.8, the PDR values are slightly worse in the case of small generation sizes.

The most interesting parameter to modify is the amount of recoded messages. Figure 6.9 summarizes the results from a simulated 5s5g scenario with a generation size of 15, a contact probability of 0.5 and an RT value of 60 %. The same parameters have been used in the 10s10g scenario. The associated results are shown in Figure 6.10.

As it can be seen, the PDR value can be heavily increased by applying network coding. If no recoded messages are added (RM = 0 %), the generation is encoded without creating additional messages. If only one message from the generation is lost, the whole generation is lost as mentioned above.

Thus, the conducted case study confirms one of the core advantages of network coding: By employing network coding, the probability of successful end-to-end bundle delivery in DTNs with lossy links can be heavily increased.

The downside of network coding in the sketched scenarios is obvious, taking the average hop count and end-to-end delays into account. As it can be seen, hop count

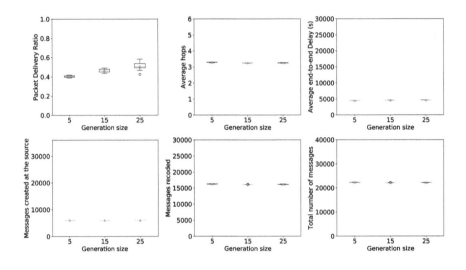

Figure 6.8: Results for scenario with 5 satellites, 5 ground stations, RM = 120 %, RT = 40 %

and delays both increase if more messages are created by recoding. The reason for this is simply the exhaustion of the best available paths. If the count of messages in the network increases, a greater portion of messages has to take less optimal routes to the destination because of the limited capacity on optimal routes. This can be verified

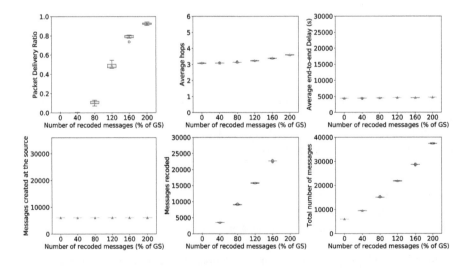

Figure 6.9: Results for scenario with 5 satellites, 5 ground stations, GS = 15, RT = 60 %

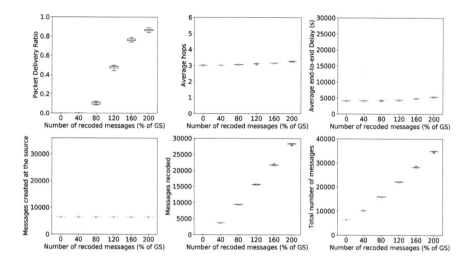

Figure 6.10: Results for scenario with 10 satellites, 10 ground stations, GS = 15, RT = 60 %

by comparing the results from the 5s5g scenario to the 10s10g scenario. Due to more nodes and more contacts being available, the 10s10g scenario offers a greater overall capacity and also a greater amount of short, low-delay paths. Thus, the impact of a greater overall message count on the hop count and end-to-end delays is smaller.

This brief discussion already shows the importance of analyzing the network in which network coding should be applied in order to determine valid parameters. Redundancy always comes with the cost of additional network resource usage. Thus, the tradeoff between these two factors has to be analyzed for the addressed scenario.

After focusing on coding within the network, coding at the physical layer and packet-level coding will be addressed in a next step.

6.3 Coding at the Physical Layer and Packet-Level Coding

6.3.1 Basics of Channel Coding with Application to CCSDS Links

Error control coding lies at the core of nearly any modern digital communication system. In fact, reliable communications over a link affected by noise cannot be established without a suitable mechanism for the recovery of transmission errors. When dealing with error correction, two main approaches shall be distinguished, i.e.

■ Forward error correction (also known as channel coding);

■ Automatic retransmission protocols.

While the latter require the presence of a feedback channel to signal to the transmitter which data units arrived corrupted at the receiver, the former attacks the error correction problem by adding (proactively) to the transmitted information a certain amount of redundancy which is then used at the receiver to detect and (possibly) correct errors. Retransmission protocols have the inherent advantage of entailing a low complexity. Nevertheless, their applicability is limited, especially in satellite and space communications, due to the large propagation delays which render buffering at the transmitter side (for future possible retransmission requests) extremely costly. In the remainder of the chapter, we will focus hence on forward error correction only.

In its most simple form a channel coding scheme is composed by an encoder, which maps an information message $\mathbf{u} = (u_1, u_2, \ldots, u_k)$ composed by k bits onto a codeword $\mathbf{x} = (x_1, x_2, \ldots, x_n)$ of n symbols, and by a decoder, which produces an estimate $\hat{\mathbf{u}}$ of the transmitted message, given the noisy observation of the transmitted codeword at the communication channel output, $\mathbf{y} = (y_1, y_2, \ldots, y_n)$ (see Figure 6.11). We shall denote the encoding function as $E(\cdot)$ and the decoding function as $D(\cdot)$. Thus, we have

$$\mathbf{x} = E(\mathbf{u}) \qquad \text{and} \qquad \hat{\mathbf{u}} = D(\mathbf{y}).$$

Two types of error measures are used to describe the effectiveness of a channel coding scheme: Block and bit error probabilities. The block error probability P_B is simply the probability that the estimated message differs from the transmitted one, whereas the bit error probability is the probability that a decoded bit differs from the transmitted value, i.e.

$$P_B = \Pr\{\hat{\mathbf{u}} \neq \mathbf{u}\} \qquad \text{and} \qquad P_b = \frac{1}{k} \sum_{\ell=1}^{k} \Pr\{\hat{u}_\ell \neq u_\ell\}.$$

Bit and block error probabilities give only a partial description on the quality of a channel code. We measure the efficiency of the channel coding scheme via the code rate $R = k/n$. The code rate summarizes the amount of redundancy introduced in the transmission: We shall expect lower error probabilities when the code rate is low (large amount of redundancy), whereas the error probability shall rise when the code rate is large (a little redundancy). It is hence obvious that code rate and error probability shall be subject of a trade-off. It is less obvious to observe that, over a given channel[10], the (bit and block) error probabilities can be made arbitrarily small if the code rate lies below the so-called *capacity* of the channel. To characterize such largest possible coding rate, we shall first detail a statistical model of the channel. The statistical model is given in the form of channel transition probability (density) $p(\mathbf{y}|\mathbf{x})$. If the channel is memory-less and time-invariant, the channel transition probability (density) can be factorized as

$$p(\mathbf{y}|\mathbf{x}) = \prod_{\ell=1}^{n} p(y_\ell|x_\ell).$$

---

[10]We refer here only to *ergodic* channels.

Figure 6.11: Model of channel coded transmission

The channel capacity is given by

$$C = \max_{Q(x)} E \left[\log_2 \frac{p(Y|X)}{\sum_x Q(x)p(Y|x)} \right] \qquad (6.4)$$

where X, Y are the random variables associated with x and y (channel input and output), $Q(x)$ is the distribution of the channel input X and the expectation if over X, Y. Shannon's channel coding theorem states that the (bit/block) error probability can be made arbitrarily small by transmitting at a rate $R < C$, and that conversely for any $R > C$ the error probability cannot be made arbitrarily small. Moreover, in absence of feedback, reliable communication at rates close to capacity can be attained only by coding over large blocks, i.e., by taking $n \to \infty$ (and $k = nR \to \infty$ for any fixed rate R).

During the past decades, a huge amount of effort has been addressed to the design of codes capable of approaching the limit (6.4) with an affordable complexity, as remarkably summarized in [91]. In fact, the conceptually simple optimum decoding strategy, referred to as maximum likelihood (ML) decoding, requires implementing the rule

$$\hat{\mathbf{x}} = \arg\max_{\mathbf{x} \in \mathscr{C}} p(\mathbf{y}|\mathbf{x}) \qquad (6.5)$$

where $\mathscr{C}$ is the codebook with size 2^k. Once $\hat{\mathbf{x}}$ is found, by inverting the encoding mapping one obtains $\mathbf{u} = E^{-1}(\mathbf{x})$. The exhaustive search in (6.5) has complexity growing exponentially with the message size, and hence it is impractical already for moderate-small k. The possibility to achieve capacity with long codes hence requires first solving efficiently the decoding problem. A breakthrough in the research of low-complexity capacity-approaching codes has been the introduction of (sub-optimum) *iterative* decoding algorithms. While iterative decoding gained public praise as an efficient method for approaching capacity only recently with the introduction of turbo codes [43], its roots lie in the seminal work on low-density parity-check (LDPC) codes by R.G. Gallager [122]. Both turbo and LDPC codes belong to the wide class of linear block codes, and are characterized by a simply graphical representation that enables channel decoding as an iterative message passing algorithm over the code's graph. The class of linear block codes, and in particular LDPC codes, will be discussed in detail in the next subsection. After that, we provide a brief overview of the channel coding options currently included in the Consultative Committee for Space Data Systems (CCSDS) recommendations.

6.3.1.1 Binary Linear Block Codes

A (n,k) binary linear block code can be conveniently described by its encoding process, which involves the multiplication of the information vector $\mathbf{u}$ by a $k \times n$ binary matrix $\mathbf{G}$, i.e.

$$\mathbf{x} = \mathbf{uG} \tag{6.6}$$

where the operations $(+,\cdot)$ are defined over the binary finite field $\mathbb{F}$,

+	0	1		·	0	1
0	0	1		0	0	0
1	1	0		1	0	1

By rewriting (6.6) as

$$\mathbf{x} = \sum_{\ell=1}^{k} u_\ell \mathbf{g}_\ell \tag{6.7}$$

where $\mathbf{g}_\ell$ is the ℓth row of $\mathbf{G}$, we may observe that the code $\mathscr{C}$ is generated as a k-dimensional linear subspace of the n-dimension vector space $\mathbb{F}_2^n$, with base vectors given by the k (linearly-independent) rows of $\mathbf{G}$. Thanks to this choice, encoding can be performed with (polynomial) complexity $\mathscr{O}(k^2)$. The matrix $\mathbf{G}$ is usually referred to as *generator matrix* of the code.

Consider next a set of $n-k$ linearly independent vectors spanning the $n-k$ subspace of $\mathbb{F}_2^n$ orthogonal to $\mathscr{C}$,

$$\mathbf{h}_1$$
$$\mathbf{h}_2$$
$$\dots$$
$$\mathbf{h}_{n-k}.$$

By definition we have

$$\mathbf{x}\mathbf{h}_\ell^T = 0 \qquad \forall \mathbf{x} \in \mathscr{C} \text{ and } \ell = 1,\dots,n-k. \tag{6.8}$$

The matrix $\mathbf{H}$ having the $n-k$ vectors $\mathbf{h}_1, \mathbf{h}_2, \dots, \mathbf{h}_{n-k}$ as rows is the *parity-check matrix* of the code. The $n-k$ equations in (6.8) are referred to as parity-check equations. By means of $\mathbf{H}$, the parity-check equations can be compactly summarized by the constraint

$$\mathbf{x}\mathbf{H}^T = \mathbf{0}. \tag{6.9}$$

A linear block code can be defined either via the generator matrix, as the set of vectors in $\mathbb{F}_2^n$ obtained by multiplying all possible k-bit vectors by $\mathbf{G}$, or via the parity-check matrix as the set of vectors in $\mathbb{F}_2^n$ satisfying the parity-check equations of (6.8).

A (n,k) code can be generated via different generator matrices, i.e., the choice of the base vectors forming the matrix $\mathbf{G}$ is not unique. In fact, starting from a given generator matrix, an alternative generator matrix can be obtained as

$$\mathbf{G}' = \mathbf{TG} \tag{6.10}$$

for any $k \times k$ non-singular binary matrix $\mathbf{T}$. The rows of the two matrices $\mathbf{G}$ and $\mathbf{G}'$ will span the same k-dimensional space, but they will eventually yield a different mapping between information vectors $\mathbf{u}$ and codewords $\mathbf{x}$. The generator matrix is referred to be in *systematic form* if

$$\mathbf{G} = (\mathbf{I}_k | \mathbf{P}) \tag{6.11}$$

with $\mathbf{I}_k$ being the $k \times k$ identity matrix, and $\mathbf{P}$ being a $(n-k) \times n$ binary matrix. For a code generated by a systematic form generator matrix, it is easy to observe that

$$\mathbf{x} = (\mathbf{u} | \mathbf{p})$$

i.e., the first k position of the codeword contains the information vector $\mathbf{u}$. The remaining part of the codeword is given by $\mathbf{p} = \mathbf{u}\mathbf{P}$ and it is referred to as *parity* or *redundancy*. Every binary linear block code admits a generator matrix in systematic form. Denote by $\mathbf{A}$ the matrix composed by the k rightmost columns of a given generator matrix. If $\mathbf{A}$ is non-singular, it is sufficient to set $\mathbf{T} = \mathbf{A}^{-1}$ in (6.10) to obtain the corresponding systematic form of the generator matrix. If on the contrary the k rightmost columns of the generator matrix are linearly dependent, a systematic form of the generator matrix can be obtained by permuting the matrix columns such that the rank of the sub-matrix $\mathbf{A}$ is full; The systematic form is then obtained as before, i.e., by setting $\mathbf{T} = \mathbf{A}^{-1}$ in (6.10).

For a binary linear block code with a generator matrix in the form of (6.11), the parity-check matrix can be easily obtained as

$$\mathbf{H} = \left(\mathbf{P}^T | \mathbf{I}_{n-k}\right)$$

with $\mathbf{I}_{n-k}$ being the $(n-k) \times (n-k)$ identity matrix. As for the generator matrix case, a binary linear block code can be defined by different (equivalent) parity-check matrices.

6.3.1.2 Channel Codes for Telemetry (Downlink)

The telemetry downlink in CCSDS admits a large set of channel coding options. Historically, the reason for this stems from the fact that the telemetry downlink is typically very power limited, demanding for the use of power error correcting codes. As a consequence, the CCSDS recommendations for telemetry links have been always quick in adopting the latest (and most powerful) channel coding schemes. At the time being, the recommended schemes include

- Rate-1/2, memory-6 convolutional codes with octal generators $(171, 133)$. The encoder is depicted in Figure 6.12. Higher rates $(2/3, 3/4, 5/6,$ and $7/8)$ can be obtained by *puncturing* (i.e., selectively removing) some of the encoder output bits according to pre-defined patterns. Decoding here can be performed by means of the Viterbi algorithm over the 64-states code trellis.

- Reed-Solomon codes over an order-256 finite field. Reed-Solomon codes belong to the class of non-binary linear block codes. Due the field order choice, the codes adopted by CCSDS work on symbols composed by 8 bits each (bytes). The code parameters are $n = 255$ and $k = 239$ symbols (with the capability of guaranteeing the correction of all patterns with 8 symbol errors or less), or $n = 255$ and $k = 223$ symbols (guaranteed error correction capability of 16 symbol errors).

■ The serial concatenation of convolutional and Reed-Solomon codes. Concatenated codes using as an outer code one of the above-mentioned Reed-Solomon codes, and as an inner code the rate-$1/2$ convolutional code (or one of its punctured versions) are recommended with block, byte- wise interleaving, via interleaving matrices of $n \times I$ bytes, with $n = 255$ and $I = 1, 2, 3, 4$ or 5. Decoding takes place in two stages: first, the inner convolutional code is decoded (e.g., via Viterbi algorithm). The convolutional code decoded stream is first de-interleaved and then input to the (algebraic) Reed-Solomon decoder, which aims at removing residual errors.

■ Turbo codes (in the parallel concatenation form). The classical turbo code construction obtained by the parallel concatenation of two recursive convolutional encoders is adopted, where the two encoders are based on memory-4 convolutional codes. Decoding may be performed by iterative soft-estimates of the information bits, which are produced in turn by soft-input soft-output Bahl-Cocke-Jelinek-Raviv (BCJR) decoders operating over the trellises of the two convolutional codes. The family of CCSDS convolutional turbo codes comprises the code rates $1/2$, $1/3$, $1/4$, and $1/6$, and input block lengths of $k = 1784, 3568, 7136, 8920$ or 16384 bits.

■ Turbo codes (in the serial concatenation form). Recently added to the CCSDS family, the serial concatenation involves two memory-2 convolutional encoders. Puncturing at the output of the inner/outer encoder may be employed to obtain different code rates. Code rates from $\approx 1/3$ up to 0.9 are foreseen, as well as block lengths from 16200 up to 48600 encoded bits. The scheme is meant to be used together with modulation formats for spectral efficient links, i.e., quadrature phase shift keying (QPSK), 8-ary phase shift keying (8-PSK), and 16, 32 or 64-point amplitude-phase shift keying (APSK). Due to this choice, the scheme is particularly suited for near-Earth missions.

■ LDPC codes. Various families have been included, i.e.

   ■ The family of LDPC codes from the Digital Video Broadcasting – Second Generation Satellite (DVB-S2) standard. Codes with block lengths $n = 64800$ and $n = 16200$ are included, with rates ranging from $\approx 1/5$ to $\approx 9/10$. Also this scheme is meant to be used together with modulation formats for spectral efficient links, i.e., QPSK, 8-PSK, and 16 or 32-point APSK.

   ■ A family of *protograph*-based LDPC codes with input block size $k = 1024, 4096$ or 16384 bits and code rates $1/2, 2/3$ and $4/5$.

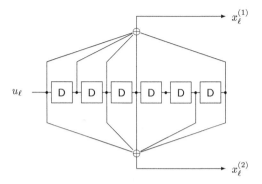

Figure 6.12: Encoder for the memory-6 convolutional codes with octal generators $(171, 133)$.

■ A single $(8160, 7136)$ LDPC code.

6.3.1.3 Channel Codes for Telecommand (Uplink)

The original telecommand uplink standard foresaw the use of a $(63, 57)$ Bose-Ray–Chaudhuri-Hocquenghem (BCH) code. The code, possessing a minimum distance of 4, allowed for the correction of all error patterns with a single bit error and the detection of all even-weight error patterns (single error correction mode), or (as error detection code only) for the detection of all even-weight error patterns together with error patterns of weight 2 (triple error detection mode). The standard has been recently amended to include more powerful codes. In particular, $(128, 64)$, $(256, 128)$, and $(512, 256)$ protograph-based LDPC codes have been included.

6.3.1.4 Channel Codes for Packet Erasures

The codes described in the Sections 6.3.1.2, 6.3.1.3 operate at the physical layer of the communication scheme, i.e., decoding takes place directly at the output of the demodulator. In the past two decades, application of error correcting codes to higher layers of the protocol stack became widespread in many wireless communication systems (see e.g. [29, 65, 187]). Here, the concept is to generate *redundant* packets at a given layer of the protocol stack, and to use the redundant packets at the receiver side to recover from packet losses. Codes that operate in such a fashion are often referred to as *erasure codes*. Recently, CCSDS investigated the inclusion of erasure codes among the recommended channel coding families, resulting in the experimental specification of [15]. We shall review next the basic concepts of erasure coding, and discuss its application to the Delay/Disruption Tolerant Network (DTN) setting.

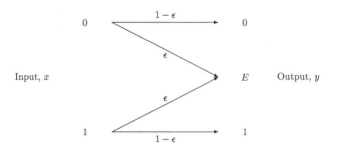

Figure 6.13: The binary erasure channel.

6.3.2 Erasure Codes

6.3.2.1 Basics

We consider now the transmission over a memoryless binary erasure channel (BEC) with erasure probability ε. The channel is represented in Figure 6.13, and it is described by the transition probabilities

$$p(y_\ell|x_\ell) = \begin{cases} 1-\varepsilon & \text{if } y_\ell = x_\ell \\ \varepsilon & \text{if } y_\ell = E \end{cases}$$

where the channel output E denotes an erasure. Thus, on the BEC a codeword bit is either correctly received (with probability $1-\varepsilon$) or erased (with probability ε). The plain application of the ML decoding rule (6.5) would require to list all codewords that are equal, in the non-erased positions, to the received vector **y**. If the list contains only one codeword, then the decoder outputs the codeword and success can be declared. If the list contains more than one codeword, then the decoder outputs a randomly-picked codeword in the list. The probability of success is given by the probability of picking the right codeword in the list, i.e., it is the inverse of the list size. By following this approach, however, would lead to a complex decoder, being the number of codewords to be checked exponential in k.

We shall observe that, if the code is linear, the decoding problem can be turned into the solution of a set of linear equations. The observation follows from the definition of linear codes from either the generator or the parity-check matrix perspective. We illustrate how ML decoding can be performed efficiently by introducing a working example. We consider the transmission with a $(7,4)$ Hamming code with generator matrix

$$\mathbf{G} = \begin{pmatrix} 1 & 0 & 0 & 0 & 1 & 1 & 1 \\ 0 & 1 & 0 & 0 & 1 & 1 & 0 \\ 0 & 0 & 1 & 0 & 1 & 0 & 1 \\ 0 & 0 & 0 & 1 & 0 & 1 & 1 \end{pmatrix}.$$

The generator matrix is in systematic form, which readily yields the parity-check matrix

$$\mathbf{H} = \begin{pmatrix} 1 & 1 & 1 & 0 & 1 & 0 & 0 \\ 1 & 1 & 0 & 1 & 0 & 1 & 0 \\ 1 & 0 & 1 & 1 & 0 & 0 & 1 \end{pmatrix}.$$

Consider the transmission over a BEC with output given by

$$\mathbf{y} = (E \quad x_2 \quad E \quad x_4 \quad E \quad x_6 \quad x_7)$$

i.e., the first, third and fifth bits are erased. We begin by showing how decoding can be performed, based on the code generator matrix. Recall that the information vector and the codeword are related via the generator matrix through $\mathbf{x} = \mathbf{uG}$. We construct a *reduced* generator matrix $\mathbf{G}_R$ by removing from $\mathbf{G}$ the columns associated to erased symbols, i.e., the first, third and fifth columns of $\mathbf{G}$ are removed leading to

$$\mathbf{G}_R = \begin{pmatrix} 0 & 0 & 1 & 1 \\ 1 & 0 & 1 & 0 \\ 0 & 0 & 0 & 1 \\ 0 & 1 & 1 & 1 \end{pmatrix}.$$

Similarly, we construct *reduced* codeword $\mathbf{x}_R$ by removing the erased symbols from $\mathbf{y}$. We have $\mathbf{x}_R = (x_2 \quad x_4 \quad x_6 \quad x_7)$, with all elements of $\mathbf{x}_R$ known (i.e., not erased) at the decoder. We observe next that

$$\mathbf{uG}_R = \mathbf{x}_R$$

where $\mathbf{u} = (u_1 \quad u_2 \quad u_3 \quad u_4)$ being a vector of unknowns, $\mathbf{G}_R$ the 4×4 system matrix, and $\mathbf{x}_R$ the known term in the linear equation. To reconstruct the information vector $\mathbf{u}$, we have hence to solve the 4×4 system of equations

$$\begin{aligned} u_2 &= x_2 \\ u_4 &= x_4 \\ u_1 + u_2 + u_4 &= x_6 \\ u_1 + u_3 + u_4 &= x_7. \end{aligned}$$

In this specific case, the matrix $\mathbf{G}_R$ has rank equal to $k = 4$, thus a unique solution exists. By applying Gauss-Jordan elimination, we get

$$\begin{aligned} u_1 &= x_2 + x_4 + x_6 \\ u_2 &= x_2 \\ u_3 &= x_7 + x_2 + x_6 \\ u_4 &= x_4. \end{aligned}$$

A similar procedure can be applied starting from the parity-check matrix. In this case, we need first to construct a *reduced* parity-check matrix $\mathbf{H}_R$ by removing from $\mathbf{H}$ all columns associated to erased bits, i.e., the columns with indexes $1, 3$ and 5, yielding

$$\mathbf{H}_R = \begin{pmatrix} 1 & 0 & 0 & 0 \\ 1 & 1 & 1 & 0 \\ 0 & 1 & 0 & 1 \end{pmatrix}.$$

The removed columns are used to construct a *complementary* matrix $\mathbf{H}_{\bar{R}}$

$$\mathbf{H}_{\bar{R}} = \begin{pmatrix} 1 & 1 & 1 \\ 1 & 0 & 0 \\ 1 & 1 & 0 \end{pmatrix}.$$

Recall that the reduced codeword $\mathbf{x}_R$ is obtained by removing the erased symbols from $\mathbf{y}$. We have $\mathbf{x}_R = (x_2 \quad x_4 \quad x_6 \quad x_7)$, with all elements of $\mathbf{x}_R$ known (i.e., not erased) at the decoder. We introduce the complementary vector $\mathbf{x}_{\bar{R}}$ whose elements are the erased codeword bits, $\mathbf{x}_{\bar{R}} = (x_1 \quad x_3 \quad x_5)$. We make use of the parity-check equations by re-writing (6.9) as

$$\mathbf{x}_{\bar{R}} \mathbf{H}_{\bar{R}}^T = \mathbf{s}$$

where

$$\mathbf{s} = \mathbf{x}_R \mathbf{H}_R^T$$
$$= (x_2 \qquad x_2 + x_4 + x_6 \qquad x_4 + x_7)$$

is the known term, $\mathbf{x}_{\bar{R}}$ is the unknown vector, and $\mathbf{H}_{\bar{R}}^T$ is the system matrix. As for the generator matrix case, the rank system matrix is equal to the number of unknowns, and the unique solution can be obtained by applying Gauss-Jordan elimination to the system

$$\begin{aligned} x_1 + x_3 + x_5 &= x_2 \\ x_1 &= x_2 + x_4 + x_6 \\ x_1 + x_3 &= x_4 + x_7 \end{aligned}$$

yielding

$$\begin{aligned} x_1 &= x_2 + x_4 + x_6 \\ x_3 &= x_2 + x_6 + x_7 \\ x_5 &= x_2 + x_4 + x_7. \end{aligned}$$

Being the code constructed with a systematic generator matrix, the information vector can be easily obtained by recognizing that $\mathbf{u} = (x_1 \quad x_2 \quad x_3 \quad x_4)$.

From the example above, we may derive some useful conclusions concerning erasure decoding for binary linear block codes, which are summarized next.

i. Decoding can be performed with complexity growing only polynomially in k via Gauss-Jordan elimination applied to a linear equation system derived from either the generator or the parity-check matrix. In particular, complexity grows with k^3, which is the complexity of Gauss-Jordan elimination for a system of equations in k unknowns.

ii. In case the generator matrix is adopted, decoding turns into solving a system of $n - n_E$ equations in k unknowns, with n_E being the number of erased bits.

iii. In case the parity-check matrix is adopted, decoding turns into solving a system of $n - k$ equations in n_E unknowns, with n_E being the number of erased bits.

iv. Decoding succeeds, in either case, if the equation admits a unique solution, i.e., if the system matrix $\mathbf{G}_R$ has rank k or (equivalently) if the system matrix $\mathbf{H}_{\bar{R}}^T$ has rank n_E.

v. From the two conditions at point (iv), it follows that a necessary condition for successful decoding is to have a number of erasures not exceeding the number of equations, i.e., to have $n - n_E \geq k$ or equivalently $n - k \geq n_E$.

vi. Codes for which decoding succeeds with probability 1 whenever $n - k \geq n_E$ are referred to as maximum distance separable (MDS). MDS codes exist only for specific combinations of the parameters n and k. For the case of binary linear block codes, only codes with $k = 1$ (repetition codes) or codes with $n = k+1$ (single-parity-check codes) are MDS. MDS codes for broader sets of parameters (n,k) can be constructed with codeword symbols defined in higher-order finite fields (with erasures taking place symbol-wise).

MDS codes can be considered as ideal erasure correcting codes. Hence, they can be used as a benchmark for existing erasure coding schemes. In particular, the block error probability attained by an (eventually, non existing) MDS code with parameters (n,k) when used over a BEC with erasure probability ε is given by the Singleton bound [221, 182]

$$P_S(n,k,\varepsilon) = \sum_{i=n-k+1}^{n} \varepsilon^i (1-\varepsilon)^{n-i} \qquad (6.12)$$

and it represents a lower bound on the block error probability achievable by any (n,k) binary linear block code. The fact that MDS codes may not exist for arbitrary n and k, anyhow, does not represent a major performance limitation: It is in fact possible to prove that, by picking a (n,k) binary linear block code at random, its block error probability is upper bounded by

$$P_B(n,k,\varepsilon) = P_S(n,k,\varepsilon) + \sum_{i=1}^{n-k} \varepsilon^i (1-\varepsilon)^{n-i} 2^{-(n-k-i)}. \qquad (6.13)$$

Equation 6.13 is referred to as Berlekamp's random coding bound for the BEC [42, 182]. The right-hand side of 6.13 is given by the Singleton bound plus a penalty term, which gets quickly small for n and k large enough. In Figure 6.14, the Berlekamp random coding bound and the Singleton bound are compared, for the case of $R = 2/3$ codes with block lengths $n = 384$ and $n = 768$. The two bounds are already very close at the shortest block length, and the gap between the two curves tend to diminish as the block length grows.

Although the result is encouraging (binary linear block codes can attain a nearly ideal erasure recovery capability with polynomial decoding complexity), the cubic (in k) cost of the decoding cost renders the use of relatively long erasure codes impractical. We shall see next how the decoding complexity can be largely reduced for a special class on binary linear block codes.

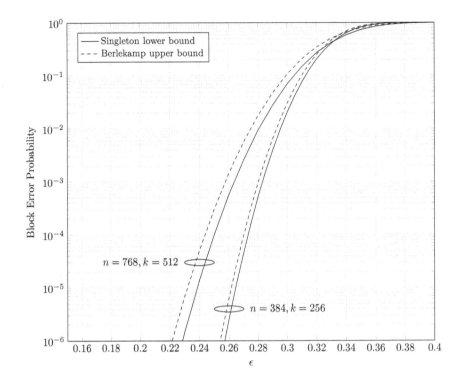

Figure 6.14: Singleton and Berlekamp bounds for rate-$2/3$ codes.

6.3.2.2 Low-Density Parity-Check Codes for Erasure Channels

An LDPC code is a binary linear block code defined by a *sparse* parity-check matrix [122]. Consider a $m \times n$ parity-check matrix $\mathbf{H}$, where $m \geq n - k$. [11] Denote by $w_{c,i}$ the Hamming weight (i.e., number of ones) of the ith column in $\mathbf{H}$, and by $w_{r,j}$ the Hamming weight (i.e., number of ones) of the jth column in $\mathbf{H}$. For an LDPC code we require the Hamming weight of each column, $w_{c,i}$ with $i = 1, \ldots, n$, to be much lower than the column length, i.e. $w_{c,i} \ll m$, and the Hamming weight of each row, $w_{r,j}$ with $j = 1, \ldots, m$, to be much lower than the column length, i.e. $w_{r,j} \ll n$.

An LDPC code is said to be *regular* if the Hamming weight of each column in $\mathbf{H}$ is fixed, $w_{c,i} = w_c$ for all i, and the Hamming weight of each row in $\mathbf{H}$ is fixed, $w_{r,j} = w_r$ for all j. A regular LDPC code is often referred to as a (w_c, w_r) LDPC

---

[11]We account here also for parity-check matrices having a redundant number of rows, i.e., where some rows linearly depend on other rows of $\mathbf{H}$.

code.[12] The code rate of a regular LDPC code is given by

$$R \geq 1 - \frac{w_c}{w_r}$$

where equality holds only if all the rows of **H** are linearly independent. An LDPC code is said to be *irregular* [188, 235] if the Hamming weight of the columns and/or rows of **H** varies.

The code rate of an irregular LDPC code is given by

$$R \geq 1 - \frac{\bar{w}_c}{\bar{w}_r} \tag{6.14}$$

where

$$\bar{w}_c = \frac{1}{n} \sum_{i=1}^{n} w_{c,i} \quad \text{and} \quad \bar{w}_r = \frac{1}{m} \sum_{j=1}^{m} w_{r,j}$$

are the average column/row weights. Also here, equality holds only if all the rows of **H** are linearly independent. LDPC codes can be well described through a bipartite (Tanner) graph $\mathcal{G}$. The graph is composed by

■ A set $\mathcal{V} = \{V_1, V_2, \ldots, V_n\}$ of n variable nodes (VNs), one for each codeword bit (i.e., one for each column of **H**).

■ A set $\mathcal{C} = \{C_1, C_2, \ldots, C_m\}$ of m check nodes (CNs), one for each parity-check equation (i.e., row) in **H**.

■ A set $\mathcal{E} = \{e_{i,j}\}$ of edges.

An edge $e_{j,i}$ connects the VN V_i to the CN C_j if and only if $h_{j,i} = 1$. It follows that the graph $\mathcal{G}$ and the parity-check matrix **H** provide an equivalent representation of the code. As an example, consider a $(3,4)$ regular LDPC code defined by the parity-check matrix

$$\mathbf{H} = \begin{pmatrix} 1 & 1 & 1 & 1 & 0 & 0 & 0 & 0 & 0 & 0 & 0 & 0 \\ 0 & 0 & 0 & 0 & 1 & 1 & 1 & 1 & 0 & 0 & 0 & 0 \\ 0 & 0 & 0 & 0 & 0 & 0 & 0 & 0 & 1 & 1 & 1 & 1 \\ 1 & 0 & 0 & 0 & 0 & 1 & 1 & 0 & 0 & 0 & 0 & 1 \\ 0 & 1 & 0 & 1 & 1 & 0 & 0 & 0 & 1 & 0 & 0 & 0 \\ 0 & 0 & 1 & 0 & 0 & 0 & 0 & 1 & 0 & 1 & 1 & 0 \\ 0 & 1 & 0 & 0 & 1 & 1 & 0 & 0 & 1 & 0 & 0 & 0 \\ 1 & 0 & 1 & 0 & 0 & 0 & 1 & 0 & 0 & 1 & 0 & 0 \\ 0 & 0 & 0 & 1 & 0 & 0 & 0 & 1 & 0 & 0 & 1 & 1 \end{pmatrix}.$$

The corresponding bipartite graph is depicted in Figure 6.15.

The *degree* of a node is the number of its adjacent edges. The *neighbors* of a

---

[12]The notation shall not be confused: Here the parameters do not refer to the block size and the code dimension, but to column and row weights.

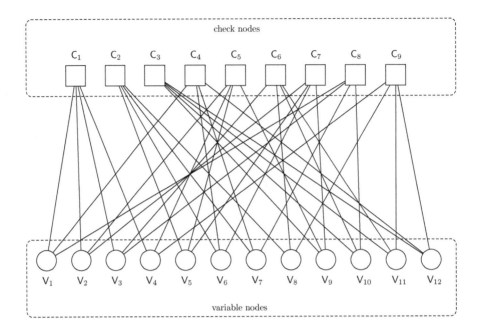

Figure 6.15: Example of an LDPC code bipartite graph.

node are the nodes connected to its adjacent edges. The set of neighbors of a VN V_i is denoted by $\mathscr{N}(V_i)$, the set of neighbors of CN C_j is $\mathscr{N}(C_j)$. Consider the graph of Figure 6.15. We have that the neighborhood of V_1 is $\mathscr{N}(V_1) = \{C_1, C_4, C_8\}$. The neighborhood of C_1 is $\mathscr{N}(C_1) = \{V_1, V_2, V_3, V_4\}$. Observe that the degree of the VN V_i is equal to the Hamming weight of the ith column of $\mathbf{H}$. Similarly, the degree of the CN C_j is equal to the Hamming weight of the jth row of $\mathbf{H}$. We can thus re-define a regular LDPC code as follows. An LDPC code is said to be *regular* if its bipartite graph has constant variable and check node degrees d_v, d_c. Being in this case $d_v = w_c$ and $d_c = w_r$, we have that

$$R \geq 1 - \frac{d_v}{d_c} \tag{6.15}$$

where equality holds only if all the rows of $\mathbf{H}$ are linearly independent. An irregular LDPC code has obviously varying degrees across its variable/check nodes. In this case, the code rate can be upper bounded as

$$R \geq 1 - \frac{\bar{d}_v}{\bar{d}_c} \tag{6.16}$$

being $\bar{d}_v, \bar{d}_c$ the average variable/check node degrees. The performance of an LDPC code are strongly related to the distribution of its node degrees, i.e., to the fraction of nodes possessing a certain degree. For this reason, we shall introduce the following definitions.

Definition 6.1 The *node-oriented variable node degree distribution* is denoted by Λ_i, $i = 1, \ldots, d_{v,\max}$ where Λ_i is the fraction of VNs with degree i and $d_{v,\max}$ is the largest VN degree.

Definition 6.2 The *node-oriented check node degree distribution* is denoted by P_j (here "P" stands for the capital greek letter *rho*), $j = 1, \ldots, d_{c,\max}$ where P_i is the fraction of CNs with degree j and $d_{c,\max}$ is the largest CN degree.

According to the definitions above, we have that $\bar{d}_v = \sum_i i\Lambda_i$ and $\bar{d}_c = \sum_j jP_j$. Often, the node degree distributions are provided in polynomial form in the dummy variable x as

$$\Lambda(x) = \sum_i \Lambda_i x^i \quad \text{and} \quad P(x) = \sum_j P_j x^j. \qquad (6.17)$$

The degree distributions can be stated from an edge perspective, too.

Definition 6.3 The *edge-oriented variable node degree distribution* is denoted by λ_i, $i = 1, \ldots, d_{v,\max}$ where λ_i is the fraction of edges connected to VNs with degree i.

Definition 6.4 The *edge-oriented check node degree distribution* is denoted by ρ_j, $j = 1, \ldots, d_{c,\max}$ where ρ_i is the fraction of edges connected to CNs with degree j.

Edge degree distributions are provided in polynomial form too,

$$\lambda(x) = \sum_i \lambda_i x^{i-1} \quad \text{and} \quad \rho(x) = \sum_j \rho_j x^{j-1}. \qquad (6.18)$$

(Note the subtle difference in the exponent definition between the node and the edge oriented polynomial degree distributions. The advantage of this choice is evident in the asymptotic analysis of LDPC codes [233].) Consider for example the LDPC code defined by the graph defined in Figure 6.16. The graph possesses $m = 5$ CNs, $n = 10$ VNs and 25 edges. We count the node degrees as follows.

VN Degree	Number of VNs
1	1
2	4
3	3
4	2

CN Degree	Number of CNs
4	1
5	2
6	2

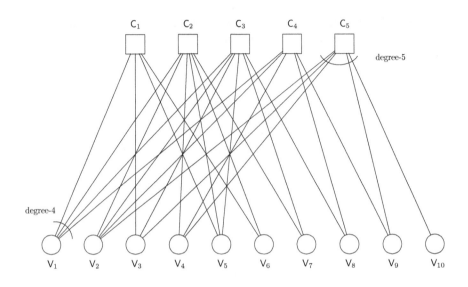

Figure 6.16: Example of bipartite graph of an irregular LDPC code.

We thus have

$$\Lambda(x) = \frac{1}{10}x + \frac{4}{10}x^2 + \frac{3}{10}x^3 + \frac{2}{10}x^4$$

$$P(x) = \frac{1}{5}x^4 + \frac{2}{5}x^5 + \frac{2}{5}x^6.$$

From the edge perspective, we have

VN Degree	Adjacent Edges
1	1
2	8
3	9
4	8

CN Degree	Adjacent Edges
4	4
5	10
6	12

We thus have

$$\lambda(x) = \frac{1}{26} + \frac{8}{26}x + \frac{9}{26}x^2 + \frac{8}{26}x^3$$

$$\rho(x) = \frac{4}{26}x^3 + \frac{10}{26}x^4 + \frac{12}{26}x^5.$$

It is possible to convert a node perspective degree distribution into an edge perspective one through

$$\lambda_i = \frac{i\Lambda_i}{\sum_l l\Lambda_l} \qquad \text{and} \qquad \rho_i = \frac{iP_i}{\sum_l lP_l}.$$

Another graph property that will condition the performance of an LDPC code under iterative decoding is the graph *girth*. A *cycle* in a bipartite graph is a sequence of edges leading to closed path where each node belonging to the cycle is visited once. The length of a cycle is given by the number of edges composing the cycle. The girth g of graph $\mathscr{G}$ is the length of its shortest cycle.

Iterative decoding of LDPC codes is based in fact on a message-passing algorithm where extrinsic information messages are exchanged along the edges of the graph. The algorithm provides correct a-posteriori estimates for the codeword bits only if the graph is cycle-free. Thus, having codes whose graphs have large cycles would allow to propagate correct extrinsic estimates for a large number of iterations, whereas short cycles tend to lead to poor iterative decoding performance. It follows that a main design principle for LDPC codes is to maximize the graph girth, for a given degree distribution pair.

The graph representation of the parity-check matrix constraints leads to a simple description of iterative erasure decoding: At each iteration, the decoder looks for CNs connected to VNs associated to erased bits (we shall refer to these VNs as *erased VNs*). If the decoder finds a CN connected to only one erased VN, then the erased VN value is recovered by summing (in the finite field) the bit values associated to the other VNs connected to the check node. Decoding stops when, during an iteration, no CNs connected to only one erased VN can be found.

Iterative decoding of LDPC codes has a complexity that grows linearly with n, allowing the use of long block lengths, and thus the possibility to operate close to the Shannon limit. The price to be paid lies in the sub-optimality of the decoding algorithm. Before clarifying this point, we shall first discuss the peculiarities of erasure patterns that cannot be resolved by ML decoding.

Let us denote by **e** the erasure pattern generated by the channel, with $e_i = 0$ if the channel does not erase the i-th codeword bit, and with $e_i = 1$ if it does. We further denote by

$$\mathscr{S}_{\mathbf{e}} = \mathrm{supp}\,(\mathbf{e}) = \{i | e_i = 1\}$$

the support set of **e**. We have the following proposition.

Proposition 6.1

*Assuming that a binary linear block code is used to transmit, an erasure pattern **e** is unresolvable under ML decoding if and only if its support set includes the support set of a non-zero codeword, i.e. if*

$$\exists \mathbf{x} \in \mathscr{C}, \mathbf{x} \neq \mathbf{0} \; s.t. \; \mathscr{S}_{\mathbf{x}} \subseteq \mathscr{S}_{\mathbf{e}},$$

*being $\mathscr{S}_{\mathbf{x}} = \mathrm{supp}\,(\mathbf{x}) = \{i | x_i = 1\}$ the support set of the codeword **x**.*

Proof 6.1   The proposition follows by observing that if the non-zero coordinates of a codeword are erased by the channel, then the resulting vector does not allow discriminating between the transmission of the codeword itself or of the all-zero codeword, leading to ambiguity.

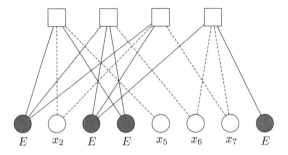

Figure 6.17: Example of stopping set.

Proposition 6.1 represents a sufficient condition for a decoding failure under iterative decoding too, being iterative decoding sub-optimal with respect to ML decoding. However, it does not represent a necessary condition. In other words, iterative decoding may fail even if the support set of an erasure pattern does not include the support set of a non-zero codeword. In fact, under iterative decoding, the decoding failure condition is given by the following proposition.

Proposition 6.2

Consider a (n,k) LDPC code with bipartite graph $\mathcal{G}$. An erasure pattern $\mathbf{e}$ is unresolvable under iterative decoding if and only if its support set includes a stopping set of $\mathcal{G}$, where a stopping set is a subset of the VNs such that any CN connected to it is connected to it at least twice.

Consider the $(n=8, k=4)$ code with graph depicted in Figure 6.17. The variable nodes V_1, V_3, V_4, V_8 represent a stopping set. In fact, all the connected check nodes "see" at least two erasures. This fact prevents iterative decoding from starting if the corresponding codeword bits are erased (recall in fact that iterative decoding on the BEC proceeds by finding at each iteration check nodes connected to only one erased variable node, whose value is then resolved). However, if ML decoding is employed, the erasure pattern could be resolved. We have that the code parity-check matrix is

$$\mathbf{H} = \begin{pmatrix} 1 & 1 & 0 & 1 & 1 & 0 & 0 & 0 \\ 1 & 0 & 1 & 1 & 0 & 1 & 0 & 0 \\ 1 & 1 & 1 & 0 & 0 & 0 & 1 & 0 \\ 0 & 0 & 1 & 0 & 0 & 1 & 1 & 1 \end{pmatrix}.$$

Assuming an erasure pattern $\mathbf{e} = (1,0,1,1,0,0,0,1)$, we have that the sub-matrix induced by the erasure pattern (first, third, fourth and last columns of **H**) is

$$\mathbf{H}_{\bar{R}} = \begin{pmatrix} 1 & 0 & 1 & 0 \\ 1 & 1 & 1 & 0 \\ 1 & 1 & 0 & 0 \\ 0 & 1 & 0 & 1 \end{pmatrix},$$

whose rank is 4.

The sparse structure of the parity-check matrix of LDPC codes enables also an efficient (though, not linear in n) decoding algorithm for LDPC codes: We refer the reader to the vast literature [234, 215, 64, 254, 212] which discuss different flavors of the efficient decoding algorithm, which all fall in the category of *structured* Gauss-Jordan elimination [171, 170]. Figure 6.18 shows the performance of a $(2048, 1024)$ LDPC code from [211], under both iterative and (efficient) ML decoding as a function of the channel erasure probability. The Singleton bound is provided as reference, too. The capability of attaining a performance close to that of an idealized MDS code

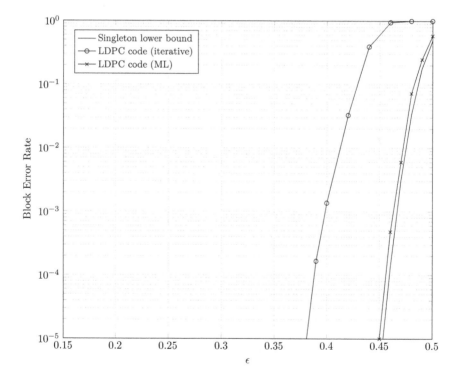

Figure 6.18: Block error rate for a $(2048, 1024)$ LDPC code under iterative and ML decoding, compared with the Singleton bound.

is evident from the ML performance curve. Under iterative decoding a performance degradation is visible, with roughly 15% of erasure correction capability lost with respect to ML decoding: At a block error rate of 10^{-6}, the iterative decoder handles erasure probabilities up to 0.38, whereas a ML decoder would allow working at a channel erasure probability close to 0.45.

So far, erasure codes have been discussed as codes capable of correcting bit erasures. However, in most of the applications where erasure codes are employed, erasures hit entire packets. This is the case, for instance, of wireless transmission where fading events may lead to packet losses. The concept presented in this and in the previous sections keeps applying, though: It is sufficient to consider the symbols involved in parity-check equations to be packets (i.e., *vectors* of bits) rather than single bits. For example, with reference to the graph in Figure 6.16, the first check node refers to a parity check equation involving the codeword symbols in position $1,3,5$ and 6, i.e., $x_1 + x_3 + x_5 + x_6 = 0$. If the symbols are vectors of bits of equal length L, then the equation can be casted as

$$\mathbf{x}_1 + \mathbf{x}_3 + \mathbf{x}_5 + \mathbf{x}_6 = \mathbf{0},$$

where all summands are vectors of L bits, and the sum is over the vector space $\mathbb{F}_2^L$.

6.3.2.3 Erasure Codes in CCSDS

The experimental CCSDS specification in [15] provides a description of a flexible family of LDPC codes able to cover a wide range of block lengths and code rates. The approach, which relies on the construction proposed in [183], is based on the concatenation of an inner LDPC code and an outer random linear block code. The LDPC codes used in [15] belong to the class of irregular repeat accumulate (IRA) codes [155]. The IRA code parity-check matrix is partitioned in two parts, i.e.

$$\mathbf{H}_{\text{IRA}} = [\mathbf{H}_u | \mathbf{H}_p].$$

Here, $\mathbf{H}_u$ is a $m \times k$ unstructured low-density (random) matrix, while $\mathbf{H}_p$ is the $m \times m$ dual-diagonal matrix,

$$\mathbf{H}_p = \begin{pmatrix} 1 & 0 & 0 & 0 & \cdots & 0 & 0 \\ 1 & 1 & 0 & 0 & \cdots & 0 & 0 \\ 0 & 1 & 1 & 0 & \cdots & 0 & 0 \\ \vdots & \vdots & \vdots & \vdots & \ddots & \vdots & \vdots \\ 0 & 0 & 0 & 0 & \cdots & 1 & 0 \\ 0 & 0 & 0 & 0 & \cdots & 1 & 1 \end{pmatrix}. \tag{6.19}$$

For a given set of parameters (n, k), the generation of the matrix $\mathbf{H}_u$ is performed through a pseudo-random algorithm which is driven by a target VN degree distribution. The algorithm is known at the transmitter and at the receiver, so is the seed used to initialize the random number generator that underlies the construction. The random construction of $\mathbf{H}_u$ enables a large flexibility in defining codes of different lengths and rates. However, due to the lack of optimization with respect to the code girth, the resulting codes are typically affected by a relatively poor performance. To enhance the code performance, an outer random code with length $n_o = k$ and information block size $k_o < k$ is employed. If we denote by $\mathbf{H}_o$ the parity-check matrix of the outer code, the overall code obtained by this concatenation would possess a parity check matrix in the form

$$\mathbf{H} = \begin{pmatrix} \mathbf{H}_o & \mathbf{0} \\ \mathbf{H}_u & \mathbf{H}_u \end{pmatrix}$$

where $\mathbf{H}_o$ has dimensions $(k - k_o) \times k$, $\mathbf{0}$ is a $(k - k_o) \times (n - k)$ zero matrix, and $\mathbf{H}_u$ and H_p have dimensions $(n - k) \times k$ and $(n - k) \times (n - k)$, respectively. The outer code is generated using again a pseudo-random algorithm shared by transmitter and receiver, and it is designed to possess a dense parity-check matrix $\mathbf{H}_o$. The dense outer code parity-check matrix allows to improve the minimum distance properties of the overall code, and hence its performance under ML decoding. By keeping the number of rows of $\mathbf{H}_o$ negligible with respect to the number of parity-equations of the inner IRA code, the overall matrix remains sparse, and the efficient ML decoders for LDPC codes can be employed with a limited increase in decoding complexity.

6.3.3 Application to DTN

6.3.3.1 Positioning of Erasure Codes in the Protocol Stack

Use of erasure codes in DTN networks is particularly appealing in the case of space networks, where the large delays therein experienced make the use of Automatic Repeat reQuest (ARQ) [88, 87] to recover from information losses particularly inefficient. This aspect is particularly evident in the case of deep-space communications where the propagation delay can be in order of several minutes or even hours, as experienced in the case of Mars-Earth (about 14 minutes in average) and Pluto-Earth (about 4.5 hours), whereby implementation of traditional retransmission schemes is actually impractical because of the large delay introduced, especially when data communications must cope with large error ratios eventually resulting in different retransmission loops. Another important consideration is that the window visibility between remote spacecraft and Earth ground stations is also limited to few hours, implying that long retransmission loops might need to be postponed to the next available contact. Overall, this results in very large latencies to accomplish image file transfers. Such a performance limitation is also coming up in the case of near-Earth or cislunar space missions, where the propagation delays are actually much lower (in the order of few seconds). In these operations conditions, the performance degradation introduced by ARQ strategies is obviously less evident than in deep-space, but still non negligible. It is therefore immediate to see that the application of erasure codes would have a twofold benefit: i) providing very reliable communication links, and ii) overall reducing the average delivery delay.

In more detail, the advantage offered by erasure codes [86] would be particularly appealing for free-space optics communications, where the presence of turbulence and scintillation can give rise to fading events of moderately short duration (i.e., $1 - 10$ ms), which can still cause important performance degradation because of the large data rate usually offered by optical communication. For instance, if we take into account that future space missions built on free-space optics technology will provide at least downlink data rate in the order of 100 Mbps, the occurrence of one fading event lasting $1 ms$ will result in the loss of 10^5 bits, roughly corresponding to 12 network packets (i.e., assuming one packet to carry 1000 bytes). Obviously as we increase the available data rate and the fading event duration too, the overall reliability of the communication systems is overall hampered, hence calling for proper recovery mechanisms.

In the perspective of future space mission planning and system design, an important question is related as to properly positioning the implementation of erasure codes. In the last decades, several studies have been done to identify the most promising approach, although no clear consensus has been reached yet, because of the different performance and system design implications that can arise. On the one hand, it would be desirable to keep the implementation of erasure codes as close as possible to the physical layer, in order to control the possible propagation of frame erasures to the upper layers, which is likely to cause the loss of large volume of data because of encapsulation and segmentation operations. On the other hand, use of erasure codes closer to the application layer is desirable as well, since it would allow defining different erasure coding strategies according to the specific requirements of the various services. For instance, telemetry services are typically expected to target a very low packet error rate, whereas transfer of image files can tolerate slightly higher packet error rate owing to the implementation of robust image coding. Ideally, erasure codes could be implemented "in the middle" of the protocol stack in order to easily exploit both features, although protocol stack engineering would be in any case needed in order to provide a necessary inter-layer signalling flow.

Another important issue is related to how erasure codes could be implemented in a protocol, whether according to a *layered* or an *integrated* approach [87]. The former consists in implementing erasure codes in a dedicated protocol layer (or sublayer) in order to keep the functions of existing protocol unaltered. On the other hand, this has the side effect that protocol interface should be in any case upgraded in order to properly exchange data with the new coding layer. Conversely, the latter (integrated) consists in implementing erasure coding directly within existing protocols, with the evident benefit that interaction between adjacent layer would result unaffected. On the other hand, however, integrating erasure codes into existing protocol would require the extension of the current functions, by means of unused protocol header fields or defining suitable extension blocks where possible. This option has in general the shortcoming of re-engineering a given protocol in order to accommodate the necessary encoding and decoding functions. From this standpoint, it is therefore arguable that the integrated approach would certainly increase the overall complexity of the system design and implementation, whereby the layered approach can be considered preferable. For this reason, the remainder of this section will only focus on the case of layered approach.

The remainder of this section is actually devoted to analyzing the implementation of erasure coding according to a layered approach in the DTN protocol stack, as proposed within CCSDS standardization framework. To this regard, the following protocol options are considered:

- CCSDS File Delivery Protocol (CFDP) [11]. Erasure coding is implemented in a shim layer placed between CFDP and Bundle Protocol (BP) protocols.

- BP [16]. Erasure coding is implemented in a shim layer placed between BP and the convergence layers.

- Licklider Transmission Protocol (LTP) [18]. Erasure coding is implemented in a shim layer placed between LTP and the CCSDS Encapsulation Service [12].

It is also worth noting that the recent specification of BP[16] for CCSDS also includes the definition of the Delay Tolerant Protocol Conditioning (DTPC), which operates on an end-to-end basis between the application layer and BP. Its main function is to provide aggregation of application data units (ADUs), priority policies, end-to-end reliability, and ordered delivery, just to cite some of the most important. Obviously this list of features could be extended so as to include also erasure coding. Although appearing as an attractive option, this approach is considered too complex according to authors' experience, since it will require adding important modifications to the core of the DTPC protocol, whose specification is already quite complex. Therefore, this option is actually not further explored in the rest of this section.

CFDP

As introduced in the previous section, the potential advantage of implementing erasure coding as close as possible to the applications is that different coding strategies could be envisioned so as to meet the different quality of service requirements of different services. In the case of CFDP, this translates into distinguishing between the contents transported in different files and accordingly map the corresponding coding strategies. For practical reasons, it would be obviously desirable to have a single code design (e.g., LDPC) and then apply different code-rates on the basis of the file content. For instance the case of telemetry files will certainly require the use of the most robust code design in order to provide delivery ratio very close to 100%. On the other hand, files demanding less stringent requirements could be those transporting images or video streams, whereby the delivery ratio could be reduced to 90% [86].

According to the layered approach considered here, erasure coding should be implemented in a shim layer positioned below CFDP and operating on top of BP. Extending the corresponding protocol interfaces of CFDP and BP is expected to be straightforward by means of primitives' overriding. On the other hand, an important point to be carefully addressed is how CFDP PDUs should be processed to be fed into the erasure coding engine. To this regard, it is worth remembering that typically erasure coding is very effective for large code-blocks, implying that a large number of input symbols should be provided to the encoding engine. To this end, it is necessary that each single CFDP PDU is big enough to fit the size of a code-block. Moreover, as erasure coding operates on symbols (i.e., independent packets of information), it will be necessary to segment each incoming CFDP PDU into $K[97, 87]$ packets, for eventually generating N-K redundancy packets according to the selected coding strategy. It can obviously happen that the size of a CFDP PDU is not sufficient to cover the size of code-block: this situation deserves some more attention since two different sub-cases need to be analyzed. In the first one, it can happen that all generated CFDP PDUs have a size lower than that of a code-block. A possible remedy is to merge multiple CFDP PDUs in a single macro-block fitting the code-block size, which will be in turn segmented into K information packets. A second case occurs when instead all the CFDP PDUs fit the size of a code-block but the last one. A possible mitigation to this event is make use of padding in order to properly and fully fill the code-block.

Finally, it can be also remarked that CFDP may provide both acknowledged (*reliable*) and unacknowledged (*unreliable*) services, the former implementing a selective ARQ strategy based on Negative Acknowledgment (NAK) issuance. Then, in case erasure coding is applied underneath CFDP, its combination with a reliable service of CFDP will correspond to implement a hybrid ARQ, hence able to ensure full delivery of data (provided that no retransmission attempts threshold is exceeded). On the contrary, the combination of erasure coding with CFDP operating an unreliable service will not ensure the full delivery of data, although a proper selection of the code family along with the code rate can guarantee very satisfactory performance (e.g., about 99% delivery rate or even higher).

BP

In the case of erasure coding operating below the bundle protocol, some more attention must be dedicated to the interaction of erasure coding itself and the other adjacent protocols. This additional analysis stems from the fact that BP can provide hop-by-hop reliability my means of custodial transfer option [319, 111] in terms of ARQ functionality. As such, the implementation of erasure coding with the BP custodial transfer [305] would work as hybrid ARQ, where the retransmission loop implemented in BP could be exploited to reach 100% delivery. Another important consideration is to take into account whether the underlying protocols already provide some reliability measures. For instance, in the case BP is operating on top of LTP (operated with red parts only) or Transmission Control Protocol (TCP) (where meaningful, i.e., in near-Earth space missions), then the reliability offered by erasure coding would be redundant since reliability should be already guaranteed by the aforementioned protocols. In this context, its use is actually not recommended, since it would not likely be able to improve the reliability and would eventually only increase the bandwidth waste.

In this line of reasoning, it could be argued as well that erasure coding could be employed in the case LTP or TCP sessions are aborted for the large packet losses or the exceeded maximum number of retransmissions. Hence, erasure coding could be used to recover the packets lost in the corresponding sessions. However, according to authors' experience, this event is likely to happen when a very large burst of packets loss is experienced and therefore it would be preferable to suspend the data transmission rather than resorting to erasure coding.

Finally, as similarly observed for CFDP, the shim layer implementing erasure coding should be capable of merging different bundles and making use of padding to overcome the problems related to the size of bundles, when they are not matching the size of a code-block.

LTP

The implementation of erasure coding within a shim layer placed below LTP layer is certainly the most effective one, since it is placed very close to the datalink layer and therefore can avoid that packet erasures can propagate to upper layers, giving rise to large loss of data. For instance, if LTP segments belonging to a green part [18] are lost, it can happen that either BP or CFDP detect the malformed corresponding PDUs and decide to drop them. This implies that the erasure coding strategy used in the shim layer implemented in both layers shall be able to recover larger losses than actually occurred. Another advantage consists in the fact that erasure coding can be applied to both green and red parts. In the first case, it helps provide a reliability measure, which usually cannot be guaranteed unless BP custodial transfer-induced retransmissions are performed. In the second case, as already observed for BP, a hybrid ARQ will be actually implemented, where the losses not recovered by the erasure coding strategy (for whatever reason) will be mitigated by the ARQ implemented with LTP red parts [18, 301, 289].

Finally as also observed for the other cases, there might be the need for merging together different bundles in order to fit the overall size of the code-block. However, differently from the other cases, this function is already provided by the LTP convergence layer adapter, so that multiple bundles can be cascaded and merged in a single and larger LTP block. In turn, this will be automatically segmented in several LTP segments, as inherent to the LTP protocol entity specification, without necessitating to implement additional segmentation/reassembly functions within the erasure coding shim layer.

An additional note must be reserved to the interaction of the shim layer with the underlying protocol, which typically is the CCSDS Encapsulation Service [12] in space missions. The proper interfacing is achieved by overriding the available primitives from the CCSDS Encapsulation Service. Moreover, the correct dispatching of packets to the shim layer will be done by exploiting the 'protocol ID' and 'protocol ID extensions' respectively [15].

6.3.3.2 Exemplary Protocol Design

As reported in the previous section, a reasonable protocol design choice for including erasure coding into the protocol stack is to position it as close as possible to the datalink layer, in order to limit the propagation of frame erasures and actually recover them as much as possible. From this standpoint, erasure coding can be considered as bound to the upper interface of the CCSDS Encapsulation Service [12] and interwork with a number of protocols operating on top of it. In other words, it is worth considering those protocols that already have an interface towards the CCSDS Encapsulation Service so that insertion of the erasure coding shim layer will come at limited cost, mostly consisting in the reuse of the existing protocol interface primitives. We have

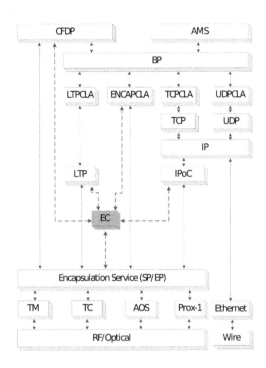

Figure 6.19: Protocol architecture for erasure coding enabled CCSDS system

already observed that this exercise is particularly attractive for the use of erasure coding with LTP, but it is not necessarily limited to it only. As a matter of fact, running on top of CCSDS Encapsulation Service is also possible for BP and CFDP, as clearly illustrated in Figure 6.19[13]. Starting from this, it is therefore possible to consider the application of erasure coding in a dedicated shim layer placed on top of the CCSDS Encapsulation Service and able to mitigate the effect of frame erasures on LTP, BP and CFDP disjointly. This is a remarkable feature for the considered CCSDS-DTN architecture, as it allows quite some flexibility in the use of erasure coding.

The overall sketch of the new architecture proposal is provided in the following figure [15].

It is however to be noticed that application of erasure coding directly on the PDUs generated by CFDP and BP requires some additional engineering (as already observed in the previous section), because of the fact that erasure coding requires a large number of symbols as input and the PDU size must fit that of the code-block.

---

[13]It can be noticed that Figure 6.19 also considers the case of the TCP/IP protocol stack running on top of CCSDS, which can be considered still meaningful in the case of missions operated for near-Earth, where performance degradations of TCP are still not critical.

Apart from these specific aspects which are more relevant from an implementation viewpoint, the main elements composing the shim layer architecture are as follows [15]:

■ Encoding/decoding engine. This is actually containing the classical functions of erasure coding, i.e. encoding and decoding, specifically depending on the code family and related construction being considered. A possible example of code design is provided in the previous sections of this chapter. This engine directly interfaces to the upper layers and then is capable of receiving the corresponding PDUs, which can be regarded as ADUs (application data units) for the sake of generality.

■ Encoding matrix. The encoding matrix is necessary to buffer a number K of the incoming PDUs from the upper layers (e.g., bundles, LTP segments, or CFDP PDUs), prior to proceeding with encoding and decoding functions. The proper implementation of such a matrix is a key-element in the implementation of the overall shim layer, since it can severely affect the performance of the considered erasure coding strategy, independently of the effectiveness of the adopted code design. For instance, the case of less than K PDUs being received from the upper layer (for encoding) or from the underlying layer (for decoding) must be properly handled by means of dedicated timers, tuned to the network characteristics. Another important aspect concerns the fact that more than an encoding matrix might be needed in order to take into account that the encoding rate could be less than the actual generation rate of the PDUs in the upper layers. Typically, correctly dimensioning the number of such matrices also depends on the bandwidth-delay product so that the upper layer protocols are able to fill the available capacity pipe. Obviously, the buffers implemented in the shim layers cannot be regarded as storage units to possibly prevent congestion events in case the generation rate of PDUs exceeds the available capacity, since this function actually belongs to the more general tasks of flow and congestion control that should be carried out by dedicated layers (e.g., LTP or BP). These implementation aspects are however not fully understood so far since no stable implementation of erasure coding for DTN is still available and therefore additional study and performance assessment is necessary before coming to a proper conclusion.

■ Protocol engine. This actually represents the protocol entity of the erasure coding shim layer, mostly dealing with appending (or removing) a dedicated protocol header and performing the corresponding protocol functions. As such, the output of the protocol engine function (in case encoding has been just performed) will be erasure coding data units, i.e. ECDUs [15]. Similarly, prior to starting decoding functions, the protocol engine will receive ECDUs from the underlying layer, i.e., the CCSDS Encapsulation Service. As to the protocol header, it should contain the necessary information to let the receiver correctly perform the decoding functions. For instance it should contain information about the used family code (i.e., the settings for K and N) as well as the coding

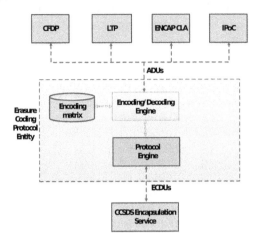

Figure 6.20: Inner architecture of erasure coding shim layer and interfaces with the other CCSDS protocol layers

block identifier and the position of the ECDU within the specific coding block. Finally it should also include the identifier of the protocol running on top of the erasure coding shim layer, in order to allow operations from different protocols (e.g., LTP, BP, and CFDP). The availability of extension blocks would be desirable too in order to make the protocol flexible enough towards future upgrade of the protocol specification itself. Proper characterization of all these elements is not given in this book, since more pertinent to a protocol specification document as illustrated in the experimental CCSDS recommendation ([reference orange book], to which the interested can refer to).

The overall architecture diagram is depicted in Figure 6.20 [15].

Please note that the interaction with BP is ensured by means of the *ENCAP CLA*, which serves as ad-hoc convergence layer adapter between BP and the erasure coding shim layer. As a matter of fact, it plays the same role as the other CLAs, with specific functions here advocated to allow the already mentioned erasure coding functions.

6.3.3.3 Performance Results

To the best of authors' knowledge, only a very limited part of the scientific literature has explored so far the advantages of erasure coding for DTN-based space missions [87], since most of the studies addressed the code design, independently of the specific protocol architecture being considered. This lack of analysis is probably due to the fact that full consensus on the use of erasure coding in space missions has been not yet reached on the one hand and a very limited set of tools are actually available on the other hand.

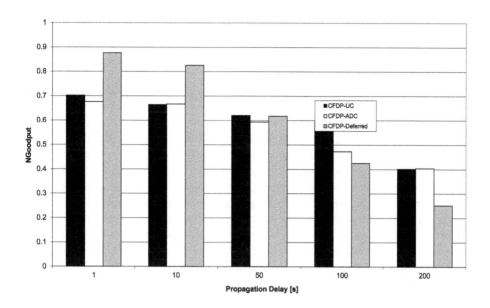

Figure 6.21: Performance analysis (normalized goodput) of CFDP under different recovery strategies, as a function of the propagation delay

Some preliminary results available from past studies actually concern the application of erasure coding to CFDP [87], assuming free-space optics transmission channel modeled as bursty erasure channel according to a 2-state discrete Markov chain. The performance analysis actually considers three variants of the CFDP protocol: *CFDP-deferred*, *CFDP-ADC*, and *CFDP-UC*. The latter two correspond to the case of CFDP running in acknowledged and unacknowledged mode with erasure coding applied, respectively. The first one refers to CFDP running in acknowledged mode without erasure coding. The tag 'deferred' refers to the fact that different recovery functions are available, 'deferred' corresponding to postponing the initiation of the possible retransmission functions once the end-of-file CFDP PDU is received. For the sake of the completeness, the performance analysis is carried out for different propagation delays, in order to analyze the effect of the latency on the overall closed loop in case of reliable service. The performance is depicted in Figure 6.21 [87], where the considered metric is the normalized throughput, also containing a correction factor for the unreliable service (CFDP-UC) in case not all CFDP PDUs are eventually recovered by the erasure coding functions, so that the actual amount of received information is normalized to the transmitted one.

From the above figure it can be seen that the "basic" CFDP service (i.e., without erasure coding) is still efficient for short or moderately large latency (i.e., propagation delay less than 50 s). Beyond that, its performance is importantly degraded by the increase of propagation delay which causes much delayed file delivery. On the contrary, the solutions making use of erasure coding (i.e., CFDP-UC and CFDP-ADC) offer better solutions, with CFDP-UC being the most performant. Finally, it can be also noticed that for much larger delays (i.e. 200 s) CFDP-ADC and CFDP-UC perform the same because the impact of additional latency introduced by retransmission loops in any case compensated by the residual error rate that erasure coding alone is not able to cut off.

Finally it can also be observed that, although the provided performance figures are related to CFDP, the aforementioned findings are of general validity, since similar ARQ strategies are also available in BP and LTP protocol specifications. Obviously, there are some important differences for what regards the size of bundles and LTP segments that have to be properly configured for efficient erasure (de)coding operations. In particular, the main lesson learnt concerns the suitability of ARQ schemes still for low-latency scenarios (i.e., Moon-Earth data communications), whereas larger latencies make the use of feedback-based strategies not particularly appropriate unless full data delivery (e.g., 100%) is required. In this latter case, still, the application of erasure coding combined to retransmission capabilities is in any case desirable since erasure coding will actually reduce the number of necessary retransmissions, ultimately reducing the overall delivery delay.

6.4   Summary

The application of coding solutions implemented at the higher layers of the protocol stack have proved to be very beneficial to increase the robustness of data communications over space networks. In particular, the present chapter has pointed out the added value in terms of performance achievement coming from the use of network coding and erasure coding.

The former comprehends a large set of coding solutions, ultimately aiming at exploiting data delivery over multiple links and therefore making a better use of the available network resources. In particular, a proper (linear) combination of packets can improve on the one hand the overall reliability of data communication, and on the other hand can help reduce the amount of link capacity actually needed, up to a theoretical bandwidth saving of 50%. In more detail, it is shown in the first part of the chapter how the concept can be efficiently applied to the case of satellite networks and then implemented in the DTN protocol stack, by proposing a suitable protocol specification supporting network coding functionalities. The performance results reported prove the validity of the proposed concept and the potentials for use in different operative conditions.

On the other hand, the second part of the chapter is devoted to erasure coding, applied to near-Earth or deep-space links, exhibiting residual packet erasures after channel decoding functions performed at the physical layer. In particular, the case of free-space optical links is taken as reference, as they introduce fades of short/moderate length, which can still result in the erasures of a large number of packets because of the offered high data-rate. Based on the experience in the course of the CCSDS standardization, it is shown how LDPC-based erasure codes can offer very good performance, actually trading off the excessive delivery latency actually resulting from the use of more traditional ARQ schemes. Similarly to the case of network coding, important notes related to the implementation of erasure coding into the DTN protocol architecture are drawn, with particular attention to the case of CFDP, BP, and LTP protocols, for which implementation preference is given to LTP in order to limit the propagation of packet erasures throughout the upper layers of the protocol stack.

According to the results shown in the two parts of the chapters, it turns out that network coding and erasure coding are very promising solutions to improve reliability and data distribution in space networks and, moreover, their suitability to the DTN protocol architecture makes their implementation in forthcoming space and satellite missions of great interest.

Chapter 7

DTN for Spacecraft

Keith Scott

The MITRE Corporation

CONTENTS

7.1 Overview

Delay / Disruption Tolerant Networking was originally conceived of by NASA as a way to both standardize and generalize the relay communications that were such a success with the Mars rovers starting in 2004. Although relay communications were just a technology demonstration when first placed on the Spirit and Opportunity rovers, it quickly became apparent that the power savings of having the rover transmit to Mars orbit rather than directly to Earth could greatly increase the science data return. This is because the received power decreases as the square of the distance between transmitter and receiver.

Essentially, given the rovers' power budgets, they could either drive to new locations or communicate with Earth on a given Sol, but not both. Relay communications have since become an integral part of Mars communications, with 85-95% of the data from Spirit and Opportunity coming through Mars Odyssey. Indeed, the power savings and reliability of relay communications led NASA to design its Phoenix lander with no direct-to-Earth science data communications.

The Mars Odyssey orbiter that performs these relay operations completed its primary science mission in August of 2004 and entered its extended mission phase, and ever since the Mars rovers have reaped the benefits of Odyssey's longevity. One of the inspirations behind the design of DTN was the notion that other spacecraft, after finishing their primary missions could, if equipped with relay communications capabilities, similarly serve to help newer missions return data to Earth. This led to the notion of constructing a 'Solar System Internet' (SSI) where spacecraft that are not otherwise occupied (e.g. past their primary missions, in cruise phase, etc.) collaborate to form a shared communications infrastructure similar in design to the Internet. The hope is that the shared infrastructure will increase data return and reliability while decreasing costs.

Because there is typically not a contemporaneous end-to-end path between landed elements on the surface of Mars and Earth, DTN must handle cases where data arrives at a router and there is no immediate link out which it can be forwarded. For example, an orbiter might receive data from a lander or rover that is on the opposite side of Mars from Earth. The orbiter would then have to wait until a communications opportunity with Earth before it could forward the data.

Because of the cost and high impact of losing a spacecraft, many of the world's civilian space agencies are cooperating through the Consultative Committee for Space Data Systems (CCSDS) to set standards for building, communicating with, and operating spacecraft. These standards result in economic savings through economies of scale and also enable cross-support of one agency's spacecraft by another agency's ground infrastructure. Cross-support enables higher data return from spacecraft, especially low-Earth-orbiting (LEO) spacecraft, since such spacecraft can typically only communicate with a given point on the ground for a few minutes or tens of minutes at a time. Cooperation, in this case, effectively increases the number of ground station antennas that can be used to return data to Earth. Cross-support also allows multiple agencies' assets to be focused on a single spacecraft in the case of an emergency. By standardizing the frequencies, coding, modulation, and data link layer communications, multiple agencies can send commands to a distressed spacecraft and/or attempt to receive telemetry from it.

Standardization of the lower layers of the communication stack began in CCSDS in 1982 with the physical and data link layers, followed shortly by standardized packet formats that applications use to encapsulate their data. These allowed for basic, low-layer interoperability but did not address the higher-layer issues of how to actually exchange information between agencies on the ground. Starting in 2002 CCSDS began standardizing what were then called Space Link Extension (SLE) services, and what have grown since into a framework and set of protocols for cross support, including a set of Cross Support Transfer Services that have replaced SLE. SLE / CSTS has evolved to provide the services needed to effect the inter-agency cooperation discussed above. The missing functionality needed to allow the in-space cross-support needed for one agency's spacecraft to relay data from another agency was addressed starting in 2007.

In June of 2007 the Interagency Operations Advisory Group (IOAG) chartered the Space Internetworking Strategy Working Group to recommend ways to introduce internetworking to space communications. The group's 2008 report (with errata from 2010) [129] recommends using IP-based communications where the environment supported them, and DTN communications in any environment. CCSDS was tasked with defining the protocols to be used. The CCSDS working group chose to adapt the DTN Bundle Protocol then specified in an experimental RFC [244] and in 2015 CCSDS finalized a profile of the IETF Bundle Protocol RFC [16].

Work is ongoing in the CCSDS to standardize routing for space missions, as well as security. Where possible, the CCSDS is working with the Internet Engineering Task Force (IETF) to standardize capabilities in the IETF first while ensuring that space mission concerns are addressed, and then to lightly adapt the IETF standards for use in space. This was done for the Licklider Transmission Protocol [224], [18] and the Bundle Protocol. Some capabilities needed for space missions have no real counterparts in terrestrial applications, and for these standards have been developed in CCSDS and are described below.

Figure 7.1 gives the timeline for development of DTN for space missions.

This chapter is organized as follows. Section 7.2 describes the current model for space missions, where spacecraft communicate at layer-2 (OSI Data Link Layer) with their mission operations centers. Section 7.3 provides an overview of the DTN model for space communications, describes the requirements for a space internetworking architecture, and a number of alternatives to realize such an architecture. Section 7.4 describes the space-specific capabilities developed to support the DTN architecture in space, including modifications to the Licklider Transmission Protocol and the Bundle Protocol made by the Consultative Committee for Space Data Systems (CCSDS). Section 7.5 describes the benefits of DTN to space missions,

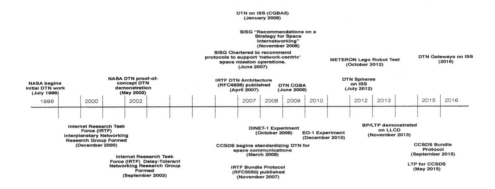

Figure 7.1: Timeline of the development of the DTN Protocols for Space.

followed by section 7.6 that discusses some standard DTN applications to support interoperability with legacy spacecraft and emergency operations. Section 7.7 discusses experiments and operational space missions demonstrating and using DTN, followed by a short section of conclusions.

7.2   Current Model for Space Missions

Current space missions use a point-to-point communications model, where each spacecraft communicates directly with its mission operations center (MOC) over a single ISO layer-2 (data link layer) connection. Since in most cases the mission operations center is not collocated with the ground station, the space data link is encapsulated and tunneled over the terrestrial network to the MOC, as shown in Figure 7.2

This model provides mission designers with a great deal of flexibility when planning space missions. All of the capabilities of the data link layer, such as special frames for hardware commanding, are available and completely configurable for each new mission. This model also ensures that the MOC has full control over the spacecraft, since the low-layer frames are built at the MOC and tunneled to the ground station for transmission. Because commands to the spacecraft have the potential to cause harm (e.g. pointing delicate instruments directly at the sun or changing the spacecraft orbit) all data sent to the spacecraft is typically highly controlled. This

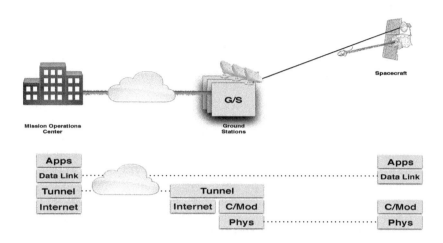

Figure 7.2: Current mission communication model with the spacecraft connected to its mission operations center via a single data link that is tunneled across the terrestrial network.

control process often includes human processing for all commands sent to the space-craft, including any commands to retransmit data that may have been lost in transit.

Current space mission operations also include the ability for different agencies to cooperate and to share terrestrial resources such as antennas. Using the terrestrial interoperability infrastructure, one agency can request service from another agency's antenna, specifying where to point, the communications parameters (e.g. frequency, modulating, coding) to use, etc. This terrestrial cross-support can greatly increase the amount of data return from space, especially for near-Earth satellites, since near-Earth satellites can typically only communicate with a small area of the Earth's surface at a time, and so ground stations might otherwise be idle for significant periods of time when they cannot communicate with any of their agency's assets.

7.3   DTN Model for Space Missions

While the current model for space communications has worked well for over 50 years, it has some disadvantages. Mission communications tend to be designed with very little automated communications, where humans are required to intervene should anything not go according to plan. For example, typical data transmissions require human intervention in order to identify portions of a data set that are not received correctly at the mission control center and to direct the spacecraft to retransmit them or to remove them from the spacecraft memory. This approach made sense in the past, where data rates were much lower and the ability to manually examine the received part of an image, for example, and to determine whether the missing parts were relevant or could be ignored was a significant savings. Advances in communications, including the use of file systems onboard spacecraft and data compression techniques that are intolerant of loss mean that partial loss of data sets is less and less desirable.

Also, while there are mechanisms for inter-agency cross-support on the ground that allow agencies to cooperate and share, for example, antenna resources, the cross-support capabilities in space were, at the time the DTN work was beginning, limited to interoperability at the data link layer of the ISO stack. What was desired was a mechanism that could provide an end-to-end, possibly relayed data delivery service that was standardized among space agencies to support in-space cross support, where not only could one spacecraft communicate with another at the data link layer, but could in fact use other spacecraft to forward data to Earth.

In June of 2007 the Inter-Agency Operations Group chartered the Space Internetworking Strategy Group (SISG) to:

"...reach international consensus on a recommended approach for transitioning the participating agencies towards a future "network centric" era of space mission operations. The group will focus on the extension of internetworked services across the Solar System, including multi-hop data transfer to and from remote space locations and local networked data interchange within and among the space end systems."

By November 2008, the SISG issued their final report recommending that the various space agencies begin developing the architecture and protocols to support in-space data relaying and cross-support based on an internetworked model of space communications. This recommendation was based on making missions more robust by extending cross-support to space, enabling more data return from missions through sharing of resources, particularly communications links, and enabling new types of science by allowing 'direct' (networked) communication between instruments on different spacecraft and between space-based and ground-based instruments.

In response to the space internetworking strategy working group, CCSDS chartered a working group to develop a set of requirements and use cases for in-space cross support and to evaluate existing and prospective protocol suites to use to implement internetworked communications against those requirements. The requirements identified by the CCSDS Green Book, "Rationale, Scenarios, and Requirements for DTN in Space," [13] are:

Network-Layer Requirements:

■ An optionally reliable end-to-end delivery service

■ End-to-end data delivery in the presence of delays and/or disruptions

■ Prioritized data delivery

■ Data link layer agility

■ Management

- Support for higher-layer services including

- File Delivery Service

- Messaging Service

- Space Packet Delivery Service

System Requirements:

- Communications via zero or more relays

- Support for general class of applications, including but not limited to file transfer

- Management information relating to data transfer shall be collected in all nodes and made available to operators

- Ability to automatically reroute data

- Security mechanisms must be provided

- Firewall capability

- Data privacy

- Data authentication

- Interoperability among ground stations (use of multiple ground stations to communicate with a single spacecraft)

- Support for simplex downlinks (some ground stations can only receive)

- Support for low-level (emergency) commanding of spacecraft via data-link-layer frames

Data Transfer Requirements

- Ability to download a single application data unit across multiple ground station contacts

- Mechanisms to support complete delivery of sequences of data

- Must function in highly asymmetric environments (e.g. 10,000:1)

7.3.1   Space Internetworking Protocol Alternatives

The CCSDS working group examined the various options available within the CCSDS suite of recommended protocols, including custom solutions, the use of CCSDS space packets, IPv4/IPv6, the use of the CCSDS File Delivery Protocol (CFDP) as an internetworking transport mechanism, and what was then the emerging DTN specification.

Custom data forwarding solutions were considered both for completeness and for the ability to optimize a custom solution to the individual scenario / environment for which it is designed. The primary drawback of custom solutions is the requirement to build, test, and maintain different solutions for each new mission, and the inability of such solutions to interoperate and leverage one another's resources. As such custom solutions would likely fail the first requirement of being able to provide an end-to-end relayed delivery service.

CCSDS Space Packets [9] were once considered as a possible data structure to support internetworking, but the addressing mechanisms available with space packets are insufficient to discriminate between intermediate (link-by-link) data transfers and final destinations without incorporating some of the addressing mechanisms of underlying layers. Space packets contain an 11-bit application process identifier (APID) that is used to identify an application much the same way a port number is used in IP, as well as sequence and length information. There have been vague notions presented of mapping the APID together with qualifying information like the spacecraft identifier to a routing token called a Label Distribution Path (LDP), but such notions are highly constrained at best. This is because the LDP notion does not support the concept of a destination address for a particular piece of data. For example, if the APID identifies a particular instrument on a spacecraft, and that information might need to be sent to different locations at different times, the entire LDP routing topology would have to change to accommodate the different destinations. At the time of evaluation there had been no standardization to date of the APID-to-LDP mapping, or of managing multi-hop paths using the LDP labels, and so Space Packets were discarded as a potential space internetworking data structure. It should be noted that space packets still play an important role in the SSI architecture, in that they identify the application associated with a particular piece of data.

CFDP was designed to provide multi-hop file delivery, and even has some features that were designed to support internetworking. CFDP can provide many of the requirements identified in the CCSDS DTN Green book, including multi-hop file delivery over piecewise-connected paths. CFDP does not however provide any of the security services (e.g. confidentiality, authentication), and the notion of using CFDP to provide a general transport service to a multitude of client applications seems forced. Finally, while there were notions of providing CFDP file delivery where the data transport was split across multiple paths, no implementations were developed to support them.

The Internet Protocol Suite (IPS), using either IPv4 or IPv6, is an attractive option for building an internetworking service in space. IP provides an end-to-end addressing structure that is independent of the underlying layers used, is very mature, and provides a range of services including quality of service (with IP differentiated services) and security (with IP Security, IPSec). The main disadvantage with the IP suite is that IP implementations, as well as some of the supporting protocols like DNS, TCP, and many of the IP routing protocols assume low latencies and continuous end-to-end connectivity that are not necessarily present in space environments.

Not surprisingly, the Bundle Protocol as specified in RFC5050 supported almost all of the requirements for internetworking in space. In particular, BP provides an optionally-reliable (via the BP custody transfer service) end-to-end data delivery service that is very similar to a modified combination of the services provided by IP and TCP while operating in environments that may have only piecewise connectivity. The main service required by CCSDS member agencies that was NOT part of the RFC5050 specification is an in-order delivery service of multiple bundles. BP, as originally conceived, was intended to support applications that put everything needed for a data transaction into a single bundle for transport and delivery. So for example, if one wanted to send multiple commands to a spacecraft, it was assumed that all of the commands would be put into a single bundle and sent. This allows the transport mechanism (BP) to operate in a more stateless manner and to not have to store data at a receiver awaiting the arrival of 'previous' bundles so that all bundles can be delivered in order. Also, the 'base' bundle protocol of RFC5050 does not support duplicate suppression; multiple copies of a given bundle that is sent may be delivered to the receiver. Both in-order delivery and duplicate suppression are services desired by CCSDS, and both are supported in CCSDS by specifying a 'shim' layer on top of BP that operates end-to-end. This is the Delay Tolerant Payload Conditioning (DTPC) service described in the BP for CCSDS book and covered in sections that follow.

Both the SISG and CCSDS working groups concluded that for network paths that might be disrupted due to the inability to form links between nodes such as ground stations and spacecraft, that the Bundle Protocol was the best candidate. For environments that can support the Internet Protocol Suite (IPS), such as in-situ communication among a group of rovers or astronaut habitats, the use of the Internet protocol suite was found to be warranted.

7.3.2   Space Internetworking CONOPS

While there is agreement in the space community that internetworked communications will improve operations, transitioning from the current operational model to one that uses DTN is complex. Not all ground stations and space agencies will implement DTN at the same time, so missions will need to be able to operate in either case.

In the simplest transition option, DTN is implemented only at the endpoints of communication (the spacecraft and at one location on the ground, probably the mission operations center). In this scenario, the use of DTN is 'self-contained' to the mission, using existing cross-support mechanisms to transport DTN bundles across the space segment to the ground. Distribution of data from the mission operations center to e.g. a science center is a mission issue. This simple DTN CONOPS is illustrated in Figure 7.3.

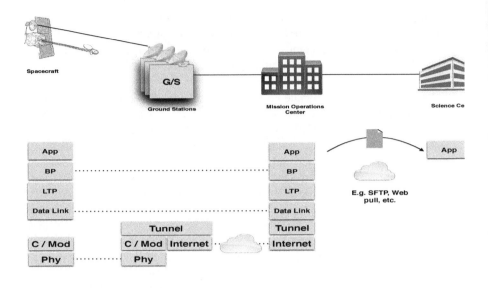

Figure 7.3: Simple CONOPS for DTN deployment with BP on the spacecraft and in the mission operations center. This scenario requires no support by the ground station(s) or ground station operations center.

Even this relatively simple change to mission operations brings several advantages. DTN automates the data transfer and data retransmission mechanisms, taking humans out of the loop. This is particularly useful in shortening the data / acknowledgment control loop for reliable end-to-end transmissions, and especially useful when those operations would take place at times that are inconvenient to operators (like the middle of the night). DTN's storage mechanisms also decouple data generation from data transmission. This decoupling is currently implemented with on-board storage that's managed by the spacecraft computer, but DTN standardizes and automates all of the data forwarding and management operations that would otherwise be re-implemented with each mission.

An intermediate scenario could place DTN in the ground station operations center, essentially replacing the mission operations center in Figure 7.3 with a ground station operations center such as JPL's Deep Space Operations Center (DSOC). This has the advantage of requiring installation in fewer locations, but foregoes many of the advantages of implementing DTN in ground stations themselves. For instance, placing DTN in a ground station operations center but not at the ground stations does not allow DTN to rate-match between space and ground links.

The end goal is to implement DTN at the ground stations and to provide in-space cross-support (relaying) for data. This has a number of advantages, including closing the space link at the ground stations so that link-layer retransmissions can be done in a timely and effective manner; BP at the ground station can serve to rate-match between the space link and the terrestrial network; and data can be distributed

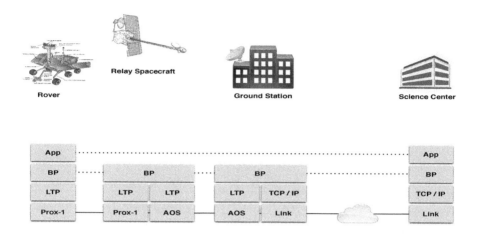

Figure 7.4: DTN Model for space communications and the Solar System Internet; internetworked communication with in-space relaying and cross-support.

directly from ground stations to destination(s) like science centers without being routed through the mission operations center. By routing data directly out of the ground station, more advanced capabilities such as efficiently multicasting multiple copies of the data to different destinations are also possible.

In-space cross-support as envisioned in Figure 7.4 will allow an agency sending spacecraft to a location of interest such as Mars to use other agencies' assets such as orbiters to relay data back to Earth. While missions to send orbiters to Mars still occasionally fail, once in orbit a spacecraft can last many years and provide relay services to a number of future landers / rovers sent later. Not sending a separate orbiter for each landed payload greatly increases the chances of mission success and greatly reduces mission cost.

It should be noted that internetworked space communications, and even the possible use of the Internet Protocol suite (IPv4/IPv6) as part of such a system, does not necessarily mean direct connectivity between spacecraft and the commercial Internet. Due to security concerns, such interconnections (where they exist) are tightly controlled and authenticated to prevent unauthorized access to space data and/or unauthorized commanding of space assets.

Table 7.1: How DTN capabilities are applied to space missions.

	Custody Transfer	Convergence-Layer Architecture	Bandwidth Efficiency Mechanisms	Enhanced QoS Mechanisms
No Contemporaneous End-to-End Path	X			
Heterogeneous Links		X		
Simplex Links			X	
Highly Asymmetric Links				X
Constrained Resources on End Systems	X		X	
Limited-Bandwidth Space Links	X			
Fine-Grained Quality of Service				X
High-Reliability Flooding of Critical Data				X

7.4 DTN Capabilities for Space Missions

This section reviews the capabilities of the CCSDS Bundle Protocol Specification that are of particular interest to space missions. The CCSDS Bundle Protocol Specification is an adaptation / extension of RFC5050, with some additional features described here. The motivations for these capabilities in the space context are summarized in Table 7.4.

7.4.1 Custody Transfer

Custody transfer provides reliable data delivery in DTN using in-network acknowledgment and retransmission of data as opposed to end-to-end acknowledgment and retransmission as in TCP, the main reliable data transfer protocol in the Internet. The rationale behind this design decision was that space links, particularly interplanetary space links, are scarce resources and retransmitting the same data across a space link should be avoided whenever possible. Thus if a bundle from Mars makes it from the surface, through an orbiter and to Earth, only to be lost within the MOC due to congestion or some sort of failure, that bundle should not have to be retransmitted by the source on the surface of Mars. In this way, custody transfer allows bundles to hold the gains that they have made, checkpointing their progress through the network.

Another advantage of custody transfer in space missions comes because the data-gathering applications typically reside on spacecraft with limited storage capacity. Using DTN and custody transfer, an application can delete bundles from its local storage once custody has been taken by another node (since the new custodian assumes

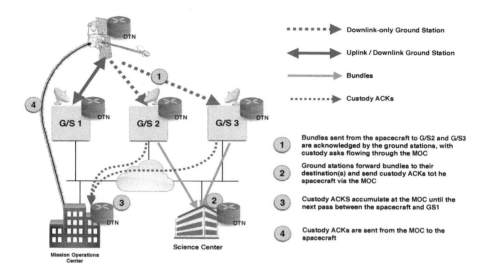

Legend:

- Downlink-only Ground Station
- Uplink / Downlink Ground Station
- Bundles
- Custody ACKs

1. Bundles sent from the spacecraft to G/S2 and G/S3 are acknowledged by the ground stations, with custody asks flowing through the MOC

2. Ground stations forward bundles to their destination(s) and send custody ACKs tot he spacecraft via the MOC

3. Custody ACKS accumulate at the MOC until the next pass between the spacecraft and GS1

4. Custody ACKs are sent from the MOC to the spacecraft

G/S 1   G/S 2   G/S 3

Mission Operations Center

Science Center

Figure 7.5: Custody transfer in an EOS example with multiple downlink-only ground stations and a single bidirectional ground station.

responsibility for keeping a copy of the data and retransmitting it as necessary in order to deliver it to its destination).

As a reliability mechanism, there is an interplay between custody transfer at the Bundle Protocol layer and the use of reliable convergence layers underneath BP. In a tightly-controlled system where all convergence layers provide reliable and timely (across individual links) data delivery underneath BP, it might be possible to rely only on the CL reliability and forego the actual custody transfer protocol. For this to work, however, every hop in the network must support a reliable convergence layer. This assumption may not hold in all situations, since some space links are simplex (unidirectional). In these cases, while there may be a network-layer path from the receiving node back to the sender over which bundles can be sent, there is no way to ensure reliable delivery of bundles using only the convergence layer. Convergence layers operating over simplex links may use mechanisms such as forward erasure coding to increase the probability of correct reception at the expense of overhead, but without acknowledgment the sender cannot know whether or not the bundle has been received.

As an example of such a system, consider an Earth-observing satellite that can make use of multiple downlink-only antennas during the course of an orbit, but with only a limited number of uplink antennas. In this case, reliable transmission (at the link layer) from the spacecraft to the downlink-only antennas is not possible. Custody transfer at the Bundle Protocol layer can, however, route custody acknowledgments via the uplink antenna to the spacecraft as shown in Figure 7.5.

7.4.2 Convergence Layer Architecture and LTP

Networks using BP as their network-layer protocol may need to function overlay paths composed of highly heterogeneous links. Taking again the example of rovers on Mars communicating with Earth, the surface-to-orbit link is relatively low delay and can support bidirectional communication. The link from Mars orbit to Earth has significant one-way delay (ranging from about 4 to about 20 minutes, depending on the positions of the planets). Connectivity from the ground station to the mission operations center is IP-based, and supports standard terrestrial protocols like TCP.

It would be extremely difficult to design a single link layer that could function in all of these environments. Instead, the DTN architecture assumes that an end-to-end path at the BP layer can be supported by different convergence layers for each transmission between DTN routers, each designed / tuned for the environment in which it has to operate.

The Licklider Transmission Protocol for CCSDS [18] is the preferred convergence layer for space links where reliability is required. The CCSDS version of LTP is a profile of the IETF LTP RFC [224] with some constraints to make LTP more amenable to space missions as well as some extra capabilities that extend the base protocol. The extra constraints of the LTP for CCSDS protocol mainly improve efficiency by ensuring that certain LTP protocol values have short representations. LTP makes use of self-defined numeric values [104] to provide efficient representation of a wide range of integers. Some of these values (e.g. the offset of data within an LTP block) must take certain values, while others (e.g. session identifiers) need only be unique over a certain span. The LTP for CCSDS Protocol restricts the values of the identifiers to reduce the header size and improve efficiency. These constraints do not impact interoperability with an RFC5326-based LTP implementation.

LTP for CCSDS also defines a Service Data Aggregation (SDA) service that is not present in RFC5326. SDA is designed to improve the efficiency of LTP when there are many small blocks to be sent by allowing an LTP sender to aggregate multiple incoming blocks into a single 'super-block' for the purposes of reliable transmission. Using service data aggregation, 10 100-byte reliable (red) LTP blocks can be aggregated into a single larger block requiring (if there is no loss) only a single report / report ACK cycle to acknowledge receipt. This is illustrated in Figure 7.6.

Service data aggregation can dramatically reduce the amount of acknowledgment traffic for streams of small blocks. Consider a stream of 150-byte bundles representing some periodic housekeeping data or science task. Assuming no loss, each LTP block will elicit a report segment from the receiving LTP engine that is on the order of 13 bytes. Especially when the asymmetry of the channel (the ratio of downlink to uplink bandwidth) is high, this can present a problem by completely consuming, and congesting, the uplink channel with acknowledgments. With SDA, a number of 150-byte blocks can be grouped together and sent as a single 'meta' block so that all of the smaller blocks are covered by a single reporting structure.

This same issue of needing a mechanism to operate efficiently over highly asymmetric channels (and, for BP, paths) arises again with respect to custody transfer at the bundle protocol layer and is discussed in the next section. For BP a slightly

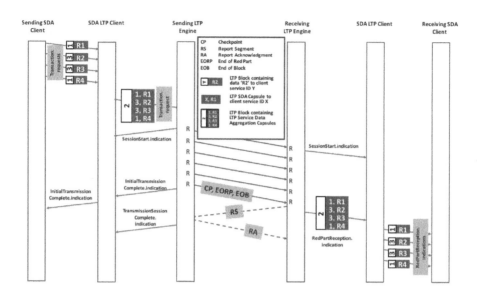

Figure 7.6: LTP Operation (from [18]).

different approach has been taken where bundles are labeled sequentially on transmission and a run-length encoding structure has been defined to allow a receiver to acknowledge multiple bundles efficiently.

7.4.3   Aggregate Custody Signaling

ACS was developed to support initial DTN experimentation on the International Space Station (ISS) [153]. In that environment, the link asymmetry was about 2800:1. This meant that for the minimum custody ACK size of 49 bytes, each data bundle needed to be about 140 kilobytes in order to not congest the acknowledgment channel. Since the bundles being generated by the end systems were significantly smaller than 140kB, something needed to be done to reduce the ACK channel bandwidth.

The idea behind ACS is to allow a node taking custody of multiple bundles to send a single custody acknowledgment with what amounts to a list of the bundles the node is taking custody of. The issue then is how to efficiently encode that list. What complicates this process is that a receiver cannot, simply from the received set of bundles, determine if there are any bundles missing. Recall that the primary block contains the source EID, the creation time of the bundle, and a sequence number that is only required to be monotonically increasing within a given second (and which may or may not be reset between seconds). To form a compact identifier space, ACS defines a custody transfer enhancement block (CTEB) that is appended to bundles by the current ACS-capable custodian. Sequence numbers in the CTEBs allow a downstream custodian to identify bundles and ranges of bundles using the CTEB identifiers.

With aggregate custody signaling, a node taking custody can signal to the current custodian that it is, for example, taking custody of bundles with CTEB identifiers: 1, 3, 5–17, and 20–25.

Delay-Tolerant Payload Conditioning, discussed below, can be used as an alternative way to address the issue of ack-channel congestion, or can be used in conjunction with aggregate custody signaling to operate efficiently at even higher asymmetries.

7.4.4 Delay-Tolerant Payload Conditioning (DTPC)

The Bundle Protocol supports most of the requirements for space missions identified by CCSDS, but not all. In particular, the data transfer requirements to deliver sequences of data both reliably and in order, and efficient support for extremely high asymmetries are not supported by the base protocol. In response, CCSDS defined an application service layer that sits immediately above the bundle protocol, between BP and applications, and that provides the following services:

- Delivery of application data items in transmission (rather than reception) order;

- Detection of reception gaps in the sequence of transmitted application data items;

- End-to-end positive acknowledgment of successfully received data;

- End-to-end retransmission of missing data, driven by timer expiration;

- Suppression of duplicate application data items;

- Aggregation of small application data items into large bundle payloads, to reduce bundle protocol overhead;

- Application-controlled elision (removal) of redundant data items in aggregated payloads, to improve link utilization.

7.4.5 DTPC and LTP Together

The combination of DTPC, BP, and LTP provides a flexible framework that allows mission designers to control the amount of overhead traffic associated with data streams with widely varying data unit sizes. Applications may generate data of whatever size is useful to the application. For example, a hyperspectral imager might generate relatively few data items, with each one being many megabytes or even gigabytes in size. Smaller instruments or housekeeping data might generate many small records at a higher rate. To reduce the overhead of transmission and of custody acknowledgments, delay-tolerant payload conditioning can aggregate small data units that are all headed to the same destination into larger bundles for transmission. This requires only a single primary bundle header to send the data, and a single custody acknowledgment to acknowledge receipt. In this way, DTPC allows a mechanism to tune the bundle-layer overhead to match the perceived end-to-end path conditions.

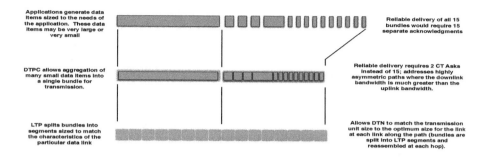

Applications generate data items sized to the needs of the application. These data items may be very large or very small

DTPC allows aggregation of many small data items into a single bundle for transmission.

LTP splits bundles into segments sized to match the characteristics of the particular data link

Reliable delivery of all 15 bundles would require 15 separate acknowledgments

Reliable delivery requires 2 CT Asks instead of 15; addresses highly asymmetric paths where the downlink bandwidth is much greater than the uplink bandwidth.

Allows DTN to match the transmission unit size to the optimum size for the link at each link along the path (bundles are split into LTP segments and reassembled at each hop).

Figure 7.7: Using the space-specific capabilities of BP and LTP to control overhead in a space environment.

DTPC operates at the endpoints and does not have knowledge of the characteristics of the individual links that make up the end-to-end path. LTP operates at each transmission/reception hop and can be tuned to the individual link characteristics using LTP's Service Data Aggregation (SDA) feature that, like DTPC can aggregate small bundles into larger LTP blocks, as well as LTP's segmentation feature that breaks larger bundles into smaller segments for transmission and acknowledgment by LTP. This is illustrated in Figure 7.7

7.4.6 Extended Class of Service

The extended class of service (ECOS) block in the CCSDS specification provides additional quality of service markers that may be useful to space missions, including:

■ An optional 255 levels of priority within the normal BP class of service 2 (expedited).

■ A mechanism to efficiently flood a 'critical bundle' over all viable paths to its destination.

■ Bits to influence the choice of convergence layers with specific capabilities, such as reliable or unreliable delivery.

■ A 'flow label' that can be used by the application layer to identify bundles that are part of a sequence.

The expanded levels of priority are intended to allow fine-grained control over the order of bundle transmission. Unlike the base RFC5050 specification, which defines the three basic priority levels but does not prescribe how they should be used, the ECOS block defines a strict priority order for bundles, with bundles of RFC5050 COS 2 and higher ECOS priorities transmitted before those with lower ECOS priorities and before bundles with lower RFC5050 COS values.

The 'critical bit' in the ECOS flags specifies that the bundle should be intelligently flooded towards the destination along all paths that have a plausible prospect of reaching the destination node without being routed back through the transmitting node. It is up to the implementation and its routing/forwarding process to decide which such paths qualify.

There are two bits in the ECOS block that influence the choice of convergence layer protocols by all DTN routers along the path. The first bit directs DTN routers to use convergence layers that do NOT retransmit data on loss. Such a choice might be suitable to support streaming voice over BP, for example, where the added latency and/or out-of-order delivery incurred by reliably delivering all bundles doesn't improve the usability of the overall data stream. The second CL-influencing bit directs intermediate DTN routers to use convergence layers that DO provide reliable data delivery. Setting both the unreliable and reliable CL bits instructs intermediate DTN routers to use a specialized convergence layer that monitors the order of bundles (by examining the bundle creation timestamps) and to forward in-order bundles using unreliable mechanisms and out-of-order bundles using reliable mechanisms. If the DTN router does not implement such a 'dual-mode' convergence layer then it must use an unreliable CL if one is available.

The purpose of setting both the 'reliable' and 'unreliable' bits in the ECOS block is to support streaming media over BP. The intent is that the destination will consume and present to users the in-order bundles as they arrive (e.g. for video streaming), and will archive the entire data stream (both in-order and out-of-order bundles) for later review. This way the receiver can implement a DVR-like capability where the most recently-received data is presented immediately and the entire data stream can be reviewed later.

7.5   Benefits of DTN to Space Missions

DTN provides a number of benefits to space missions beyond the ability to provide end-to-end information transfer in environments with only piecewise-connected paths.

7.5.1   Automated Data Transfer

One of benefits of using DTN for space missions is that it automates the data transfer process, even for missions that use 'traditional' single-hop architectures. Instead of commanding each data transfer operation, DTN allows mission operators to set up the parameters of communication such as which links will be available when, along with the link characteristics such as data rate and latency. Once the system has been set up in this way, data sources can submit data to the network whenever it becomes available, and the network will automatically ensure that the data reaches its destination without further human intervention.

Automated data transfer will be even more important for multi-hop missions because the protocol will be able to act autonomously on local information rather than having to wait for information about received data to be sent to Earth before commands to transfer it across the next hop can be generated. As with the single-hop case, DTN allows mission designers to set up the parameters of communication, but now those parameters can enable multi-hop communications (if there is a link at time T coming in to a node, and a link at time T+X leaving the node, for example).

Allowing BP's custody transfer mechanisms to automate the process of reliable data transfer removes the need for humans in the loop to examine the data and generate and check commands to retransmit missing portions. Automated and reliable data transfer also allows missions to use portions of communication windows that they might not otherwise use. Typically the very beginning and the very end of a communication window are not used because the quality of the link is questionable. Automated reliable data transfer, assisted by prioritization mechanisms that ensure that more important data is transferred first, allows missions to use the 'edges' of communications passes that might otherwise go unused.

7.5.2   Rate-Matching

With the move to higher frequencies, including optical communications, many Earth-observing missions are able to transmit data to the ground station faster than the ground station can forward the data to the Mission Operations Center [94]. In these cases, DTN can provide more efficient reliable data transfer if there is a DTN router at the ground station. A DTN router at the ground station can acknowledge bundles to the spacecraft immediately when those bundles reach the ground station. If data is lost or corrupted on the space downlink (due to weather or clouds, e.g.), the lack of a custody acknowledgment by the ground station can elicit retransmission by the spacecraft well before the data loss could be detected by the MOC. The custody transfer mechanism, where the ground station can acknowledge receipt of data and ensure its reliable delivery onwards, allows the spacecraft to remove the data from its buffers and to make that space available for new data. This can provide a significant savings in spacecraft resources.

Because the current ground architecture requires the forward space link to be constructed at the MOC, the ground station cannot simply transmit custody acknowledgments to the spacecraft. Instead, the custody acknowledgments must be forwarded through the MOC for uplink to the spacecraft as in Figure 7.8. By prioritizing custody acknowledgments forwarded via the MOC to the spacecraft over data sent to the MOC, this architecture still allows the ground station to receive and acknowledge data faster than it can be forwarded over the terrestrial network.

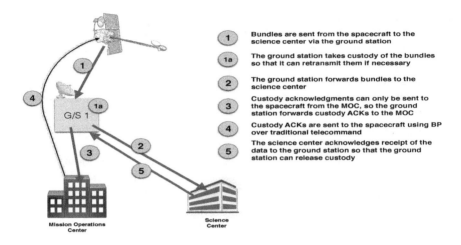

Figure 7.8: DTN routing and custody acknowledgments in missions with a mix of bi-directional and simplex links where all data uplinked to the spacecraft must pass through the mission operations center.

7.5.3 In-Space Cross-Support

CCSDS is currently refining and space agencies are implementing terrestrial cross-support mechanisms that allow an agency to request service from another agency's ground infrastructure. Practically speaking, this means that AgencyA can, using standard mechanisms and protocols, request that an antenna belonging to AgencyB track and exchange data with a spacecraft belonging to AgencyA, forwarding the data to AgencyA over the Internet. In-space cross-support extends this notion to allowing an agency to request data forwarding services, using DTN, from another agency's spacecraft. One benefit of this mechanism is that in locations with multiple spacecraft, there will be more opportunities to relay data back to Earth, and thus the ability to return more science data. In addition, spacecraft that are in relatively close proximity (two spacecraft in orbit around Mars, for example) can communicate much better with each other than with Earth. If something goes wrong with one of the spacecraft, the other may be able to get diagnostic information from it and to send it commands to fix the problem more easily than if the operation has to be conducted from Earth.

7.6 Standard SSI Applications

Space missions are very expensive and considerable care is usually taken to ensure that if something goes wrong, the mission can be recovered and continued. The low-

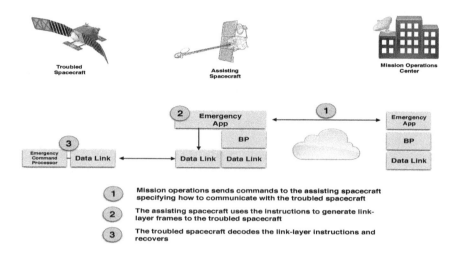

Figure 7.9: First-hop last-hop model.

est levels of fault recovery typically attempt to reduce the number of spacecraft systems that need to be operating in order to diagnose and recover from errors. In addition, part of the failure mode may be that the spacecraft has lost attitude control and is spinning or tumbling, and as a result may not be able to receive long and complicated sequences of instructions. What is typically done is to include a simple set of fault recovery commands that can be transmitted directly in data-link-layer frames to the spacecraft. The commands are typically set up to be decoded in hardware by the spacecraft radio, so that they do not rely on a function computer on board. In the current model for space communications, such commands are sent directly from Earth to the spacecraft.

In the networked model for space communication, it may be that a spacecraft doesn't have a direct-to-Earth link at all. If the only means of communication is over the network, a spacecraft fault that takes down the networking system would essentially disable the entire spacecraft. Thus even in a networked mode of operations, we would still like to be able to provide the same type of link-layer emergency commanding capability that is available with current missions.

The current plan is to provide both emergency commanding (forward) and emergency telemetry (return) capabilities as standardized network applications. These applications would reside on all network-enabled spacecraft (or at least on all such spacecraft that might be used as relays) and would support the sending of link-layer commands and receipt of link-layer telemetry to/from a spacecraft in distress as shown in Figure 7.9.

If mission operators need to send emergency commands to a spacecraft that has only network connectivity, they will send an application-layer set of commands to a networked spacecraft that can reach the distressed spacecraft over a single data link. The application commands will include information such as the data link layer frames to emit together with timing, frequency, modulation, and coding information. These commands can be sent to the assisting spacecraft over the network, since presumably the assisting spacecraft is still functioning normally. The assisting spacecraft will then emit the frames towards the distressed spacecraft according to the instructions. Effecting these transmissions will probably also involve operator configuration of the assisting spacecraft, as the emergency transmissions will likely fall outside of normal communications windows.

When invoked, an 'emergency receive' mode will trigger reception of data link layer frames from the distressed spacecraft to be encapsulated into BP bundles by the assisting spacecraft for forwarding to Earth. The intent of the emergency receive mode is to allow low-layer telemetry from the distressed spacecraft to be sent to Earth for analysis.

7.7   DTN on Space Missions

DTN has flown as an experiment on a number of space missions, and more recently has transitioned to operational use by scientists and flight personnel.

7.7.1   DINET

In October and November of 2008, NASA's first Deep Impact Network Experiment [296] tested DTN using the Deep Impact spacecraft operated by the EPOXI project. DINET demonstrated DTN's internetworking capabilities by using the spacecraft as a DTN router connecting multiple nodes on the ground at NASA's Jet Propulsion Laboratory in Pasadena, CA that acted as assets on Earth, and on and around Mars. Over the course of the experiment, some 300 images were uplinked to the spacecraft, which at the time was about 15 million miles (about 80 light seconds) from Earth. To communicate with the spacecraft, NASA used its Deep Space Network, and used the Licklider Transmission Protocol as the convergence layer protocol underneath BP. In addition to demonstrating DTN's routing, forwarding, and quality of service functions, the DINET experiment also demonstrated reliable data delivery in the face of intentionally dropped LTP segments, and unexpected lapses in service from the DSN stations servicing the satellite. In both cases, the reliable data transmission mechanisms automatically detected the losses and retransmitted the needed segments and bundles.

7.7.2   DTN on the International Space Station

In June of 2009 DTN was deployed onto the Commercial Generic Bioprocessing Apparatus (CGBA) unit #5 on board the International Space Station (ISS) [153]. This was one of the first cases of deploying DTN into an operational space mission, and it both validated the utility of DTN and required the development of additional capabilities to meet the stringent requirements of the mission.

Prior to using DTN, the CGBA5 data transfer system was not allowed to automatically acknowledge correct receipt of data. To increase the reliability of data transfer, the system inserted new data into circular buffer 'carousel' and simply transmitted the carousel over and over. If new data wasn't being generated quickly, this could result in the same data being transmitted to the ground hundreds or even thousands of times. As part of the DTN testing, NASA engineers agreed to allow custody ACKs to be transmitted to the spacecraft automatically (after strict automated format checking and with the requirement that they could be disabled by operators). This resulted in a reduction in the number of transmissions used to downlink data from around three thousand to just one or two. This means that using DTN, scientists can plan on getting roughly 1,500 times more data down than without it. Arguably this is not a benefit of DTN per se, but of allowing automated acknowledgment and retransmission of data. DTN, however, provides a standard mechanism for acknowledgment around which tools and procedures can be built by mission operations so that they can have the level of confidence required to allow automated uplink of acknowledgments to the spacecraft.

It was the asymmetry of the communication channel allocated to the CGBA's (roughly 400kbps down and 150 bps up) that drove the development of the aggregate custody signaling mechanism described above.

7.7.3 EO-1, IRIS, and TDRSS Demonstrations

To gain practical experience with the DTN protocols and to determine their benefits to space missions in operation, NASA's Goddard Space Flight Center conducted a number of flight experiments using DTN [94]. The first of these used the Earth Observing Mission 1 (EO-1) in late 2010 / early 2011. During the EO-1 experiments, DTN showed the following: • The ability to decouple the communication schedule from command and telemetry generation. Like most previous and contemporary missions, standard EO-1 operations required that commanding of the spacecraft, including commanding it to dump data to Earth, take place in 'real-time' when the spacecraft link to Earth was established. Using DTN and the ability to store data in the network, commands could be 'sent' at any time and stored in the network until the link with the spacecraft was available. This included commands to the spacecraft to transmit data back to Earth. • Recovery of some housekeeping data that was not being sent due to a hardware failure on board. The EO-1 suffered a hardware failure that prevented the storage of routing housekeeping data (not its primary science data) between contacts with Earth. While this data could have been recovered in other ways as well, by sending the data via DTN, the data was stored in DTN's storage (which was allocated out of the still functional storage pool) and forwarded to the ground once a link was established. • Rate-matching between the satellite link and the terrestrial network. The EO-1 DTN experiment used DTN's ability to receive data at a very high rate from the spacecraft and to automatically and robustly buffer that data before trickling it out to the terrestrial network at a rate the terrestrial network could handle. The same functionality is usually achieved by recording the data at the ground station and shipping it later, sometimes via physical media such as magnetic tapes. This is an important area where the DTN concept of operation provides greatly improved performance over current methods, since the DTN data is decoded and routed from the ground station, the ground station can immediately detect errors or lost data and request that they be retransmitted from the satellite. Without this, the satellite must hold on to old data for a long time until it can be verified to be correct (using precious on-board storage) or lost/corrupted data is simply lost.

The second GSFC experiment leveraged the Communications, Standards, and Technology Lab (CSTL) to emulate a collaborative sensor web experiment that extended the communications path from EO-1 across an IP network comprised of the terrestrial NASA network and an in-space IP router (the Internet Router In Space, or IRIS) on board Intelsat 14. This experiment demonstrated the viability of 'standard' IP routing operating on-board a space platform as well as DTN's ability to ride on top of different underlying transport mechanisms, including leveraging IP multicast to deliver the same data to multiple destinations simultaneously. The combination of DTN store-and-forward and in-space routing reduced the latency between simulated sensor events and their reception at mission / science operations centers. This was one of the first demonstrations to show a multicast / pub/sub model directly from the instrument to multiple recipients.

Finally, GSFC tested DTN over the Tracking and Data Relay Satellite System (TDRSS) to ensure that DTN could operate correctly and efficiently in the presence of asymmetric links, including simplex links. This experiment simulated a spacecraft in ground equipment at GSFC connected to a mission control center over the TDRSS system, with DTN nodes at the simulated spacecraft and MOC, as well as at the TDRSS ground station at White Sands, NM. This test leveraged DTN's ability to use convergence layers appropriate to the local conditions by using BP over LTP and a CCSDS data link for the spacecraft-to-ground station link, and LTP over UDP/IP for the ground station to mission operations center link. DTN was able to use the simplex downlink that was available throughout the experiment, and to batch responses to the spacecraft until the uplink became available.

The three EO-1 tests demonstrated DTN's ability to provide benefits to near-earth space missions by simplifying mission operations, providing a more timely data transfer mechanism, and making use of both a wide variety of links and heterogeneous links along the path from spacecraft to mission/science operations.

7.7.4 LLCD

NASA also demonstrated DTN on the Lunar Laser Communication Demonstration (LLCD) on the Lunar Atmosphere and Dust Environment Explorer (LADEE) mission [149]. Optical communication is of increasing interest to the space community because of the higher data rates that can be achieved with relatively low power. An issue in optical communications is loss due to obstruction from clouds or other, shorter-term events. Here the DTN stack, especially when DTN is implemented at ground stations, can help. By closing LTP at the ground station, any loss can be detected immediately and a retransmission requested. For near-Earth missions, such retransmissions can be requested and fulfilled within the current communications pass; for deep-space missions this may not be the case, and LTP may have to buffer segments until a subsequent pass.

In the DTN demonstration on LLCD, BP over LTP was used to transmit data to the spacecraft which relayed the data back to Earth. In this experiment there was no DTN node on the spacecraft itself, but the link did have to go from the Earth to lunar orbit and back. The experiment showed the benefits of the automated retransmission capabilities of BP and LTP, as during one of the experiments the link was disrupted for about two minutes during which time communications were lost. At the end of the disruption, DTN picked up and automatically retransmitted the lost bundles.

7.7.5   ISS DTN Gateways

The International Space Station (ISS) is nominally continuously connected to the ground infrastructure via NASA's Space Network of TDRSS satellites. Due to the exact placement of the TDRS satellites, there is a small zone of exclusion the prevents 100% coverage using the high-bandwidth Ka-band communication system. This means that not infrequently, the ISS goes through loss-of-signal (LOS) / Acquisition of Signal (AOS) events where data transfers are interrupted. These interruptions would cause normal internet protocols like TCP to slow their transmission rates and start exponentially increasing the time between when they probe to determine if the connection is available again. While DTN with the LTP convergence layer cannot make a link exist where there is none, they can either be aware of the outage, including planning for it, when the outage appears in the contact plan of a scheduled routing algorithm, or they can simply use LTP's loss detection and retransmission to 'ride out' the outages.

After its success with the CGBAs on board ISS, DTN was adopted as a standard service as part of the Telescience Resource Toolkit (TReK) produced by NASA's Marshall Space Flight Center (MSFC). The TReK toolkit provides investigators with a set of software tools to both control and receive data from their experiments on board the International Space Station (ISS). As part of the DTN deployment two DTN routers have been deployed onto the ISS, one on the payload LAN and one on the operations LAN, and a DTN node has been installed in the Huntsville Operations Science Center (HOSC) at MSFC. By using the DTN infrastructure to send and receive data, scientists can more easily decouple their schedules and the schedules of their experiments from the communication schedule with the ISS. For the HOSC, DTN's storage and retransmission services buffer data when PI machines are not connected to the network and automatically send and receive data when the machines are connected. This reduces the operational load on the personnel of the HOSC, as it translates into fewer requests to retransmit data that was already downlinked from the ISS.

The DTN router on the operations LAN is being tested for use by everyday operation of the ISS, including functions like file uploads and downloads.

7.8 Summary

For the past 30 years, space design has been moving relentlessly from new point designs for each mission to common standards and protocols that enable cross-support among agencies on the ground, and is now extending that interoperability to space. The international interest in Mars and Lunar exploration form loci of interest where multiple orbiters and landers will be in relatively close proximity and where it will be physically possible to share communication resources in order to get more science faster and more reliably. The idea of a Solar System Internetwork formed by the cooperation of multiple spacecraft in different mission phases and bringing the benefits of networking to space missions promises to extend our reach into the solar system and to make missions less expensive and more robust. We have already seen the real benefits of these concepts in the way data is being returned from Mars, and they are rapidly being extended to near-Earth missions such as the International Space Station.

Chapter 8

Delay-Tolerant Security

Edward Birrane

The Johns Hopkins Applied Physics Laboratory (APL)

CONTENTS

8.1 Introduction

Securing network communications remains a challenging and evolving field. New threats to networks are continuously emerging as desktop computational capabilities increase and exploitation tools grow in both availability and sophistication. A proliferation of poor security practices across many corporate enterprises leave systems open to even well-understood attacks. The increasing need for security services is evidenced by the requirement that new protocols must explain how they secure their operation in a variety of hostile conditions. International standards bodies such as the Consultative Committee for Space Data Systems (CCSDS) and the Internet Engineering Task Force (IETF) require security mechanisms be addressed in all of their new specifications.

A natural question to ask is whether this existing body of security capabilities, most often deployed across the Internet or corporate intranets, can be extended as-is to address the security needs of a DTN. Put a slightly different way, what are the common challenges of securing the traffic into and out of (1) a corporate data center, (2) a set of energy-harvesting sensor nodes, and (3) an interplanetary spacecraft providing a communications link to Mars? Are these three networks similar enough that the same protocols, algorithms, and tools can work without modification? As DTNs mature from a specialty in the field of challenged network research, do we need to research new kinds of security mechanisms to successfully engineer and deploy these networks safely?

Most modern security research focuses on protecting Internet-style networks because they represent the majority of operationally deployed networks in the world today. However, deployments of DTN-style networks are becoming increasingly more frequent to enable new classes of networked sensing concepts. While the fundamental nature of security (services, algorithms, cipher suites) may be unchanged based on the type of network deployment, the physical implementation of security mechanisms may be impacted by the capabilities of the network. The tools, data sets, and protocols used to secure networks will need to be different between the Internet and a DTN deployment because our current tools are often unable to scale to accommodate the high signal propagation delays, frequent link disruptions, and lack of state synchronization that define a DTN.

This chapter explores how to translate the familiar application-level services and concepts of end-to-end Internet security to a DTN. Since the mechanisms by which DTN services are implemented and maintained will be driven by the assumptions, requirements, and constraints of the DTN operating environment, this chapter will define common security services, examine what challenges are unique to a DTN, explain how those challenges lead to the creation of a new security model, and review some emerging security protocols to implement that model. Finally, a discussion of best practices, policies, and configurations for securing networks are provided.

8.2 Common Security Services

Network security can be categorized into services relating to confidentiality, integrity, authentication, and availability[1]. The definition of these services remains the same regardless of whether the network is well-resourced or challenged. Table 8.1 enumerates these services and briefly discusses the benefit they bring to a network. The remainder of this section describes these services in greater detail.

---

[1] The relatively modern term "CIA Triad", which refers to the services of confidentiality, integrity, and availability, has undocumented origin. These fundamental concepts have ideological origin in antiquity.

Table 8.1: Common security functions as applied to a DTN.

Security Service	Benefit
Authentication	- Determines who originated a message. - Determines who was the last person to forward a message. - Determines permissions or rights for a message in the network.
Integrity	- Determines that protected parts of a message were not changed in transit from a source to a destination.
Confidentiality	- Prevents a third party from viewing protected parts of a message as it is in transit from a source to a destination. - Provide similar protection when a message is "at rest" waiting for its next hop in a DTN.
Availability	- Reduces the likelihood that the network will be unavailable to its users.

8.2.1 Confidentiality

A Confidentiality service obfuscates protected traffic in a network such that only authorized recipients of the information can determine its original content. Confidentiality is usually implemented through the processes of encryption and decryption, as illustrated in Figure 8.1.

In this figure, user data is provided in an unprotected form (called plaintext[2]) and must be permuted in accordance with one of a variety of mathematical algorithms (called ciphers) to produce a new, unintelligible version of the user data (called cipher text). This process is called enciphering or encryption, because it passes the user data and other information through a cipher and captures the output of that cipher as the version of the data to be passed across the network. In addition to the user data, another critical input to a cipher is some secret information (called a key) that is understood by both the message sender and the message receiver. Keys located at the sender and destination nodes may be identical (symmetric keys) or they may be different but entangled in some mathematical way (asymmetric keys). When cipher text is received at a message destination, the encryption process is reversed. The cipher text and key information are provided as input to the cipher which then outputs the plain text which should now represent the original message generated at the sender. This process is called deciphering or decryption.

The Internet provides multiple examples of why confidentiality is a necessary component of any type of deployed network. Consider transmitting medical records, bank account numbers, or other personal information without any assurance that the information would be protected from unauthorized viewing. The consequences of

---

[2]Typically, the input to a confidentiality suite is information that has not already been encrypted, and is therefore considered plain text and not cipher text. There are situations in which the output of one confidentiality suite may be given as input into another confidentiality suite. In this case the second cipher suite would accept cipher text and produce new cipher text.

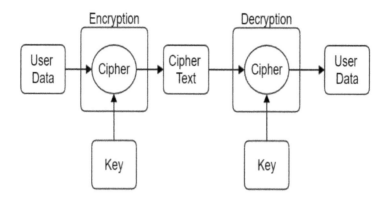

Figure 8.1: Confidentiality hides data from the point of encryption to the point of decryption.

information getting into the "wrong hands" ranges from simple embarrassment to financial loss to significant changes in company or government policy.

The "strength" of an encryption scheme addresses how well the scheme prevents unintended observers from accessing information. The "stronger" the scheme, the harder it is for such observers to gain access to the original plaintext information being protected. As such, strength grows proportionally with the difficulty of deciphering data without proper credentials. There is no standard unit of measurement for strength across all encryption schemes and no standard mechanism for measuring such strength across the myriad of methods someone could use to overcome various encryption schemes. However, since a key is typically the necessary credential to decipher some ciphertext, and since guessing longer keys is harder than guessing shorter keys, encryption strength is typically expressed in terms of the length of the key. Put another way, all things being equal, a cipher using a large key value is considered more secure than the same cipher suite using a shorter key value. For example, the Advanced Encryption Standard (AES) [209] supports three key lengths: 128, 192, and 256 bits. AES-256 is considered to be stronger than AES-128 due to key length. In 2011, a key recover attack on AES performed by Bogdanov, Khovratovich, and Rechberger [48] showed the ability to recover a 128-bit AES key in $2^{189.9}$ operations whereas $2^{254.3}$ operations are required to recover an AES 256-bit key. [272]

The strength of the confidentiality service of a network, however, is more than the strength of its encryption scheme. In addition to a strong encryption scheme, information about the configuration, policy, and management of the cipher and cipher text message traffic must also be considered to say that the network has a strong confidentiality service. For example, a very long key does not protect cipher text if the key can easily be stolen from an unsecured server or otherwise retrieved by social engineering attacks. Alternatively, certain encryption schemes can be compromised if the plaintext is known beforehand and can be compared to cipher text observed on the network, such as is the case when transmitting a well-known status packet or requesting a web page in a certain way.

8.2.2   Integrity

An Integrity service ensures that data are not modified from the time they are generated by a sender until the time they are received at their destination. Integrity is usually achieved by sending an abbreviated[3] representation of the data to the receiver along with the data themselves. This abbreviated representation is almost always a hash value and so the term "hash value" will be used in lieu of the term "alternative representation" for this discussion. The generation of the hash value is performed by a hashing function which accepts the data to be summarized and produces the relatively unique hash value of that data. A secure hashing function is a special type of hashing function that also accepts a security key and produces a hash value that is a relatively unique mathematical combination of both the user data and the security key.

One constructed, the hash value is sent with the original data as a tuple. At any point along the message path, and certainly at the message destination, a receiving node may calculate a local hash value[4] from the received data and compare it to the received hash value. If the received hash matches the locally calculated hash, then (with confidence) the received data has not changed since the original hash was calculated. Exotic schemes may emerge where a hash value may be transmitted more intelligently than simply bundling it with the data. However, all integrity mechanisms work in the same way: the data and some calculated summarization of that data are both presented to a receiver who recalculates a local representation of the summary data and compares it to the received summary to determine integrity. The general integrity procedure, including secure hashing functions, is illustrated in Figure 8.2.

The consequences of running a network without an integrity service is that there is no way to understand whether a message was modified between the sender and the receiver. The concept of integrity is fundamental to networking of any kind and, therefore, integrity checks are built into multiple levels of protocol stacks. The reason for these multiple layers of integrity checking is that a message may be altered not by a malicious attacker, but by the environment itself. Very long runs of Ethernet cable, noisy WiFi environments, or even ionizing radiation can affect the mechanisms that translate physical effects into digital 1s and 0s for a variety of communications media. In some cases of integrity services, the alternate representation of the user data is verbose enough to allow for small errors in the data to be both detected and corrected.

The "strength" of an integrity service addresses the likelihood that a change in

---

[3] Abbreviated versions of the user data are preferred as they reduce the overall size of the message and associated bandwidth costs.

[4] When using a secure hashing function, a receiving node must also have an appropriate key to produce a proper hash value. This key may be identical to the key used when originally creating the hash value (when using symmetric keys) or it may be mathematically entangled with the original key (when using asymmetric keys).

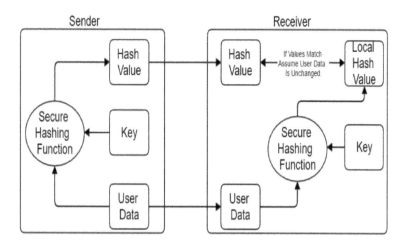

Figure 8.2: Integrity ensures that a message has not changed while in the network.

the user data would result in a message recipient calculating a different local hash value than the one received with the message. Similar to encryption algorithms, this is often correlated to the size of the hash value. A larger number of bits to represent a hash value indicates a larger computational space to hold hash values and reduces the likelihood of having hash function collisions. Assuming an integrity function that calculates hashes uniformly across this computational space, it becomes less likely that two different sets of user data (such as original user data and a slightly modified version of that user data) would hash to the same value.

8.2.3 Authentication

An Authentication service confirms that a received message was sent by the presumed message sender. Authentication is typically accomplished by using an integrity service to generate a secure hash value over the combination of the data-to-be-authenticated and some value that uniquely identifies the sender (such as the sender's private security key). The authentication hash is bundled with the data and sent as a tuple. Similar to an integrity service, any node along the message path could authenticate data by calculating its own local authentication hash based on the user data and the local node's understanding of the sender's identity (such as the sender's public security key). If the authentication hash value in the message matches that which was computed locally, then the user data was both unmodified in transit and was generated by the appropriate sender.

Authentication may be applied to an entire message, only the user data portions of a message, only portions of the user data portions of the message, or to certain meta-data in a message (such as a primary header). Determining what information to include in the has value is based on the desired goal of enabling authentication in the network. In cases where data traffic is trusted if it is sourced from a particular node and always kept on a private network, then authenticating the portions of the message (such as a primary header) that confirm the source of the message may be sufficient. In many networks, the entirety of a message must be authenticated to prevent malicious users from altering data that is asserted to come from a known source. In certain networks, user data must be authenticated, but secondary headers carrying ephemeral and otherwise annotative information do not require authentication.

The concepts of authentication and integrity are distinct. One way to differentiate these cases is to understand that an integrity service detects (and sometimes corrects) an incoherent change to a message – random changes to a message that are not designed to appear syntactically correct. An authentication service detects (and does not necessarily correct) coherent change to a message – nonrandom, intentional manipulation of a message to change meaning while appearing syntactically correct.

The Internet provides several cautionary tales of the consequences when authentication is not present in a network. Consider the cases of changed legal contracts, medical direction, business directives, or personal correspondences that could be altered en-route in a network without detection.

8.2.4   Availability

Network Availability refers to the combination of other services to ensure the network can perform its intended purpose. Most networks are built to carry specific data in specific circumstances. Corporate networks carry company traffic and internal websites. Sensor networks report measurements back to a central data analysis center. Telecommunications networks carry voice and data to subscribers in accordance with pay-for-access schemes. If a network becomes congested, misconfigured, or otherwise unable to support its intended traffic, then the network may as well not exist. The family of attacks that attempt to prevent the intended function of a network by overloading its resources is called "Denial of Service" (DOS) attacks. Modern networks typically have sufficient resources that a single malicious computer could not overwhelm them. When malicious users arrange for multiple computers to coordinate attacks on a network this is referred to as a Distributed Denial of Service Attack (DDOS).

In cases such as a DDOS where the network is slowed or stopped due to an influx of legitimate but overwhelming set of properly formed messages, authentication services can be used at network boundaries to limit access to internal network resources to certain privileged sets of users. In cases where the network becomes congested with retransmission requests, integrity services can be used to check the correctness of data early and often to minimize the number and severity of corrupted messages. Generally, networks that require high levels of confidentiality make it difficult for unintended message traffic to be understood making the construction of large quantities of well-formed messages difficult.

8.2.5 Cipher Suites

Ciphers are the core algorithms that implement the aforementioned security services. Rarely does a single cipher provide all of the mechanisms necessary to implement security for a network; often configuration and other tools are necessary to provide ancillary inputs to these ciphers. Also, for certain networks, the selection of a cipher for a service such as confidentiality might favor using a related cipher for integrity services. The term "cipher suite" refers to a series of cryptographic algorithms that can be bundled together to provide a series of compatible security operations. The concept was formalized in the Transport Layer Security (TLS) / Secure Sockets Layer (SSL) network protocols [102] that defined a cipher suite as a set of four algorithms: integrity, encryption, authentication, and key exchange[5].

Today, a cipher suite refers to any such bundling of ciphers and associated configurations and tools to implement security services in a network. They are powerful ways of describing best practices for combining the myriad of security algorithms together to accomplish specific security goals. For example, a cipher suite may explain how to use the Hash-Based Message Authentication Code (HMAC) to generate both authentication signatures and integrity signatures while using the Advanced Encryption Standard (AES) for applying confidentiality.

The cipher suite, as a series of complementary algorithms and best practices for the configuration and application, has become a needed tool in a frequently changing security field. A famous set of bundled cryptographic algorithms is published by the United States National Security Agency as "Suite B Cryptography" [210][205] in 2005 which described the use of AES and a series of ciphers based on the properties of elliptic curves. As advancements have been made in the fields of quantum cryptography (and how it endangers the properties of elliptic curves) the Suite B bundle is being replaced by the Commercial National Security Algorithm (CNSA) suite. Many decades from now, the CSNA may, itself, be replaced by another cipher suite.

8.3 Security Challenges Specific to DTNs

Even well-resourced, highly available networks, such as the Internet, struggle to effectively secure communications. In many ways, securing a DTN requires overcoming all of the standard network security challenges as well as several new ones. New security challenges unique to DTNs can be identified by analyzing the assumptions users make about how secured networks operate, the types of features users expect to be present in a network, and what capabilities may reasonably be expected by nodes in the network.

---

[5]The 1.3 version of the TLS specification seeks to limit the scope of a cipher suite definition to encryption and hash-based authentication/integrity algorithms.

Table 8.2: Security implementation assume certain network features.

Id	Title	Summary
A1	Rapid, Round-Trip Communications	Nodes can negotiate security in real time.
A2	Naming and Addressing	Nodes can name items end-to-end.
A3	Stateful, session-based data exchanges	Endpoints help verify security operations.
A4	Homogeneous end-to-end data representations.	The format or encoding of user data does not change in the network.

8.3.1   Security Networking Assumptions

Individual algorithms, protocols, and software implementations of security mechanisms each have their own specific sets of assumptions necessary for their proper use. However, all Internet-style security mechanisms make certain common, fundamental assumptions relating to the underlying structure and operation of the network. To the extent that existing security mechanisms assume network capabilities, there is value in listing these assumptions and determining whether they hold when migrating from an Internet-style deployment to a DTN. Table 8.2 lists commonly assumed network features and provides examples of security service functions enabled by those features.

8.3.1.1   Rapid, Round-Trip Communications

Internet-style communications assume that nodes will be able to transmit information to and from peer nodes in some operationally relevant timeframe. For security services, these timeframes must be exceedingly short (milliseconds). Many trust-based computing patterns require that a network node validate information necessary to implement secure algorithms sequentially as part of sending a message. If the transmission of a message in the network is held up waiting for contact with some trust-based service (such as checking to make sure keys have not expired) then that round-trip communication delay to that service must be very short.

The following are examples of secure mechanisms that assume rapid access to network peers.

- **Key establishment.** A common case is to negotiate an ephemeral key (a session key) that can be used for a communications session and then discarded. The initial negotiation is protected by a long-term key and, at the point in which the ephemeral key is established, the long-term key is no longer used.

- **Key Authentication.** The Public Key Infrastructure (PKI), used for most secure transactions on the Internet, uses the concept of a published public key and a secret private key. A public key is represented by a public key certificate that is published by a Certificate Authority (CA) [231]. The CA is the authoritative source for such certificates and can let a user on the network know if a public key that they wish to use is correct, or whether it has expired, been revoked, or otherwise changed. This validation happens every time a new security session is established.

8.3.1.2   Naming and Addressing

Data and nodes involved in secure communications are reliably named and addressed so that cipher suites and security policies can be applied uniformly throughout the network.

The following are examples of security mechanisms that assume common naming and addressing.

- **Authentication.** When signing information so that it may be authenticated as being from the sender by the receiver, key selection becomes important at the receive so that the appropriate key can be used as proper input to the receiver-side cipher suite. If the receiver uses a different naming or addressing scheme than the sender, then this function may result in selecting an incorrect key and therefore an inability to authenticate the message.

- **Security Policy Enforcement.** Security policies, which cover what types of security services should be applied to messages, are often expressed as a function of the message sender and message receiver (e.g., when sending from node A to node B, always encrypt the data). If nodes do not support the same naming and addressing as the security policy mechanisms it becomes difficult or impossible to interpret policy across the network.

8.3.1.3   Stateful, Session-Based Data Exchange

Nodes in a network contain sufficient resources to establish and synchronize common connection-related information necessary for monitored data exchange. Secure communications on the Internet between two network nodes typically requires that each node store some local state capturing the configuration and expectations of the communication. This set of state information synchronized between nodes is considered a session, and the session persists for as long as the state information may be usefully synchronized or until the secure communication has completed.

The following are examples of security mechanisms that assume stateful, session-based data exchange.

- **Time Synchronization.** Synchronizing time between nodes is used to prevent replay attacks in a secure communications model. If individual messages are encoded with information regarding the time at which the message was generated, and a copy of the message is received later, the message may be seen as a duplicate (or no longer applicable) and discarded. Sometimes, these time windows are as short as fractions of a second.

■ **Nonce Generation.** SSL uses nonce values to prevent replay attacks. However, this assumes that nonce values are randomly generated and not re-used (otherwise an older nonce could be relevant again making identifying message replay attack more difficult).

8.3.1.4   Homogenous Data Representation

Data in a network must be resolvable to a uniform representation so that it can be operated upon at different nodes and yield the same result each time. For example, consider a 32-bit unsigned integer with the value 0x1. Some encoding schemes might compress this value into a single byte (0x01) whereas other encoding schemes may represent the entire 32 bits (0x00000001). While the logical value is the same, the representation of the value on the wire is different.

The following are examples of security mechanisms that assume a homogenous data representation.

■ **Cipher Inputs.** A cipher accepts data and keys and produces a cipher output. When encrypting, the inputs to the cipher are plain text and a key. When decrypting, the inputs to the cipher are the cipher text and a key. For decryption to succeed, the cipher text must be, bit-for-bit, the same as the output of the cipher from the encryption process. If an underlying encoding swaps the value 0x01 for 0x00000001, the ciphertext has changed and will not successfully decrypt.

■ **Key Exchange.** Key establishment algorithms require two nodes to either directly or indirectly exchange some shared secret. If nodes do not use the same encodings for capturing or representing this shared secret, then the generated keys will not function mathematically when fed to a cipher or other cipher suite algorithm, like the cipher inputs example above.

8.3.2   Relevant DTN Characteristics

By their nature, DTNs are deployed in challenged networking environments that cannot provide the rich set of Internet-style services assumed by most modern security protocols and policies. There are cases, such as when a DTN overlays an otherwise well-connected network, where Internet-model assumptions may still apply. However, these special circumstances are not normal conditions and a DTN security model must not rely on special circumstances for successful operation.

Table 8.3 lists four common characteristics of DTNs that have the largest impact on a security model and the security assumptions from Section 8.3.1 that they violate.

Table 8.3: DTN Characteristics violate common security assumptions.

DTN Characteristics	Common Security Assumption			
	A1	A2	A3	A4
Link Characteristics - Store-and-Forward Operations - Propagation Delay and Disruptions - Asynchronous transmission	X		X	
Internetworking Characteristics - Overlay - Encapsulation		X		X

8.3.2.1 Link Characteristics

The three link characteristics of a DTN that affect a DTN security model are listed as follows.

Propagation Delays and Disruptions. DTNs may be deployed across vast distances relative to the propagation of signals in their medium. For example, RF signals traveling at the speed of light still take minutes or hours to reach destinations that are located at interplanetary distances within our solar system. Alternatively, the relatively slow propagation of signals from acoustic modems under water introduce significant signal propagation delays in underwater networks. Additionally, DTNs may be deployed in areas where link disruption is frequent (or even considered normal operation), such as when the network undergoes rapid topological change, experiences administrative access limitations, or whose links are challenged by natural or man-made sources.

Asynchronous Transmission. Where DTNs are deployed on wireless and power-constrained networks, transmission may be unidirectional or otherwise asynchronous. For example, certain spacecraft may transmit at one frequency and receive at another, using different antenna and having different data rates. In certain cases, the transmit and receive capabilities may be significantly different in their supported data rate by several orders of magnitude.

Store-and-Forward Operation. Messages in a DTN must support a store-and-forward capability because an end-to-end pathway from a message source to a message destination may not exist at any given point in time. In cases where the DTN is partitioned, messages will migrate towards their destination until they reach a point at which they must wait for a future contact opportunity for further transmission.

This characteristic violates the following security assumptions.

■ **A1: Timely, Round-Trip Data Exchange.** When operationally-relevant time-frames are on the orders of milliseconds to seconds, a DTN may not have a path from a sender to receiver (or from a sender to a helper node such as a certificate authority). Waiting for an end-to-end path to be established, or for a message to store-and-forward its way to a destination could greatly exceed practical timeframes for security operations.

■ **A3: Stateful, Session-Based Data Exchanges.** Without timely, round-trip data exchange, establishing a session may not be practical and, once established, keeping the endpoints of a session synchronized may also be impractical. Further, even if synchronization between nodes is possible, if the round-trip exchange takes too long (minutes, hours, or days) there will likely be a backlog of queued data coming over the session at the same time as any synchronization messages, making it impractical to impossible to differentiate data sent by the source and queued by a node as part of a backlog versus replay attacks added to a node's message backlog at a later time.

8.3.2.2 Internetworking Characteristics

The two internetworking characteristics of a DTN that affect a DTN security model are listed as follows.

Overlay. A network overlay is a virtual network built on top of a physical network. The most popular example of an overlay is a Virtual Private Network (VPN) in which a secure, small network (e.g., your private network) is created as a subset of a larger network (e.g., the Internet). The overlay may have its own naming scheme, addressing scheme, and security capabilities and policies. A smaller niche of overlay is the act of establishing a virtual network that sits on top of multiple, different physical networks.

Encapsulation. Encapsulation is the process of placing one network message inside of another network message for the purpose of sending the encapsulated message through a network or network layer controlled by the encapsulating message protocol. The most popular example of encapsulation is the TCP/IP networking stack made famous by the Internet. In this case, a TCP message is placed as the payload of an IP message which is then placed inside of a frame for communication (e.g., an Ethernet frame). When a DTN spans multiple different networks, DTN messages traverse the network by being encapsulated into the native message structures for those networks.

DTNs deployments complicate the overlay/encapsulation space by operating across multiple networks with different application, transport, or even link layers. Further, the DTN may sit at different networking layers in different networks, existing at the application layer in one network and at the transport layer in a different network. This creates a difficulty in the naming and addressing of DTN messages – the messages from the point of view of the overlay have a certain set of naming conventions and addresses. The messages as they are encapsulated for transport over individual networks have names and addresses specific to that network. The bindings between them may need to occur at different layers in different networks.

In some circumstances, the address bindings for the DTN may be static and long-lived. In other cases, the address binding may change rapidly and happen just-in-time for the network. In either case, the addressing schemes may be significantly different from one another. Consider the case where a spacecraft talks to a ground station. The ground station and uplink dishes may talk to each other over the Internet using IP addresses, but the RF uplink to the spacecraft may conform to a CCSDS standard which does not use IP Addresses, and the components on the spacecraft may use a MILSTD 1553 or SpaceWire bus for internal communications that also do not use IP Addresses. If the DTN spans the ground station computer, the uplink dish, the spacecraft radio, the spacecraft data bus, and the individual payloads and instruments on the spacecraft, then it must handle a multitude of naming and addressing schemes.

These characteristics violate the following security assumptions.

- **A2: Naming and Addressing.** Security services where user data are encrypted or signed at a node in one network for decryption or validation in another network must be careful which version of naming and addressing is used for the security operations. Security services cannot operate if they use the physical name spaces and addresses of a particular network if the message receiver will be in another network or using another scheme.

- **A4: Homogenous Data Representation.** Just as different networks may have different naming schemes, they may also have differing default encodings. Care must be taken to ensure that cipher text, key information, and other security-related data remain unchanged from generation by one cipher suite algorithm to its consumption at the receiver by another cipher suite algorithm. This may require an additional layer of encapsulation to prevent message translation from occurring at network boundaries.

8.3.3 Security Constraints

A security constraint represents a restriction on the design of a DTN security model identified by considering the assumptions made by common security mechanisms and where DTN characteristics violate those assumptions. Constraints are an important aspect of verifying the scalability of a security model with increasing signal propagation delays and increasingly frequent link disruptions. In cases where DTNs can support assumptions, algorithms and protocols can be reused. In cases where DTNs cannot support these assumptions, new algorithms and protocols are required to implement security services.

Table 8.4: Security Challenges

Constraint	Violated Assumptions	DTN Characteristic
Unreliable	- Rapid, Round-Trip Communication	■ Store-and-Forward Operations ■ Propagation Delays and Disruptions ■ Asynchronous Transmission
Logical Security Policies	■ Naming and Addressing ■ Homogeneous End-to-End Data Representations	■ Overlay ■ Encapsulation ■ Asynchronous Transmission
Unreliable/ Insecure Link Layers	■ Stateful, Session-Based Data Exchange ■ Homogeneous End-to-End Data Representations	■ Asynchronous Transmission ■ Propagation Delays and Disruptions ■ Overlay
Multiple Data Representations	■ Naming and Addressing ■ Homogeneous End-to-End Data Representations	■ Overlay ■ Encapsulation

8.3.3.1 Unreliable Access to Oracles

An oracle, in this context, refers to any global source of common knowledge in the network, such as key information (e.g., certificate authorities), policy information, or time/state configuration (e.g., DNS servers)[6]. These oracles are necessary when a network service requires configuration information to perform its function effectively and/or securely. For example, a network security service running on a network node should verify that a RSA public key is valid before using it to verify a signature. The node may have a sense of whether the public key is valid based on local past history, but that does not imply that the public key is currently still valid, or will continue to be valid in the future. In a well-connected, highly-available network such as the Internet, this problem is solved by shifting this validating responsibility to some "key validating oracle". Whether implemented as a single service running on a single server or load-balanced across globally distributed servers, oracles represent easy access to "common knowledge" that can be distributed and used before it changes again.

DTNs cannot support this kind of architecture. As previously mentioned, such networks cannot provide reliable, round-trip communications to such an oracle. Also, DTNs may comprise nodes (such as environmental sensors) that cannot support accurate clocks, bulk storage, memory, or other resources necessary to meaningfully interact with an external oracle.

When access to oracles is limited, security mechanisms must substitute local information in lieu of external information whenever possible. Therefore, decentralized algorithms are preferred over centralized algorithms.

8.3.3.2 Logical Security Policies

A DTN deployment is rarely a single overlay of a single physical network. More often, a DTN will span multiple networks each having their own naming and addressing schemes. In these situations, each node participating in the DTN will have (at least) two responsibilities: (1) as a member of a network with a physical network address and (2) as a member of the DTN with a logical DTN address[7]. Each logical address of a DTN node must be bound to the physical address of the node, and this binding may be unique for the different types of physical addresses used by the variety of encapsulating physical networks. This binding may be static when the relationship between DTN node and physical network node will never change (such as when a DTN node is deployed on a specific sensor or spacecraft) or it may be dynamic when a DTN node may be relocated onto a different physical node as a matter of network configuration.

Since DTN traffic may experience multiple types of physical addresses, and since the relationship between DTN nodes and physical nodes may change over the lifetime of the network, security policies must not rely on physical addresses. Security mechanisms used to configure policy must use the logical addressing used within the

---

[6]Updating such oracles is not instantaneous (e.g., a new DNS entry can take a day to propagate globally), but these delays are relatively small and occur only on changes, which are infrequent.

[7]Including networks that do not use the Internet Protocol (IP).

DTN and accept that logical names and addresses may be bound to different physical entities in a physical network at different times.

8.3.3.3 Unreliable/Insecure Link Layers

Establishing trust within a DTN is complicated by the aforementioned constraints. Without ready access to oracles a DTN node can be delayed in its understanding of updates to security credentials. When the DTN spans multiple physical networks each participating network may have a different trust model, different administrative policies, and different risk models. Therefore, some links supporting the DTN may have less (or no) trust relative to other links supporting the DTN. In cases where the DTN supports true unidirectional links, a "forward path" from node 1 to node 2 will be different from the "return path" from node 2 back to node 1. In these cases, roundtrip communications may either be impossible (when dealing with a transmit-only function), impractical (where latency asymmetries are large), or undesirable (when an alternate return path cannot itself be secured).

For these reasons, DTN security mechanisms cannot build a "chain of trust" at the link layer to establish end-to-end security across the DTN.

8.3.3.4 Multiple Data Representations

As previously mentioned, DTN security mechanisms rely on logical addresses because supporting physical networks may use different addressing schemes. Similarly, different physical networks might use different data encodings. Certain physical networks may use a custom-packed binary format (such as in highly resourced constrained space networks) whereas other networks may use a more verbose encoding such as XML.

There are generally two ways of communicating data from one network encoding to another: encapsulation and re-encoding. When encapsulating data the payload from one network is wrapped, unchanged, for transport in the next network. This can result in inefficient network utilization in cases where the original encoding is much more verbose than the native encoding of the next network. For example, encapsulating an uncompressed XML payload may stress a resource-constrained network that uses a more efficient binary encoding of its information. Since the encapsulating network does not understand the original payload encoding it cannot apply security and administrative policy. When re-encoding data between two networks the original payload encoding is parsed and a new encoding of the payload, preserving semantic meaning, is generated and sent through the next network. This requires significant processing resources at the boundaries between networks. Also, certain end-to-end functions may be confused if the format of the payload as generated by some source network is different than the format of the payload at some destination network. These options are shown in Figure 8.3.

The decision to encapsulate or encode is often decided as part of network architecture and design and may be difficult to reverse-engineer into deployed systems. For example, making the decision to adopt a re-encoding strategy is difficult if ex-

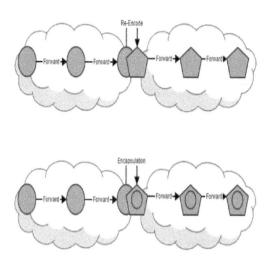

Figure 8.3: There exist multiple ways to join different networks.

isting border nodes do not have sufficient resources and new nodes cannot easily be deployed in the network.

The addition of end-to-end security adds a special concern into this decision as it relates to the way in which information is presented to the cipher suites that implement security services. When attempting to verify the integrity of a message, the payload provided to the integrity cipher suite when generating the signature must be identical to the payload provided to the integrity cipher suite when verifying the signature. If the payload encodings are different, then the cipher suites will generate different signatures and the integrity check of the system will fail. Similarly, when a payload is provided to a confidentiality cipher suite the resulting cipher text must be provided, without modification, to the confidentiality cipher suite used for decoding, otherwise the original plain text cannot be recovered.

To address this concern, security services must define a format for the representation of information that is independent of how the message may or may not be represented physically in any one network. When this format is also used as the on-the-wire encoding scheme and encapsulated across the DTN the problem of cipher suite inputs it solved: the original encoding persists end-to-end in the system. When using a re-encoding scheme, payloads received by a node in a particular encoding scheme must be re-encoded into a shared format prior to being input into a cipher suite.

8.4   The End-to-End Security Model

An effective security model provides needed security services within the context of the constraints imposed by the DTN architecture.

8.4.1   A Multi-Layered Approach

Layering is a well-understood and frequently used mechanism for separating concerns in computer architectures. Networking stacks are often visualized using a layered model that separates concerns relating to user data delivery, resource control/planning, and physical infrastructure. For example, the Internet uses packetization as a strategy for data resiliency by separating end-to-end data delivery from per-packet routing decisions[8]. In this way, a network can be layered in accordance with many models, such as the four layer Internet Model (RFC1122), the OSI seven layer OSI model, the five layer TCP/IP model and so on.

Similarly, a multi-layered model can separate security concerns. This is particularly the case where a DTN may span multiple constituent networks with different security capabilities. The security layers associated with the DTN security model are identified in Figure 8.4.

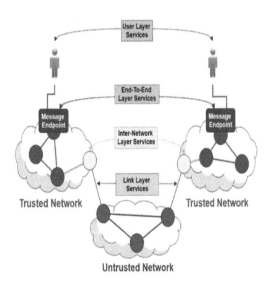

Figure 8.4: Securing a network happens at multiple layers.

---

[8]There are several instances, particularly relating to video streaming and Quality of Service, where virtual circuits are established in a packetized network to trade path diversity with a higher probability for timely, sequential packet delivery.

As illustrated in Figure 8.4, the DTN security model can be decomposed into four layers of security processing. The link layer is responsible for forming the backbone of any communications network, and the per-networking-device negotiation and security represents the common function of the network: data cannot flow if a link is not present. The policy layer is responsible for how individual networks encapsulate/re-encode data and apply additional security services when crossing network boundaries (such as when encapsulating over untrusted networks). The application layer is responsible for the end-to-end data exchange across the DTN. Finally, the user layer is responsible for any security services that are performed outside of the context of the network. For example, a user may choose to pass their own pre-encrypted, pre-signed information to the network in lieu of unprotected data.

Notably, the security responsibilities of each layer are independent, not additive. User security does not rely upon the presence of end-to-end security, which does not rely upon the presence of inter-network security, and so on. By not assuming (or trusting) an underlying security capability, the overall security stance of the network can be improved because a compromise of one security layer does not imply a compromise of another security layer.

The resource constraints and sparse connectivity of some DTN deployments, such as interplanetary spacecraft communications and energy-harvesting environmental sensors, historically apply security only at the user and the link layer. The difficulty of establishing a link in these scenarios, and the difficulty of complex processing on constrained nodes, may make a single-layer security solution appropriate for certain networks. However, as these isolated networks mature and/or become replaced by internetworks, new layers of security must be implemented.

8.4.1.1 Link Layer

The link layer has two primary security responsibilities that protected against both planned and unplanned challenges to the data exchange.

1. Point-to-point data exchange protected from data corruption.

2. Link-specific security mechanisms at both the physical and data layers.

Prevent Data Corruption. Unplanned challenges to data exchange involve changes to the data based on environmental factors that perturb the physical exchange of information. In these cases, applying unsigned integrity mechanisms such as checksums and cyclic redundancy checks (CRCs) are sufficient to detect that data was corrupted in transit (and in some cases recover that data).

Apply Link Security. Planned challenges to data exchanges encompass the larger concerns relating to standardized threats such as altered and injected data into the links. In this case, authentication, integrity, and confidentiality services may be used at this layer. Additionally, where links are subject to jamming a variety of physical approaches to communication, such as spread spectrum techniques in wireless networks, can also be used as part of the overall security and resiliency of the link.

However, link-layer security is insufficient to protect end-to-end data exchange.

The three fundamental limitations to link-layer security are coordination, shared access, and administrative domains, described as follows.

1. **Link coordination.** To securely send an end-to-end message using only link-layer security, a network must verify that every link along the message path is trusted by the sender and supports the appropriate security services and security policies. This approach might be possible in sparsely populated or single-owner networks. Coordination becomes exponentially more difficult as the number of messages that comprise the user data set grows and the number of different links used by these message paths increases. Even assuming a single user in the network at a time, this coordination can easily become intractable in a DTN.

2. **Multiple link access.** Link layer security applies to all messages transmitted over that link, including messages with different sources, different trust levels, different qualities of service, and different security policies. If security services were restricted to the link layer, all users would have credentials for the link and use the same security services available on the link. In this scenario, the link would also have no mechanism for finely-tuning the security settings per message, or for providing data-at-rest for DTN messages which may be persistently stored on a node waiting for their next transmission opportunity.

3. **Administrative domains.** A physical network typically has a set of privileged users responsible for its configuration. Sets of such users represent administrative domains for portions of the network. When a DTN spans multiple physical networks, each controlled by different administrative domains, information sharing amongst networks may be difficult or impossible. Business policies may prevent sharing information necessary to establish a chain-of-trust. Security policies in different networks may support incompatible security services or configure those services differently.

The literature is densely populated with mechanisms for securing individual links, many of which are tightly coupled to the physical aspects of the link being secured. However, the primary consideration is that link-layer security, alone, is not sufficient to secure end-to-end traffic in a DTN.

8.4.1.2 The Policy Layer

The policy layer has two primary security responsibilities when exchanging information across security boundaries in a DTN.

1. Apply security policies at the boundaries of trust between nodes and/or networks covered by the DTN.

2. Abstract differences from other networks by encapsulating or re-encoding information prior to entry or exit from the network.

Manage Trust Boundaries. Trust boundaries exist within a network whenever data is translated from one administrative domain to another. In traditional networking it is common for a trusted network (such as a home network) to communicate with an untrusted network (such as the Internet). Tools such as Virtual Private Networks (VPNs) are commonly used to apply special security to data when leaving a trusted network and being transported over an untrusted network. Using the VPN example, many businesses use servers directly connected to the Internet to act as endpoints of a VPN rather than allowing employees to directly tunnel to the computers at their desks accessed over the company intranet.

Abstract Physical Differences. Where security boundaries coincide with boundaries between physically different networks, the policy layer must also address the decision as to whether data is encapsulated or re-encoded for transition to the next network. Given the security implications of each approach, this results in the application of a security policy at boundaries that is different than the application of security policy within the trusted part of a network.

It is important to note that the inter-networking layer is a conceptual layer in the security model and may not have a distinct physical manifestation in a DTN. For example, some DTNs may overlay a single trusted network (or a series of networks with identical properties and trust) in which case there is no need for a policy layer separate from other layers. However, when there exist trust boundaries within the network separate from the end-to-end boundaries along a message path, the policy layer represents how nodes at these boundaries behave differently from other nodes in the DTN.

8.4.1.3   End-to-End Layer Considerations

The end-to-end layer has two primary security responsibilities when exchanging information across physically different networks spanned by the DTN.

1. Apply security services to DTN messages that are independent from encodings or mechanisms deployed on any individual network or link.

2. Propagate security policy and processing mechanisms independent of the path taken by any particular message from source to destination.

Apply Security Services. The end-to-end layer is the primary layer for implementing DTN security services, as the application of such services almost always is in the service of data exchange between a message source and a message destination.

Propagate Security Policy and Configuration. Security policies that exist at the DTN layer may have use policies and configurations that are separate from the policies and configurations of underlying physical networks. The end-to-end layer must understand how policies are applied regardless of the underlying network topology, link layers, and (in some cases) network boundaries.

8.4.1.4   User Layer Considerations

The user layer has a single security responsibility as it related to secure data exchange in a DTN.

1. Apply security independent of the network by passing already-secured data to the network as opaque user payload.

Apply Network Agnostic Security. DTN security services apply only to the data that has been passed into the DTN. Users/applications may choose to apply their own level of security to their data prior to its injection into the DTN. In this case, the user-secured data is considered opaque by the network and may be further secured by underlying layers or have no security applied to it.

One standard for applying user-layer security is the Cryptographic Message Syntax (CMS) which provides a common way of representing multiple types of security operation (including authentication, integrity, and confidentiality) for all or parts of a message payload.

8.4.2   A Multi-Component Approach

DTN security should not be represented by a single, monolithic security model as cipher suite specifications, network threats, and other capabilities change over time. Additionally, some DTN deployments have different resources, risk-tolerance, and characteristics than others, increasing the chances that a monolithic model becomes overly complex and overly prescriptive. Therefore, a multi-component approach is necessary to provide both specificity and flexibility: when some aspect of security best practice changes, that change may be reflected in modifications to a single security component rather than changing the entire security model.

There are five types of components in a DTN security model, as illustrated in Figure 8.5.

Best Practices Best practices capture common services, interfaces, libraries and other security-related mechanisms. Several examples of best practices are covered in this chapter, to include defining security services and special constraints and characteristics of DTNs. Several of the references at the end of this chapter provide information on other best practices. These practices justify the policies used in a system and determine what kinds of protocols are required for secure data exchange.

What is considered best practice evolves as protocols and policies are deployed, customized, and used in networks. Periodically, best practices must be revisited when faced with new computing techniques and resources, evolving threat models, and new user operational concepts.

Protocols Protocols characterize the data structures and state machines associated with data exchange. Security protocols should be designed with singular purpose such that they are highly cohesive and loosely coupled to other protocols.

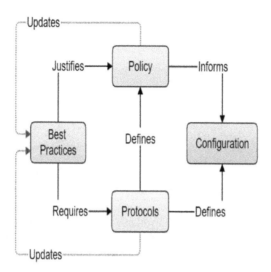

Figure 8.5: A security model is comprised of multiple components.

Scoping protocols is very important. If a protocol is not cohesive there will be many reasons to want to change it over time, leading to many protocol revisions and confusion regarding what is being deployed in a network and how to ensure interoperability of the network over time.

For this reason, a protocol specification often identifies areas for user customization in the areas of policy and configuration. This allows the actual policies and configurations that affect the protocol to be specified in separate documents that can be made specific to an individual network and avoid hard-coding values that may change.

Policy Policy refers to customization that details how the state machines and processing orders comprising the protocol should react to time and/or state on implementing nodes. This includes discussions on how implementing nodes will handle error conditions, logging and reporting, and management.

Configuration Configuration refers to customization that details how portions of the data structures comprising the protocol should be populated according to time and/or state on implementing nodes. This includes setting data values, enumerations, threshold values, and magic numbers.

8.4.3 DTN Model

A general security model cannot be specified external to the types of networks to which the model will be applied. This is because the best practices, policies, and configurations components of a security model cannot be discussed in a generic way. As previously discussed, the reason that protocols support points of customization is because there is no "one size fits all" way to configure and manage the security services of a network.

A DTN security model must document the following.

■ List the protocols that should be deployed to implement security at various security layers.

■ Provide guidance on how to set configuration and policies in various situations.

Often, a DTN will span one or more existing networks that implement link layer and inter-network layer security. For example, in cases where a DTN spans nodes on the Internet, Internet security protocols such as Transport Link Security (TLS) and the Secure Socket Layer (SSL) are available. Similarly, when the end-to-end layer also runs over the Internet protocols such as IP Security (IPSec) and the Secure Hypertext Transport Protocol (S-HTTP), can be used. At the user layer, any method may be used to generate secure payloads, including custom or proprietary mechanisms.

In the cases where a DTN spans networks with no inherent security and/or those that are unable to adopt Internet-style security mechanisms, end-to-end DTN-specific security protocols must be defined. For the DTN Bundle Protocol (BP), that protocol is identified as BP Security (BPSec).

8.5   Securing the Bundle Protocol

8.5.1   Bundle Protocol Extension Mechanisms

The Bundle Protocol (BP) is used for end-to-end data exchange in a DTN. The protocol data unit (PDU) of the BP is the Bundle and it represents the data that must be secured to implement end-to-end security in a DTN. A bundle is a set of blocks, with each block holding specific information related to the bundle and its data. Certain blocks are required in every bundle: the Primary Block (which acts as a message header) and the Payload Block (which contains user data). BP allows other blocks to be added to the bundle, at any time and at any node, which provides a built in self-extensibility mechanism to the protocol. When a new feature or capability is added to a network, that feature can be represented by defining a new block in a bundle, rather than making changes to the BP specification itself. The relationship of blocks in a bundle is illustrated in Figure 8.6.

Since a DTN cannot guarantee continuous end-to-end connectivity, concepts such as session-establishment, timing analysis, and other approaches that require endpoints to save state cannot be part of a security mechanism. This requires Bundles to carry not only user information (in the payload block), but also annotate themselves and their payload (using extension blocks) with information that cannot otherwise be maintained solely at endpoints.

Extension blocks are particularly important for securing BP, as security-related

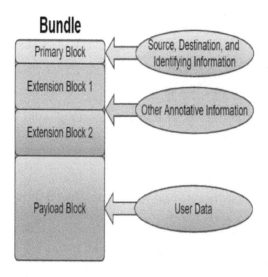

Figure 8.6: Bundle extensions can be used to extend the base protocol.

extension blocks can be used to carry key and cipher suite information that would, otherwise, be kept in session state by an endpoint node.

8.5.2 Key Properties

As illustrated by the challenges of securing a DTN, network architects cannot always deploy Internet-like security mechanisms because network characteristics do not support the underlying assumptions of those mechanisms. Where new mechanisms must be developed, or existing mechanisms must be deployed in a new way, a set of key properties can be developed to guide delay-tolerant security services.

Properties of an end-to-end security service are tied to the mechanisms used to provide the end-to-end data transport. The Bundle Protocol defines the structure of messages (e.g., Bundles) exchanged in a DTN and the behavior of nodes when receiving, storing, and transmitting. As such, the action of securing endpoint services in a DTN is the action of securing the Bundle Protocol and the following key properties assume the design and implementation features and constraints associated with Bundles.

8.5.2.1 Block-Level Granularity

Applying a single level and type of security across the entire Bundle fails to recognize that blocks represent different types of information with different security needs. Endpoint security services must recognize the different types of information represented within a Bundle and apply different security services to different Blocks.

For example, a payload block might be encrypted to protect its contents and an extension block containing summary information related to the payload might be integrity signed but unencrypted to provide waypoints access to payload-related data without providing access to the payload itself.

8.5.2.2 Multiple Security Sources

A security source in this context is a Bundle Protocol Agent that adds security service to a Bundle, either because it is creating a Bundle or forwarding a Bundle it has received. When a waypoint adds a security service to the bundle, the waypoint is the security source for that service. Therefore, there can be as many security sources as there are security services in a Bundle.

For example, a node representing a boundary between a trusted part of the network and an untrusted part of the network might want to encrypt the payload of any Bundle leaving the trusted portion of the network. The node performing the encryption would be the source for the payload encryption service. That source information may be necessary when the Bundle reaches its destination and the payload must be decrypted. In such cases, the key and/or cipher suites used to perform the decryption could be selected as a function of which node performed the encryption.

8.5.2.3 Dynamic Security Policy

Since Bundles carry their own security information with them, the enforcement of security services is a function of the nodes that transmit, forward, and ultimately receive that Bundle. Waypoints in a network may have different behaviors as a function of their configured security policy. Example security behaviors for waypoints in a DTN are as follows.

- **No Security.** Nodes across the DTN may not all have the same security capabilities, particularly in cases where certain nodes have very limited processor, memory, or storage resources. Therefore, some waypoints will not have security capabilities and cannot process security services embedded in a received Bundle.

- **Frequent Inspection.** Some waypoints may have security policies that require evaluating security services even if they are not the Bundle destination or the final intended destination of the service. For example, a waypoint may choose to verify the integrity of a block in the Bundle even though the waypoint is not the Bundle destination and the integrity service will be needed by other waypoints along the Bundle's path. The value of frequent inspection is that bad data may be detected and dealt with in the network quickly.

- **Enforcing Security Endpoints.** Some waypoints will determine, through policy, that they are the intended endpoint of a security service and terminate the security service in the Bundle. For example, a gateway node may determine that, even though it is not the destination of the Bundle, it should verify and remove a particular integrity service or attempt to decrypt a confidentiality service, before forwarding the Bundle along its path.

■ **Destination Nodes.** All security services in a Bundle must be processed at the Bundle destination, assuming the destination is security aware.

All of these are valid use cases based on the particulars of the Bundle contents, the network architecture, and the configured security policies.

8.5.2.4   Multiple Cipher Suites

Different cipher suites are used to provide different security services of various cryptologic strength and of various needs for processing resources. Any security mechanism for a DTN must be able to use different cipher suites to implement security services based on the configuration of the security sources. For example, some networks might require a SHA-256 based hash for integrity whereas others may require a SHA-384 hash instead.

There are two kinds of cipher suites that may be specified for use in a particular network.

■ **Inter-operability cipher suites.** These cipher suites are usually specified in standards documents and used to ensure that different implementations of a security protocol can correctly exchange data. Cipher suites selected for this kind of testing are typically simple to understand and implement so as to not distract from the time and effort necessary to verify proper protocol operation. As such, these cipher suites might not be suitable for deployment in an operational network.

■ **Operational cipher suites.** These cipher suites represent the algorithms and implementations that implement security in a particular network, selected as a function of the risk tolerance and computational ability of the network. Operational cipher suites may differ from one DTN to another.

8.5.2.5   Deterministic Processing

Whenever a node determines that it must process more than one security service in a received Bundle (either because the policy at a waypoint states that it should process a security service or because the node is the Bundle destination) the order in which security services are processed must be deterministic. All nodes must impose this same deterministic processing order for all security services. This specification provides determinism in the application and evaluation of security services, even when doing so results in a loss of flexibility.

8.5.3   The Bundle Security Protocol

The "Bundle Protocol Security" (BPSec) defines a set of extension blocks for use within Bundles to apply security services to the contents of a Bundle.

Table 8.5: Architecture that encodes security in application-layer messages introduce new terminology.

Terminology	Definition
Security Service	The security features that can be applied in a bundle: integrity and confidentiality.
Security Source	The node in the network that adds a security service to a bundle.
Security Target	The block that receives a security service as part of a security operation.
Security Block	An extension block within a bundle that implements a security service.
Security Operation	The application of a security service to a specific security-target. A security operation may be noted as $OP < service, target >$, for example: $OP < confidentiality, payload >$.

8.5.3.1 Terminology

Table 8.5 defines the terminology necessary to describe how security is applied to blocks within a bundle using the BPSec protocol.

8.5.3.2 Security Blocks

BPSec uses security blocks within a bundle to implement security operations. Since there are two security services supported by the protocol (integrity and confidentiality) there are two types of security block: the Block Integrity Block (BIB) and the Block Confidentiality Block (BCB).

A BIB ensures integrity by holding key information and digital signatures for each of its security target(s). This integrity information may be verified by any node in between the BIB security source and the bundle destination. Security-aware waypoints may add or remove BIBs from bundles in accordance with their security policy.

A BCB encrypts the contents of its security target(s), in whole or in part, at the BCB security source in order to protect their content while in transit. The BCB may be decrypted by security-aware nodes in the network, up to and including the bundle destination, as a matter of security policy.

The general relationship between security blocks, security targets, and cipher suites is given in Figure 8.7.

In this figure, a cipher suite accepts content from the target block and configuration from the local node to generate some set of cipher results. When performing an

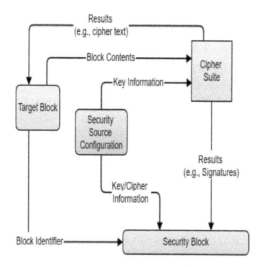

Figure 8.7: Generating a security block using BPSec and cipher suites.

integrity service, the result is a signature which can be placed in the security block representing that service. When performing a confidentiality service, the result is cipher text which replaces the plain text in the target block as well as optional additional results that can be placed in the security block representing that service. In addition to holding the cipher results, the security block also holds information relating to the target block identity and the configuration used by the cipher suite, to include key information.

Uniqueness

To prevent ambiguity in security processing, all security operations provided by BPSec must be unique. That means that the same security service cannot be applied to the same security target more than once in a bundle. The following are examples of combinations of security operations and whether they are allowed or disallowed by the BPSec protocol.

Encrypting the payload twice. The two operations OP(confidentiality, payload) and (confidentiality, payload) imply that an extension block has been encrypted two times and, therefore, must be decrypted in proper order. However, there are no reliable mechanisms for determining which of these operations happened first[9] leading to ambiguity in how to order decryption. Even in cases where the key information and cipher suite selection are different for the two operations,

---

[9]Even assuming the addition of a space-inefficient timestamp, many nodes in a DTN have inaccurate or otherwise unsynchronized clocks, making it very difficult to do time comparisons.

they still represent the same service applied to the same target and are, therefore, disallowed in the BPSec protocol. The same logic applies to encrypting any block twice, or to signing any block twice.

Signing different blocks. The two operations OP(integrity, payload) and OP (integrity, extension_block_1) are not redundant and both may be present in the same bundle at the same time. Similarly, the two operations OP(integrity, extension_block_1) and OP(integrity,extension_block_2) are also not redundant and may both be present in the bundle at the same time. The same logic applied to encrypting different blocks.

Signing and encrypting the same block. Since integrity and confidentiality are different services, they can be applied to the same target. For example, OP(integrity, payload) and OP(confidentiality, payload). In this case, the order of operations matters: if the integrity service is applied first, then it has signed plain text. If the integrity services happens second, then it has signed cipher text. However, since the services of integrity and confidentiality are unique, order can be prescribed by the protocol to prevent ambiguity.

These requirements for uniqueness determine whether a security block can be added to a bundle. If adding a security block to a bundle would cause any security operation to be duplicated in that bundle, then the block must not be added.

Multiplicity

The BPSec provides an optimization in cases where the same security service (with the same configuration) is applied to a set of blocks. The optimization is to have a single security block represent all of the security operations to avoid the inefficiency of repeating key and cipher suite information in multiple security blocks. This optimization can only be used when all the following are true.

- The security operations all apply the same security service (e.g., all integrity or all confidentiality).

- The same key is to be used for all operations captured by the security block.

- The same cipher suite parameters are to be used by all operations captured by the security block.

- All security operations in the security block were added by the same security source (e.g., at the same node in the network).

- None of the security operations in the security block are duplicates with any other operations in the security block or already present in the bundle.

The concept of multiple security targets is illustrated in Figure 8.8.

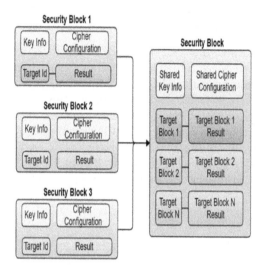

Figure 8.8: BPsec target multiplicity prevents repeating redundant information in a bundle.

Block Processing Rules

A key property of the BPSec protocol is to ensure deterministic processing in a DTN where regular methods of determining processing order (such as timestamps and session state) may not be available to the network. To avoid potential ambiguities related to security block processing in the network, the following block processing rules are enforced by the protocol.

- **Order Services.** Integrity services must always be applied before confidentiality services. If a block is already the target of a BCB block, then a BIB cannot be added. If both a BIB and a BCB share the same security target, it means, without ambiguity, that the BIB was added first and the BCB was added second.

- **Encrypt Coupled Blocks Together.** If confidentiality is being applied to a target block that also has an integrity service on it, then the confidentiality must also protect the integrity service. Put another way, when adding a BCB which targets the same block as an existing BIB, the BCB must also target that BIB.

- **Always Decrypt First.** An integrity service in a bundle cannot be evaluated if it has been encrypted by a confidentiality service. The confidentiality service must be handled first to decrypt the blocks representing the integrity service, and then the integrity service can be handled. In the terminology of security blocks, a BIB cannot be evaluated if it is the target of a BCB and(therefore) encrypted. This implies that BCBs must always be handled before BIBs.

■ **No Infinite Loops.** A constraint is made on how integrity and confidentiality services can reference each other. A BCB block may have as its target a BIB block, if it needs to encrypt the BIB block for some reason (such as mentions in "encrypt coupled blocks together" above). However, if a BIB were also allowed to target a BCB block, an infinite recursion would be possible in the protocol where a BCB could target a BIB which targets a BCB and so on. This would endanger the determinism of processing order. Therefore, the BPSec protocol prevents this kind of nesting by disallowing any BIB from targeting a BCB in any circumstance. A BCB block cannot be integrity signed by a BIB block[10].

8.5.3.3 BPSec Example

Consider a bundle that is created at a source node (N1) and traverses a network to a destination node (N4) passing through a series of security-aware waypoints along the way (N2 and N3). At N1 the bundle consists of simply a header and a payload. N2 adds integrity services and an additional extension block related to priority. N3 adds confidentiality services for the payload only. Finally, the bundle arrives at N4 with a series of extension blocks representing the evolution of security services in the bundle along its journey. This progression of blocks added to the bundle is shown in Figure 8.9.

8.5.4 Bundle-in-Bundle Encapsulation

While the BPSec protocol provides unambiguous and block-level security within a bundle for end-to-end DTN security, its simplifying constraints prevent the implementation of some complex security behaviors. In these cases, DTN supports a bundle encapsulation mechanism whereby a bundle may be made the payload of a second bundle. This process, called Bundle-in-Bundle Encapsulation (BIBE) is illustrated in Figure 8.10.

BIBE allows security combinations not otherwise possible using the BPSec, such as the following.

■ **Whole bundle integrity or confidentiality.** When an entire bundle must be integrity signed or encrypted, it can be encapsulated into the payload of another bundle. In that case, integrity and/or confidentiality can be applied to the payload of the encapsulating bundle. This is particularly useful when the primary block of a bundle should be hidden while traversing portions of an untrusted network.

---

[10]If an integrity signature of a BCB block is desired, one can be generated by the confidentiality cipher suite and kept with the BCB block itself.

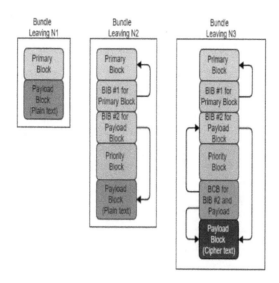

Figure 8.9: The evolution of security services in a bundle over time.

- **Multiple Encryption.** In certain cases, an encrypted payload might be encrypted a second time through portions of an untrusted network. Since the BPSec disallows encrypting the payload twice, BIBE provides a mechanism for accomplishing this by encapsulating the entire bundle and then encrypting it. If certain blocks of the encapsulated bundle need to be available in the encapsulating bundle, they can be copied into the encapsulating bundle as necessary.

- **Security Waypoints.** To ensure that a bundle arrives at a particular waypoint in the network that is not the bundle's destination (such as when establishing a VPN), BIBE can be used to generate a virtual tunnel. The entrance to the tunnel being the node applying the encapsulation, and the exit to the tunnel being the waypoint that must be visited by the bundle. These endpoints become the source and destination of the encapsulating bundle and can be kept separate from the source and destination of the encapsulated bundle.

8.6 Special Security Threats

DTN constraints affect the manifestation of security threats in much the same way they affect the implementation of security services. Because DTNs are different than other networks in this respect, common security threats must be re-evaluated in this

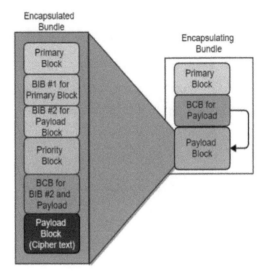

Figure 8.10: Using encapsulation as a security mechanism.

new context. Threats will be discussed in the context of BPSec and BIBE as these two protocols most directly affect end-to-end security across a DTN.

8.6.1 Attack Capabilities and Objectives

Threat analysis is layered for the same reason as network protocols and security models are layered: to help define and distribute responsibilities. Threat analyses may focus on a variety of responsibilities including vulnerabilities in hardware, operating systems, software implementations (e.g., openssl), and physical access to contents of memory and disk space. Network threat analysis focuses on the format of data structures as defined in secured protocols (such as BPSec) and the behaviors of nodes that process these protocols. For that purpose, it is assumed that all other aspects of the system operate correctly and without other vulnerability.

Even with the narrower focus on network threats there is a myriad of information for how to think about security using the Internet model. Attacks to networks can be characterized as passive or active [232].

Passive attacks rely on simply listening to information traversing the network. Attacks never need to actually write information back onto the network for the attack to occur. Examples of such attacks include the following.

■ Confidentiality violations (reading data in plain text sent over the network)

■ Password Sniffing (reading passwords sent in plain text over a network from legacy protocols)

- Offline Cryptographic attacks (reading cipher text and breaking the encryption offline)

Active attacks result in an alteration of message traffic in the network. This can be accomplished by adding messages into the network, removing or rerouting existing messages, or simply deleting messages. Examples of such attacks include the following.

- Replay Attacks (sending the same message multiple times)

- Message Insertion (creating new, potentially invalid, messages in the network)

- Message Deletion (removing messages from the network by not forwarding them)

- Message Modification (altering portions of the message to change their meaning)

- Man-in-the-Middle (a special combination of the above where a node masquerades as a message source to fool a message destination)

Given the delayed nature of communications in a DTN, Man-in-the-Middle (MITM) attacks can become much more problematic. Recall that, in a DTN, bundles can exist in the network for days/weeks/months, nodes might not have synchronized clocks, and there is no session state at endpoints. Each of these constraints preclude the types of MITM protections used on the Internet today. BPSec, as a DTN end-to-end security protocol, is specifically designed to protect against these kinds of attacks in a DTN.

8.6.2 Terminology

A comment terminology in security discussions is to name certain nodes in a network. Three nodes are used in MITM discussions:

- **Alice** and **Bob.** Alice and Bob refer to two nodes in a network attempting to communicate with each other. Typically, Alice is sending a message to Bob (e.g., node A sending to node B).

- **Mallory.** Mallory refers to a non-cooperative node existing between Alice and Bob that can receive bundles, examine bundles, modify bundles, forward bundles, and generate bundles at will in order to compromise the confidentiality or integrity of data within the DTN.

In order to inject messages into a network, Mallory must be a node on the network. Therefore, terminology is required to discuss the permissions of the Mallory node with respect to the network. It is assumed that Alice and Bob are trusted nodes in the network and do not need their own hierarchy of access. Mallory can exist on one of the following kinds of nodes.

■ **Unprivileged Node:** Mallory has no credentials for the network and can only see unencrypted data and publicly shared cryptographic material (such as keys).

■ **Legitimate Node:** Mallory has a personal set of security credentials for the network. She has her own private key(s) (KEYmallory) and other nodes may have her public key. In this sense, Mallory appears similar to any other non-administrative or non-privileged node in the network.

■ **Privileged Node:** Mallory is an administrative node in the network and has access to all public and private information across the network. This includes private keys for Alice and Bob (KEYalice and KEYbob).

When using asymmetric cipher suites there is a difference between private keys and public keys and, therefore, a difference between a legitimate node (which would know Alice's public key) and a privileged node (which would know Alice's private key). However, when using symmetric cipher suites, the public and private keys are the same. In this case, every legitimate node is also a privileged node because it has both private and public key information (because the keys are the same).

8.6.3   Special Cases

There are two special cases of network attack for which there are no good mitigations. The first special case is when an administrative node in the network is compromised. The second is when Mallory is one of the endpoints in the message path, and not actually in the middle.

Compromised privileged node. If Mallory exists on a privileged node in the network, then all private security information known to that node (including security information private to Alice and Bob) are known to Mallory. In this case, Mallory may sign information in a way that is indistinguishable from Alice and Bob because Mallory will have their private keys. Detecting and mitigating attacks in this scenario is extremely difficult (and in many situations impossible).

Mallory is an Endpoint. If Mallory is the same node as either Alice or Bob then she has access to Alice's (or Bob's) private key because they are, in this instance, Alice (or Bob).

8.6.4   Attacks and Mitigations

The BPSec protocol must provide reasonable mitigations for MITM-related attacks that are uniquely enabled or exacerbated by the properties of a DTN. Certain attacks, such as message insertion, message replay, and message deletion can be handled in a straightforward way.

■ **Message Deletion.** If Mallory always deletes messages, then Alice must re-send messages through multiple paths, hoping that one of the paths does not go through Mallory.

■ **Message Insertion.** If Mallory inserts messages into the network, then security policy in the network must be configured to require signatures on all data. Mallory can only sign data as herself and therefore cannot insert messages into the network as if from Alice.

■ **Message Replay.** The nature of a DTN requires that nodes be prepared to accept multiple copies of a bundle. In this scenario, then, replay attacks are inherently harmless in the network because the network assumes that multiple copies of a message will be received over time. Replay attacks are often used to perturb synchronization mechanisms, which do not affect DTNs where there is no concept of end-to-end synchronization.

Other attacks, such as eavesdropping and modification attacks require some additional thoughts on how to prevent.

8.6.4.1 Eavesdropping Attacks

Mallory may eavesdrop on any bundle that is sent to her. In this context, eavesdrop refers to storing a copy of the bundle and attempting to recover the data from that bundle. Sometimes this recovery attempt happens in near-real-time. Sometimes, the data may be stored and analyzed over a period of days, weeks, or longer.

The standard mitigation for these attacks is to apply confidentiality to data in the bundle; to encrypt the contents that should not be viewed by the public (including Mallory). This is accomplished in the BPSec protocol through one or more BCB extension blocks which encrypt their security target.

On the Internet there is, in some circumstances, a time limit within which the stored messages must be accessed. Certain kinds of Internet data (such as session information, keys, and some passwords) expire far faster than reasonable encryption can be compromised. However, this is less the case in a DTN where messages may have lifetimes of days, weeks, months, or longer. In such a case the useful lifetime of information in a message may be as long (or longer) than the time it takes to compromise encryption.

8.6.4.2 Modification Attacks

When Mallory receives a bundle from Alice to Bob, in addition to storing the bundle for offline analysis, she may also attempt to modify some aspect of the bundle and send the modified bundle to Bob. This modified bundle could have changed routing information, priority information, new extension blocks added, existing extension blocks removed, and changes to status fields within the block headers themselves.

The standard mitigation for these attacks is the application of an integrity service. Alice may digitally sign some or all of the blocks in the bundle prior to sending to Bob. Were Mallory to attempt to modify the bundle, her modifications could not be mistaken for Alice's data because Mallory does not have Alice's private key information[11]. This is accomplished in the BPSec protocol through the use of one or more BIB blocks which carry digital signatures for their security targets.

However, certain modifications involve Mallory removing blocks from a bundle. An integrity signature (BIB) on a block has no beneficial effect if Mallory removes the BIB and the extension block that it targets. Mallory could remove any block in a bundle (other than the primary block and the payload block) and/or any associated BIB and BCB blocks to alter the content and security stance of information in the bundle. In these scenarios there is no residual information in the bundle that those blocks ever existed and one of two mitigations must be deployed to tell Bob what to require be in a received bundle: an in-bundle inventory or an external-to-the-bundle policy.

- **Inventory Blocks.** A bundle, when created by Alice, may have a special inventory block added to the bundle which captures the contents of the bundle at the time of its creation. This inventory block must be signed by Alice to ensure that it is not modified en route. When the bundle is received by Bob, Bob can check that the inventory block exists and that the bundle contains (at least) the content noted in the inventory. In this way, Bob can detect whether or not Mallory has removed important blocks from the bundle.

- **Inventory Policy.** Pre-configured security policies between Alice and Bob could require that certain blocks be present in a bundle. For example, a security policy could enforce that all payloads have integrity services applied to them, such that if Mallory drops a BIB whose security target is the payload (and then modifies the payload), Bob can detect this condition since he would receive a bundle with a payload and no integrity on the payload provided by Alice.

There is a final concern relating to Mallory modifying information from Alice. If Alice applies integrity to a block in a bundle Mallory could simply delete Alice's signature, modify information in the bundle, and then sign the modification with Mallory's own signature. In cases where Mallory is a legitimate node in the network and Bob doesn't care that the signature is not from Alice (but just from any legitimate node in the network) then Bob is at risk of using modified information from Mallory.

This scenario can be prevented by using both confidentiality and integrity at the same time. Alice can create a BIB to sign some block in the bundle, and then use a BCB to encrypt both the BIB target and the BIB itself so that only Bob can see it. Since Mallory cannot decrypt information, she cannot sign the original block (because she can only see the block's cipher text). Further, Bob must decrypt Alice's original BIB to see the signature for the target block's plain text so Mallory cannot drop Alice's encrypted BIB without Bob knowing that there has been a modification attack.

---

[11]Note that the case where Mallory is a privileged node is a special case covered earlier in this section and not applicable to this analysis.

8.7 Policies Considerations

The configuration of a security policy in a network is a way to pre-configure and then asynchronously manage node behavior in a variety of scenarios. This configure and manage approach is necessary in a DTN, where nodes may not have timely connectivity with network operators.

There are several important events that will be encountered by a node that must be given consideration and covered by the network security policy.

1. **Bundle received with less security than required.** When the network requires a certain level of security, such as encrypted payloads or authenticated message exchange and a message is received without this information, the network must handle this in a uniform way. Most policies require not forwarding the message, but the level of logging, error messaging, and updates to local configurations should be discussed as a matter of policy.

2. **Bundle received with more security than required.** Similarly, when messages are received that contain authentication, integrity, or confidentiality when they should not, a decision must be made as to whether these services will be honored by the network.

3. **Security Evaluation in Transit.** Some security services may be evaluated at a node, even when the node is not the bundle destination or a security destination. For example, a node may choose to validate an integrity signature of a bundle block. If an integrity check fails to validate, the intermediate node may choose to ignore the error, remove the offending block, or remove the entire bundle.

4. **Fragmentation.** Policy must determine how security blocks are distributed amongst the new bundle fragments, so as to allow received fragments to be validated at downstream nodes.

5. **Block and Bundle Severability.** Distinct from fragmentation, nodes must decide whether a security error associated with a block implies a larger security error associated with the bundle. If blocks and bundles are considered severable, then an offending block may be omitted from the bundle. Otherwise, a bundle should be discarded whenever any of its constituent blocks are discarded.

8.8 Summary

Securing modern computer networks remains a difficult task due to the asymmetric nature of network threats. As computing and networking resources become more powerful, less expensive, and more densely connected, the resources available to attack a network quickly outpace the resources available to defend the network. This situation is exacerbated when securing a DTN because network defenses may not be in contact with a network node while it is under attack. Therefore securing DTNs requires careful engineering and the pragmatic application of new algorithms, protocols, and applications rather than fully adopting what has worked on highly available terrestrial networks.

The careful engineering of DTN security requires an understanding of how non-DTN security works; the assumptions made on the underlying network infrastructure that make existing algorithms, protocols, and applications function. In many cases, a DTN cannot provide the high-availability, low-latency connectivity enabling traditional layered and compartmentalized security services. Understanding the unique intersection of security assumptions and DTN constraints provides a useful way to determine what existing security mechanisms can be re-used and where the community must construct new capabilities.

Similar to other security models, the DTN security model decomposes into multiple layers and multiple components. DTN security model layers differ from traditional networking layers because DTNs span multiple networks each with their own networking stack. Where DTNs span multiple transports and/or links, the security layering defines areas of trust within or across the component networks. This is conceptually similar to the concept of secured subnetworks, except these boundaries of trust may include whole networks, multiple networks, or multiple subnetworks of networks. Within these layers there exists a series of components, including protocols, configurations, and policies. Protocols provide the most generic expression of capability for any secure DTN and configurations and policies provide information for specific instances of a secure DTN.

Since the Bundle Protocol represents the common denominator for any DTN deployment, securing bundles across the DTN represents the fundamental service of a secure DTN. The Bundle Protocol Security (BPSec) protocol uses the extension block mechanism within the BP to define security information carried within the bundle itself to implement security services. Carrying security information with the data it secures solves an important problem encountered in DTNs: nodes have no timely access to external oracles. Sending security data with the data it secures dramatically reduces the need to refer to oracles external to the node. The BPSec protocol provides integrity and confidentiality services for individual blocks within a Bundle and, combined with encapsulation mechanisms, a variety of authentication approaches can be used.

The DTN security model exists in the context of the security threats most likely to be encountered in a DTN. In some ways, attackers suffer from the same constraints as defenders in the network: sparsely connected networks limit opportunities to attack nodes and combine resources for distributed denial of service attacks. Similarly, attacker nodes deployed in a challenged environment may, themselves, have limited access to external oracles necessary for their operation (e.g., malicious software that checks dedicated Internet addresses for ransom payments, new configurations, key updates, or other commanding). Alternatively, the DTN architecture uniquely enables some attacks, such as man-in-the-middle attacks. Best security practices use mechanisms related to timing analysis and state synchronization to defeat these attacks, but both approaches become difficult or impossible to implement as delays and disruptions grow in the network.

BPSec cannot place all security information in a bundle, lest an attacker modify the bundle in transit. A DTN must deploy some state synchronization, in the form of security configuration and policy, to establish required security services. To determine configuration settings and policy considerations, network architects must answer important questions relating to the operation of the network. These universal questions address two fundamental aspects of network operation: What security services should security-aware bundle node expect and how should such a node react when a bundle fails those expectations? While these questions must be asked for every DTN deployment the answers depend on the specifics of individual DTNs.

The future of secure DTNs holds significant promise with advances in embedded, low-power devices and the accompanying desire to sense/instrument environments without ready access to power and communications opportunities. This chapter defines a rich, challenging problem space that the DTN community will spend the next several decades filling with new and novel algorithms, cipher suites, applications, and operational concepts.

8.9　Problems

1. Explain three challenges to securing an interplanetary network that are not challenges in securing a web server on the Internet.

2. Explain three ways in which securing an interplanetary network would be easier than securing a web server on the Internet.

3. Confidentiality, Integrity, and Authentication are considered the fundamental set of security services. Do these services protect against replay attack in a non-DTN network? In a secured DTN network? What additional service or services might be needed if these services do not adequately protect against replay attacks in a DTN?

4. Describe a situation where a network might want to deploy an integrity service but not a confidentiality service.

5. Explain in your own words the concept of a coherent and incoherent change to data while in transit in the network and how integrity and authentication services may protect against each.

6. Consider that you are architecting a network that must have high availability. Why is security important for achieving that goal, and what services are necessary to keep the network available?

7. Identify a security application, protocol, or other mechanism that does not make one or more of the security assumptions listed in Table 8.2. Would your selected security mechanism work when deployed in a DTN? Why or why not?

8. Section 8.3.1.1 identified two examples of security-related mechanisms that require rapid, round-trip communications to other nodes in the network to function. Can you list and defend a third example?

9. Explain how end-to-end authentication in a network is jeopardized if different nodes in the network use different naming and addressing schemes? Explain at least one way to solve this problem.

10. If secure communication endpoints in a network cannot synchronize on time, explain one way to determine if a message is being replayed by an Attacker at a later time.

11. Compare and contrast the concepts of a network overlay and network encapsulation. What security considerations must be made when choosing to use one approach over the other?

12. Describe a theoretical DTN deployment (such as a space network or distributed sensor network) and use that example to explain examples of data exchanges at the link, inter-network, end-to-end, and user layer.

13. Can a single security protocol work to secure the network? Why or why not?

14. Describe how a man-in-the-middle attack would succeed in a DTN using BPSec if there is no way to configure policy expectations in the network.

15. Assume that BPSec has added security-related blocks to a bundle in transit. Describe a mechanism by which the bundle destination node can determine if all necessary security services are present in the bundle. Your mechanism must not rely on any synchronized state information beyond the initial configuration of the node when it is deployed on the network. Your mechanism must adapt to changes in the network topology as new nodes are added to the DTN and existing nodes leave the DTN.

16. Explain three reasons why it is important to have different security for different blocks in a bundle.

17. Within BPSec, a security block may only include multiple targets if all targets have the same security source. Explain how the security of a bundle could be compromised if a single BPSec security block could be modified by multiple security sources.

18. Explain one benefit and one drawback of using bundle-in-bundle encapsulation to create a security tunnel for a bundle.

19. Explain three reasons why a DTN enables man-in-the-middle attacks more than a non-DTN network.

20. Define a security policy for a DTN comprising a network of ocean buoys. This policy must address the five policy considerations identified in Section 8.7.

Chapter 9

DTN of Things

Juan A. Fraire

Universidad Nacional de Córdoba - CONICET, Córdoba, Argentina

Jorge M. Finochietto

Universidad Nacional de Córdoba - CONICET, Córdoba, Argentina

CONTENTS

This chapter introduces **DTN of Things**, a networking paradigm that extends current capability of **the Internet of Things** (IoT). While IoT solutions assume persistent end-to-end connectivity, DTN of Things do not. As a result, highly disrupted scenarios (with high node mobility or long power saving periods) can also be considered. Both paradigms shape the concept of **Network of Things**, a growing idea of connecting objects. The network of things has and will have significant impact in day to day life, ranging from healthcare, smart house and buildings, transportation, industry, and even cities.

The chapter is divided in three main sections. In the first section, we introduce the concept of network of things, its applications, architecture and services based on existing literature. The second section provides an overview of existing infrastructure and application communication protocols developed for the IoT part of the network of things. DTN of Things is introduced in the third section as an architecture to cope with link disruptions. In this section we motivate the application scenarios, survey existing experiences on applying DTN to the network of things and discuss persistent challenges toward deploying large-scale DTN of Things networks.

9.1   Introduction

We are entering an era of connected things or more specifically, the era of the Network of Things. As we move to this new networking paradigm, network traffic is and will no longer be generated by human users, but also by things such as passive objects, sensors and actuators placed in our cities, industries, vehicles, homes, clothes and even body. The research community has made a significant effort in building the network of things by integrating them with the infrastructure of the most extensive and flexible network ever created, the Internet. Thus, the term "Internet of Things", abbreviated as IoT, is currently an active and attractive research topic.

Manifold definitions of IoT are traceable within the research community, which testify to the strong interest in the issue and the vivacity of the debates on it. However, by browsing the literature, an interested reader might experience a real difficulty in understanding what IoT really means, which social and economical implications the full deployment of IoT will have, and most importantly, how the networking technology and protocols need to evolve in order to efficiently connect things in the future [35]. In this chapter, we will argue that the difficulty in converging in a unique infrastructure model and set of IoT protocols lies not only in the heterogeneity of the things, but in the attempt of integrating them via Internet. After providing a thorough overview of what the community has been expecting and developing for the IoT, we will focus on one of the most important assumptions inherited of Internet protocols: end-to-end connectivity. Despite any effort in optimizing and fine-tuning IoT solutions, the mobility and resource-constrained nature of connected things will hinder (and at some point directly forbid) to keep up such a stable connectivity. In fact, we will highlight that many of the IoT protocols features are actually an attempt to temporarily disconnect the devices in such a controlled fashion that the rest of the network can believe the node is continuously connected. Instead of masking and controlling disruptions, the DTN architecture supply accommodates them in normal network operations based on store and forward principles. In the second part of this chapter we introduce and describe Delay-Tolerant Network of Things (DTN of Things) as an alternative to the existing families of IoT protocols based on the Internet paradigm. Although both IoT and DTN of Things are different approaches to tackle the network of things, they can be considered complementary. We will explore which topologies and scenarios of connected things can benefit from the DTN of Things approach, which can better rely on classical IoT solutions and which are good candidates for hybrid approaches.

9.2   Network of Things

9.2.1   *Overview*

The Network of Things is a concept that aims at connecting objects. The principal networking paradigm studied for the Network of Things is based in Internet-like continuous connectivity and has been called Internet of Things (IoT). Later in this chapter, we introduce DTN of Things as an alternative store-and-forward approach. Both paradigms aim at connecting things, but using different approaches.

Figure 9.1 illustrates how these two paradigms relate in the context of network of things. Traditional IoT literature and related technology has assumed a continuous connectivity between the source and the destination node. Whenever a disruption in the link happens (because of mobility, power saving or erroneous behavior), data remains stored in the source node until the end-to-end connection is reestablished. We will call this data handling approach as single-hop store and forward and is typically tackled at the application level in existing IoT protocol stack.

On the other hand, DTN of Things proposes to address disruption in lower protocol layers. Since network partition is considered a natural property of the network, there is no need for applications to deal with it. Therefore, store and forward functionality is taken to lower layers on the protocol stack, and can now happen not only in the source node but also in intermediate nodes using the DTN protocol stack. This means that when several disruptions happen along the data path, the data can advance step-by-step towards the destination, being stored in intermediate nodes when necessary. We name this scheme as multi-hop store and forward. Multi-hop store and

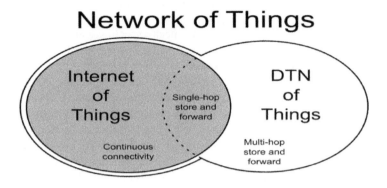

Figure 9.1: Network of Things ecosystem addressed by IoT and DTN of Things.

forward has been addressed in the DTN architecture and can be tackled by standardized DTN protocols. It is interesting to note that single-hop store and forward and continuous connectivity are a particular case of multi-hop store and forward where storage happens once or does not happen at all. As a result, a DTN of Things protocol stack will also work in networks without disruptions such as those addressed by IoT.

9.2.2   Internet of Things

The IoT term has been defined by different authors in many different ways. Among them, Vermesan et al. [281] define the IoT as simply an interaction between the physical and digital worlds. The digital world interacts with the physical world using a plethora of passive objects and smart sensors and actuators. Another definition by Peña-Lopez [202] defines the Internet of Things as a paradigm in which computing and networking capabilities are embedded in any kind of conceivable object. In general, the basic idea of the IoT concept is the pervasive presence around us of a variety of interconnected objects that cooperate to reach common goals [126]. The terms Machine-to-Machine (M2M) and Device-to-Device (D2D) communications have also been used to describe tightly coupled application logic, device and sensors without human intervention.

In general, the term Internet of Things can be semantically interpreted as "a world-wide network of interconnected objects uniquely addressable, based on standard communication protocols". As a result, the IoT enables physical objects to see, hear, think and perform jobs by having them "talk" together, to share information and to coordinate decisions. Connecting things allows to consider a wide range of application and services which are expected to grow in a significant way in the near future. In this chapter, the IoT is considered as one of the possible paradigms to build the network of things.

9.2.3 Applications and Services

There are several domains and environments in which the network things in general can play a remarkable role and improve the quality of our lives.

Smart home: connected things contribute to enhancing the personal lifestyle by making it easier and more convenient to monitor and operate home appliances and systems (e.g., air conditioner, heating systems, energy consumption meters, etc.) remotely [25], [90]. For example, a smart home can automatically close the windows and lower the blinds of upstairs windows based on the weather forecast.

Building Automation Systems (BAS): networked things allow to control and manage different building devices using sensors and actuators such as HVAC (heating, ventilation, and air conditioning), lighting and shading, security, safety, entertainment, etc. BAS can help to enhance energy consumption and maintenance of buildings. For example, a blinking dishwasher or cooling/heating system can provide indications when there is a problem that needs to be checked and solved. Thus, maintenance requests can be sent out to a contracted company without any human intervention at all.

Intelligent transportation systems (ITS): connected things allow to monitor and control transportation networks, whether public or private [269]. ITS aims to achieve better reliability, efficiency, availability and safety of the transportation infrastructure. Based on vehicles equipped with GPS, RFID or communication that can connect with road-side equipments (a.k.a. vehicle-to-infrastructure or V2I), relevant position and other data can be made available to optimize and take efficient decisions. Moreover, connected vehicles (a.k.a. vehicle-to-vehicle or V2V) are becoming more important with the aim to make driving more reliable, safe, enjoyable and efficient [173].

Industrial automation: By computerizing robotic devices, manufacturing tasks can be completed with a minimal human involvement [286]. The network of things is utilized in industrial automation to control and monitor production machines' operations, functionalities, and productivity rate through the Internet. For instance, if a particular production machine encounters a sudden issue, a connected things system can send a maintenance request with all necessary information immediately to the maintenance department to quickly handle the fix. In general, connecting things increases productivity industries by analyzing production data, timing and causes of production issues.

Smart healthcare: Connected devices can play a significant role in healthcare applications through embedding sensors and actuators in patients and their medicine for monitoring and tracking purposes. The network of things is already used by clinical care to monitor physiological statuses of patients through sensors by collecting and analyzing their information and then sending analyzed patient data remotely to processing centers to make suitable actions [198].

Smart grids: The network of things can be utilized to improve and enhance the energy consumption of houses and buildings [299]. For example, smart grids use the IoT to connect millions or billions of buildings' meters to the network of energy providers. These meters are used to collect, analyze, control, monitor, and manage energy consumption. Also, utilizing the IoT in the smart grid reduces the potential failures, increases efficiency and improves quality of services.

Smart city: Smart city is a connected things concept that aims to improve the quality of life in the city by making it easier and more convenient for the residents to find information of interest [156]. In a smart city environment, various systems based on smart technologies are interconnected to provide required services to the citizens (health, utilities, transportation, government, homes and buildings, etc.). The large area of smart cities deployments is probably the distinguishing feature of smart cities networks.

Overall, these applications can be built on top of basic services which can be categorized under four different classes or levels [125].

1. Identity-related Services: The most basic and important services that are used by other types of more complex services. Every application that needs to bring real world objects to the virtual world has to identify those objects and have a minimal addressing functionality to reach them through the network.

2. Information Aggregation Services: Collecting and summarizing raw sensory measurements is the second level to provide value to a network of connected things. Information needs to be aggregated in order to be processed and reported to the final IoT application or user.

3. Collaborative-Aware Services: Acting on top of Information Aggregation Services, the obtained and summarized data can be used to make decisions and react as per specific set of rules. This third level allows for automated and rapid response with minimal human intervention.

4. Ubiquitous Services: The fourth level is aimed to provide Collaborative-Aware Services anytime they are needed to anyone who needs them anywhere. While abstract and ambitious, this level service is considered the ultimate goal for all network of things applications.

Ubiquitous services are not achievable easily since there are a lot of difficulties and challenges that have to be addressed. Most of the existing applications provide identity related, information aggregation, and collaborative-aware services [25]. Smart healthcare and smart grids fall into the information aggregation category and smart home, smart buildings, intelligent transportation systems, and industrial automation are closer to the collaborative-aware category. The smart city concept can be considered closer to the ubiquitous service type. It is expected that further growth and interest in the network of things market will eventually push other applications to ubiquitous service characterization.

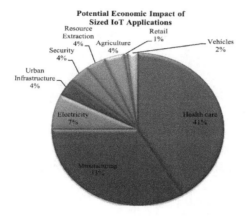

Figure 9.2: Projected market share of dominant network of things applications by 2025.

9.2.4 *Projection*

The size of the network of connected objects is expected to reach 212 billion entities deployed globally by the end of 2020 [123]. Furthermore, by 2022, traffic flows between objects are expected to constitute up to 45% of the whole Internet traffic [106]. Beyond these predictions, McKinsey Global Institute reported that the number of already connected machines (units) has grown 300% over the last 5 years [193]. Traffic monitoring of a cellular network in the U.S. also showed an increase of 250% for device to device traffic volume in 2011 [249]. Indeed, machine-to-machine (M2M) traffic is taking and will take a significant role in the future Internet which will connect not only humans but also things.

Economic growth of connected things services is also considerable for businesses. Healthcare and manufacturing applications are projected to form the biggest economic impact. The whole annual economic impact caused by the network of things is estimated to be in range of $2.7 trillion to $6.2 trillion by 2025 [193]. Figure 9.2 shows the projected market share of dominant network of things applications. All these statistics point to a potentially significant and fast-pace growth of connected things in the near future, related industries and services. This progression provides a unique opportunity for traditional equipment and appliance manufacturers to transform their products into "smart things."

9.2.5 *Layered Architecture*

The network of things should be capable of interconnecting billions or trillions of heterogeneous objects, so there is a critical need for a flexible layered architecture. However, the ever increasing number of proposed architectures for the IoT has not

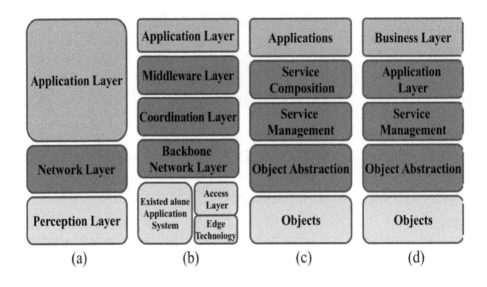

Figure 9.3: Layered architectures for the Network of Things.

yet converged to a reference model [193] [165]. Moreover, the introduction of the DTN of Things might further complicate the endeavor.

From the pool of proposed models, the basic model is a 3-layer architecture consisting of the Application, Network, and Perception Layers [162]. In the recent literature, however, some other models have been proposed that add more abstraction to the IoT architecture [302] [271] [77]. Figure 9.3 illustrates some common architectures which were surveyed in [25]. Below, we summarize the 5 layer architecture in Figure 9.3 d) and comment how DTN of Things would modify the behavior of each layer.

- **Objects Layer:** The Objects (devices) or perception layer, represents the physical sensors of the network of things that aim to collect information, process it, and eventually act upon it. Querying functionality is expected such as identification, location, temperature, weight, motion, vibration, acceleration, humidity, etc. The perception layer digitizes and transfers data to the Object Abstraction layer through secure channels. When in presence of a large quantity of objects, we can say that this layer is the initiator of the Big Data process [302].

- **Object Abstraction Layer:** Object Abstraction presents and transfers data produced by the Object's layer to the Service Management layer typically through wireless links such as RFID, 3G, GSM, Bluetooth, ZigBee, and others reviewed later in this chapter. In this layer, data obtained from sensing and device status is formatted and forwarded to the sink or gateway node [302]. In this layer, IoT and DTN of Things will differ on the data handling. While IoT will forward and assume an instantaneous reception of the data on the destination, DTN of Things must consider the possibility of data being temporarily stored in an intermediate node.

- **Service Management Layer:** Service Management or Middleware (pairing) layer pairs a service with its requester based on addresses and names. This layer enables the network of things application programmers to work with heterogeneous objects without consideration to a specific hardware platform [298]. In the case of DTN of Things, this layer cannot assume that the available information of the network nodes represents the latest status of the devices.

- **Application Layer:** The application layer provides the services requested by customers. For instance, the application layer can provide temperature and air humidity measurements to the customer who asks for that data [271]. As based on the service management layer, the application based on DTN cannot assume the provided information represents the current measurement of the consulted objects. However, applications might know at which time the provided information was measured to successfully build a history if necessary.

- **Business Layer:** The business (management) layer manages the overall network of things system activities and services. The responsibilities of this layer are to build a business model, graphs, flowcharts, etc. based on the received data from the Application layer. It is also supposed to design, analyze, implement, evaluate, monitor, and further develop the elements of the connected things system. The Business Layer makes it possible to support decision-making processes based on Big Data analysis [25].

9.3 Existing Protocols

The network of things term is syntactically composed of two terms. The first one pushes towards a **network** oriented vision of how things should be connected, while the second one moves the focus on which type of **objects** shall be integrated into the network of things. Indeed, the properties of the objects will define what kind of network protocols they can and have to implement. In this section, after describing the things part, we overview the existing protocols in the network part.

9.3.1 The Things Part

The original term of IoT is, in fact, attributed to The Auto-ID Labs, a world-wide network of academic research laboratories in the field of networked RFID (Radio Frequency Identification) and emerging sensing technologies [35]. These institutions, since their establishment, have been targeted to architect the IoT with a primary focus on spreading use of RFID in worldwide modern trading networks.

RFID systems [112] are composed of one or more reader(s) and several RFID tags. From a physical point of view a RFID tag is a small microchip attached to an antenna in a package which usually is similar to an adhesive sticker. The chip is used both for receiving the reader signal and transmitting the tag ID while harvesting the power of the received signal (Tags are generally passive). Dimensions can be very low (for example, Hitachi has developed a tag with dimensions $0.4mm \times 0.4mm \times 0.15mm$) and signals can reflect in short ranges ($10cm$ to $200m$) [290]. Tags are characterized by a unique identifier and are applied to objects (even persons or animals). On the other side, readers trigger the tag transmission by generating an appropriate signal, which represents a query for the possible presence of tags in the surrounding area and for the reception of their IDs. Accordingly, RFID systems can be used to monitor objects in real-time, without the need of being in line-of-sight; this allows for mapping the real world into the virtual world.

Indeed, item **identity-related services** which allow for object traceability and adressability are natively provided by RFID objects and are considered a basic feature of any connected things application. Indeed, RFID maturity, low cost, and strong support from the business community has pushed its adoption in an incredibly wide range of application scenarios, spanning from logistics to e-health and security. However, while the main aspects stressed by RFID technology shall definitely be addressed by the network of things, alternative, and somehow more complete visions recognize that the term network of things implies a much wider vision than the idea of a mere objects identification.

Autonomous and **proactive behavior**, **context awareness**, **collaborative communications** and **elaboration** are just some required capabilities of a broader category of things known as smart objects. In this case, we are no longer talking about passive tags, but complex sensing nodes equipped with autonomous energy supply, capable wireless communication (not only gateway-based but also multi-hop), memory, and elaboration capabilities. Note that we broadly define the term sensor; a mobile phone or even a microwave oven can count as a sensor as long as it provides inputs about its current state (internal state + environment). Furthermore, an actuator is a device that is used to effect a change in the environment such as the temperature controller of an air conditioner. Indeed, a smart object not only provides identification such as an RFID tag, but it can also interact with the environment by providing relevant sensed information and acting upon it. Specifically, smart nodes report the results of their sensing or acting to a small number (in most cases, only one) of special nodes called sinks or gateways using a single or multi-hop communication. Table 9.1 summarizes the key properties of the connected things.

9.3.2 Computation

Various hardware platforms were developed to run the network of things applications in smart devices powered by an Arduino, FriendlyARM, Intel Galileo, Raspberry PI, BeagleBone, Cubieboard, WiSense and others. Embedded processing units can range from microcontrollers, microprocessors, Systems on Chips (SOCs) and Field Programmable Gate Array (FPGAs) which can run one of the many software

Table 9.1: Type of "Things" in the network of things.

Object Type	Processing	Sense/Act	Power	Lifetime	Size
RFID	No	No	Harvested	Indefinite	Very small
Smart	Yes	Yes	Battery	3 − 10 years	small

platforms to provide functionalities on these devices. Among these platforms, Operating Systems (OS) are vital since they run for the whole activation time of a device. There are several Real-Time Operating Systems (RTOS) that are good candidates for the development of RTOS-based IoT applications such as Contiki RTOS [290]. TinyOS [174], LiteOS [71] and Riot OS [39] also offer lightweight OS designed for IoT environments. Their characteristics are listed in Table 9.2. Together, the hardware and software represent the "brain" of devices in network of things [25].

Table 9.2: Common Operating Systems (OS) in the network of things [25].

OS	Language	Minimum Memory (KB)	Event-based Programming	Multithreading	Dynamic Memory
TinyOS	C	1	Yes	Partial	Yes
Contiki	C	2	Yes	Yes	Yes
LiteOS	C	4	Yes	Yes	Yes
RiotOS	C/C++	1.5	No	Yes	Yes
Android	Java	-	Yes	Yes	Yes

On the network side, the data provided from the arbitrarily high number of devices need to be received, processed, stored and acted upon. This is where the "Big Data" is generated [191]. Indeed, cloud platforms form another important computational part of the network of things. These platforms, surveyed in [25], provide facilities for smart objects to send their data to the cloud, for big data to be processed in real-time, and eventually for end-users to benefit from the knowledge extracted from the collected big data.

9.4 The Network Part

In general, the design objectives of network of things protocols are energy efficiency (which is the scarcest resource in most of the scenarios involving sensor networks), scalability (the number of nodes can be very high), reliability (the network may be used to report urgent alarm events), and robustness (sensor nodes are likely to be subject to failures for several reasons). Altogether, these technologies will become "the atomic components that will link the real world with the digital world" [281].

9.4.1 Infrastructure Protocols

Infrastructure protocols are those that allow objects to reach a server located at the core of the network, typically connected to the Internet. While TCP/IP is the protocol set used in the Internet part, different infrastructure protocols have been developed to wirelessly connect resource-constrained devices at short, medium and long range.

9.4.1.1 Short Range

The most common communication technologies for short range low power communication protocols are RFID and NFC (Near Field Communication). In these types of protocols, the object proximity enables energy harvesting by passive devices.

RFID: Introduced in the previous section, RFID is a system where two-way radio transmitter-receivers called interrogators or readers send a signal to a tag and read its response containing an identification code. RFID is standardized in ISO/IEC 18000, an international standard that describes a series of diverse RFID technologies, each using a unique frequency range [40]. The standard defines the air interface required to provide efficient item identification. The standard is a seven part standard with the first part describing the general parameters while the other six parts tackle specifics on each of the possible RFID frequencies (below 135 kHz or low-frequency, 13.56 MHz or high-frequency, 2.45 GHz, 860 MHz to 960 MHz or ultra high-frequency and 433 MHz).

NFC: On the other hand, NFC is a specific branch of high-frequency RFID standardized in ISO/IEC 18092, that operates at the 13.56 MHz frequency. NFC devices have taken advantage of the short read range limitations of this radio frequency. Because NFC devices must be in close proximity to each other, usually no more than a few centimeters, it has become a popular choice for secure communication between consumer devices such as smartphones. However, in contrast with RFID, an NFC device is capable of being both an NFC reader and an NFC tag. This unique feature allows NFC devices to communicate peer-to-peer, an ability that has made NFC a popular choice for contactless payment. Also, NFC devices can read passive NFC tags, and some NFC devices are able to read passive HF RFID. The data on these tags can contain commands for the device such as opening a specific mobile application and can be seen in advertisements, posters and signs.

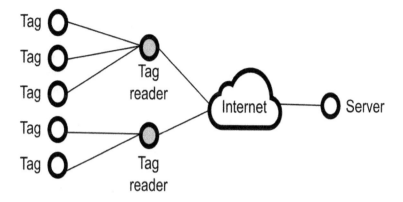

Figure 9.4: Typical short-range infrastructure protocol topology.

The resulting networking topology of RFID and NFC based low-range networks is a star where the reader is the central node that relays data to a centralized server typically connected to the Internet. Even when using NFC from our phone to read an advertisement, we are acting as a tag reader to then connect to Internet and retrieve the advertised web site. The resulting topology is illustrated in Figure 9.4.

9.4.1.2 Medium Range

A large scientific literature has been produced on medium-range Wireless Sensor (and actuator) Networks (WSN) in the recent past, addressing several problems at all layers of the protocol stack [203]. Today, most of commercial wireless sensor network solutions are based on the IEEE 802.15.4 standard [237] or Bluetooth Low Energy (BLE) and are typically addressed to smart buildings or homes also known as wireless personal area networks (WPANs).

IEEE 802.15.4: the IEEE 802.15.4 standard defines the physical (PHY) and MAC layers for low-power, low bit rate communications in WPAN [298].

- ■ PHY: The basic framework of IEEE 802.15.4 conceives a 10-meter communications range with a transfer rate of 250 kbit/s, but tradeoffs are possible to favor more radically embedded devices with even lower power requirements, through the definition of not one, but several physical layers. One is the low-band (868/915 MHz), that uses binary phase shift key modulation and another is the high band (2.4GHz) that uses offset quadrature phase shift keying modulation, both unlicenced. There are 27 available channels, 11 in low band and 16 in high band. The raw bit rates are 20-40 kbps in low band and 250 kbps in high band. By using these modulations and spreading techniques, the transmissions are robust and noise resistant [239].

■ MAC:On the MAC layer, important features include real-time suitability by reservation of guaranteed time slots, collision avoidance through CS-MA/CA and integrated support for secure communications. Devices also include power management functions such as link quality and energy detection [213]. In the MAC frame, the first two bytes indicate the type of the frame. There are four types of frames defined in the standard [113]. The first ones, beacon frames are used by coordinators to describe the way the channels can be accessed. Then, there are data frames and acknowledgments sent as responses for control and data packets. The last, control frames are used for network management, such as associations or disassociations [213]. The next flag, the sequence number, is used by acknowledgments and refers to the received frame. Then, there are addressing pieces of information, related to source, destination addresses and security. The last two bytes represent the checksum, used for data integrity.

■ Device roles: The devices are classified into two categories: full function (FFD) and reduced function devices (RFD). An RFD can communicate only with an FFD, while an FFD can communicate with both types. The latter have, generally, the role of coordinators. They keep a routing table and are responsible with network creation and maintenance. FFD and RFD can form different topologies including star, mesh and tree structures.

Bluetooth Low Energy (BLE): Bluetooth Low Energy, also known as "Bluetooth Smart," was developed by the Bluetooth Special Interest Group . The BLE protocol stack is similar to the stack used in classic Bluetooth technology, but it supports quicker transfer of small packets of data (packet size is small) with a data rate of 1Mbps. BLE supports 40 channels with 2 MHz channel bandwidth (double of classic Bluetooth) and 1 million symbols/s data rate. BLE supports low duty cycle requirements as its packet size is small and the time taken to transmit the smallest packet is as small as 80 μs. As shown in Figure 9.5, BLE uses the same spectrum as IEEE 802.15.4, but have their own modulation scheme, bit rate, channel map and channel spacing, and upper layers as stated in their comparison in [204]. Unlike IEEE 802.15.4, which is restricted to the physical and MAC layers, BLE is a full protocol stack. However, BLE can only form star topologies and not mesh such as IEEE 802.15.4. Despite different protocols, devices supporting both already exist. For example, Redpine Signal's RS9113 also supports 2.4 GHz/5GHz 802.11b/g/n WiFi in addition to a dual mode Bluetooth 4.0 and 802.15.4-based ZigBee.

IEEE 802.15.4 does not include specifications on the higher layers of the protocol stack, and considering IP is not so straightforward. The reason is that the largest physical layer packet in IEEE 802.15.4 has 127 bytes, which implies the maximum frame size at the media access control layer is 102 octets, which may further decrease based on the link layer security algorithm utilized. Such sizes are too small when

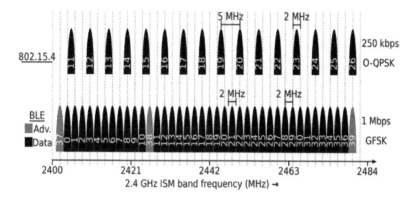

Figure 9.5: The BLE and 802.15.4 channel maps share the 2.4 GHz ISM spectrum [204].

compared to typical IP packet sizes. As a result, other standards such as ZigBee, 6LoWPAN, and others have extended the standard by developing the upper layers.

6LowPAN: 6LowPAN is developed by an IETF working group especially for small networks and embedded devices interconnected by IEEE 802.15.4 in RFC4944 [38]. It maintains an IPv6 network, but with compressed headers. A new layer is added, between the network and data link, responsible with fragmentation, reassembly, header compression and data link layer routing for multi-hop.

ZigBee: Zigbee builds on the physical layer and media access control defined in IEEE standard 802.15.4. The specification addressed the rest of the upper layers by defining four additional key components: network layer, application layer, Zigbee Device Objects (ZDOs) and manufacturer-defined application objects. ZDOs are responsible for some tasks, including keeping track of device roles, managing requests to join a network, as well as device discovery and security.

The resulting topologies of IEEE 802.15.4-based medium-range networks include star, with a central FFD and several RFDs that communicate with it, peer-to-peer (mesh), where there is a coordinator, several FFDs and RFDs and cluster, having a coordinator, a FFD and several RFDs. BLE only supports star topologies in a similar fashion as short range infrastructure protocols. Whichever the topology, the data is typically relayed to a sink node typically connected to the Internet and to a centralized server storing and acting upon the collected data. This is illustrated in Figure 9.6.

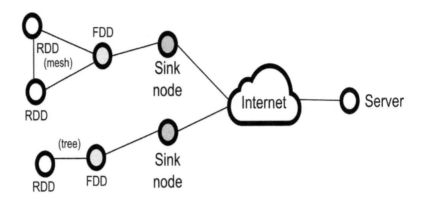

Figure 9.6: Typical medium-range infrastructure protocol topology.

9.4.1.3 Long Range

For the long range protocols, different technologies for Low-Power Wide-Area (LPWA) network have recently been proposed. LPWA technologies are designed for a wide area coverage and an excellent signal propagation to hard-to-reach indoor places such as basements. Quantitatively, a +20 dB gain over legacy cellular systems is targeted. This allows the end-devices to connect to the base stations at a distance ranging from a few to tens of kilometers.

In LPWA, low power operation is achieved by opportunistically turning off power hungry components of the devices such as the data transceiver. Radio duty cycling allows LPWA end devices to turn off their transceivers, when not required. Only when the data is to be transmitted or received, the transceiver is turned on. As a result, LPWAs are suitable for smart cities or smart agriculture projects, where applications need to send little amount of data over large distances.

As recently as early 2013, the term 'LPWA' did not even exist [28]. However, as the network of things market rapidly expanded, LPWA became one of the faster growing spaces. Many of the LPWA technologies have arisen in both licensed and unlicensed markets. Among them, LoRa and NB-IoT are the two leading emergent technologies [257]. A side-by-side comparison of LoRa and NB-IoT is provided in [257], while SigFox is overviewed in [230].

LoRa / LoRaWAN: LoRa is an emerging technology in the current market, which operates in a non-licensed band below 1 GHz for long range communication link operation. LoRa is a proprietary spread spectrum modulation scheme. LoRa is its first low-cost implementation for commercial usage. The name LoRa comes from its advantage of long-range capability. LoRaWAN defines the communication protocol and the system architecture, while LoRa defines the physical layer. On the one hand, LoRa features low power operation (around 10 years of battery lifetime), low data rate (27 kbps with spreading

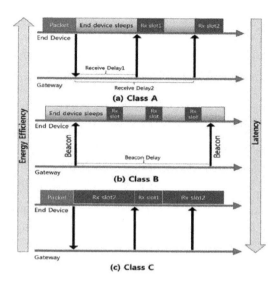

Figure 9.7: Device classes in LoRa.

factor 7 and 500 kHz channel or 50 kbps with FSK) and long communication range (2-5 km in urban areas and 15 km in suburban areas). The payload of the packet has a maximum size between 51 Bytes and 222 Bytes, depending on the configured duty cycle [20]. On the other hand, LoRaWAN uses long range star architecture in which gateways are used to relay the messages between end devices and a central core network. In a LoRaWAN network, nodes are not associated with a specific gateway. Instead, data transmitted by a node is typically received by multiple gateways. The network server has the required intelligence for filtering the duplicate packets from different gateways. End nodes in the LoRaWAN network can be divided into three different device classes according to the trade-off between network downlink communication latency versus battery life. As illustrated in Figure 9.7, three different MAC protocols were designed for these three device classes.

- **Class-A** end-devices are battery powered sensors. It has maximum battery lifetime and must be supported by all LoRa devices. All end-devices start and join the network as end devices of Class A and can then decide to switch to other classes. Class A devices use pure ALOHA access for the uplink. After sending a frame, a Class A device listens for a response during two downlink receive windows. Downlink transmission is only allowed after a successful uplink transmission [20].

- **Class-B** end-devices are battery powered actuators. Class B devices are designed for applications with additional downlink traffic needs [20]. These devices are synchronized using periodic beacons sent by the gateway to allow the schedule of additional receive windows for downlink traffic.

■ **Class C** end-devices are the main powered actuators. It has the minimum latency in downlink communication compared to the other two classes. Class C end-devices not only open two receive windows as Class A but also open a continuous receive window until the end of transmission.

NB-IoT: NB-IoT is a new IoT technology set up by 3GPP as a part of Release 13; therefore, it is a technology that will likely be offered by established mobile service providers. Although it is integrated into the LTE standard, it can be regarded as a new air interface [257]. It uses the licensed frequency bands, which are the same frequency numbers used in LTE, and employs QPSK modulation. There are 12 subcarriers of 15 kHz in downlink using OFDM and 3.75/ 15 kHz in uplink using SC-FDMA. NB-IoT core network is based on the evolved packet system (EPS), the core network of the LTE system. Thus, for NB-IoT, the existing evolved UMTS terrestrial radio access network (E-UTRAN) architecture and the backbone can be reused (the protocol stack for NB-IoT is the general fundamental protocol stack of LTE). However, the NB-IoT is based on a new air interface. In the NB-IoT air interface, the maximum packet size is 1600 Bytes.

SigFox: SigFox is a proprietary and patented technology [230]. Exclusive SigFox Network Operators (SNOs) deploy the proprietary base stations equipped with cognitive software-defined radios and connect them to the backend servers using an IP-based network. The end devices connect to these base stations using Binary Phase Shift Keying (BPSK) modulation in an ultra narrow (100Hz) SUB-GHZ ISM band carrier. It uses free sections of the radio spectrum (ISM band) to transmit its data. Sigfox focuses on using very long waves. Thus, the range can increase to a 1000 kms. The cost is bandwidth. Indeed, all these benefits come at an expense of maximum throughput of only 100 bps. It can only transmit 12 bytes per message, and a device is limited to 140 messages per day. Also, SigFox initially supported only uplink communication but later evolved into a bidirectional technology, although with a significant link asymmetry [230]. The downlink communication can only precede uplink communication after which the end device should wait to listen for a response from the base station.

Table 9.3 summarizes and compare the key properties of the existing long-range infrastructure protocols for the network of things.

Although a very different application domain, the topologies in LPWA long-range communications networks are not very different than the ones used in short-range infrastructure protocols. The reason for the latter, is that the existing long-range protocols do not allow devices to communicate on a peer-to-peer fashion. As a result, devices and gateways form a star topology where data is relayed to and from a centralized server via the gateways. This is illustrated in Figure 9.8.

Table 9.3: Long range infrastructure protocols.

Parameters	LoRa	NB-IoT	SigFox
Deployment Control	High (the user is responsible for deploying and operating the gateways)	Medium (the mobile service provider will probably have the base stations deployed)	Medium (exclusive SigFox operators are responsible for deploying the gateways)
Spectrum	Unlicensed	Licensed LTE	Unlicensed
Modulation	CSS	QPSK	BPSK (UL), FSK (DL)
Bandwidth	500 Hz - 125 kHz	180 kHz	100 Hz
MAC	unslotted ALOHA	LTE-based	unslotted ALOHA
Peak Data Rate	290 bps/ 50 kbps (UL/DL)	204/234 kbps (UL/DL)	100/600 bps (UL/DL)
Payload Size	250 bytes	1600 bytes	12 bytes (UL), 8 bytes (DL)
Power Efficiency	Very high	Medium high	Very high

Table 9.4 summarizes the key properties of the existing infrastructure protocols for the network of things.

Table 9.4: Infrastructure protocols in the network of things.

Object Type	Technology	Comms	Range	Lifetime	Standard
Passivel	RFID, NFC	Asymmetric (tag reader)	10, 1m	Indefinite, Indefinite	ISO 18000, ISO/IEC 18092
Smart	WSN	Peer-to-Peer (sink node)	100m	3 years	IEEE 802.15.4, 6LoWPAN, BLE
Smart	LPWAN	Asymmetric (gateway node)	15 km	10 years	LoRa, NB-IoT, SigFox

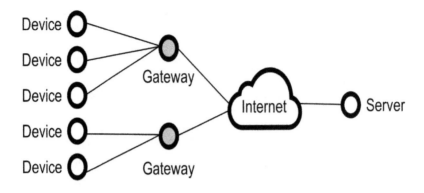

Figure 9.8: Typical long-range infrastructure protocol topology.

9.4.1.4 Application Protocols

Different application protocols have also been specified to be run typically over medium and long-range infrastructure protocols. While applications protocols target sensing, acting, and management commands, RFID systems typically transmit simple binary data encoding an identifier and do not require major intelligence on the device side.

Constrained Application Protocol (CoAP): The IETF Constrained RESTful Environments (CoRE) working group created CoAP, which is an application layer protocol for network of things applications [251] [49]. CoAP is defined in RFC 7252 and aims to enable tiny devices with low power, computation and communication capabilities to utilize RESTful (client/server) interactions. The protocol defines a web transfer protocol on top of HTTP functionalities. In general, REST represents a simpler way to exchange data between clients and servers over HTTP. REST can be seen as a cacheable connection protocol that relies on stateless client-server architecture while enabling clients and servers to expose and consume web services. It is used within mobile and social network applications and it eliminates ambiguity by using HTTP get, post, put, and delete methods. Unlike REST, CoAP is bound to UDP (not TCP) by default which makes it more suitable for the network of things applications. CoAP detects duplications and provides reliable communication over the UDP transport layer using exponential backoff since UDP does not have a built-in error recovery mechanism. Furthermore, CoAP modifies some HTTP functionalities to meet the network of things requirements such as low power consumption and operation in the presence of lossy and noisy links. CoAP utilizes four types of messages: confirmable, non-confirmable, reset and acknowledgment. In CoAP's non-confirmable response mode, the client sends data without waiting for an ACK message. CoAP uses a simple and small format to encode messages. The first and fixed part of each message is four bytes of header. Then a token value may appear whose length ranges from zero to eight bytes. The token value is used for correlating requests and responses. The options and payload are the next optional fields. A typical CoAP message can be between 10 to 20 bytes [89].

Message Queue Telemetry Transport (MQTT): MQTT is a messaging protocol aimed at connecting embedded devices and networks with applications and middleware [185]. Instead of a client/server model, MQTT utilizes the publish/subscribe scheme to provide transition flexibility and simplicity of implementation. MQTT is built on top of the TCP protocol and delivers messages through simple routing mechanisms (one-to-one, one-to-many, many-to-many) and three different levels of QoS. MQTT simply consists of three components, subscriber, publisher, and broker. An interested device would register as a subscriber for specific topics in order for it to be informed by the broker when publishers publish topics of interest. The publisher acts as a generator of interesting data. After that, the publisher transmits the information to the interested entities (subscribers) through the broker. The maximum packet size is 256MB, but small packets less than 127 bytes have a 1 byte packet length field which is recommended for IoT applications. An alternate version of MQTT has been developed for sensor networks and is of particular interest for the network of things. MQTT-SN (SN for Sensors Network) offers a few advantages over MQTT for embedded devices such as reduced header in negotiation phase and UDP as underlying layer. The disadvantage is that a gateway is necessary.

Extensible Messaging and Presence Protocol (XMPP): XMPP is an IETF instant messaging (IM) standard that is used for multi-party chatting, voice and video calling and telepresence [238]. XMPP was developed by the Jabber open source community to support an open, secure and decentralized messaging protocol. XMPP allows IM applications to achieve authentication, access control, privacy measurement, hop-by-hop and end-to-end encryption, and compatibility with other protocols. Many XMPP features make it a preferred protocol by most IM applications and relevant within the scope of the network of things. It runs over a variety of Internet-based platforms in a decentralized fashion.

Advanced Message Queuing Protocol (AMQP): AMQP is an open standard application layer protocol for the network of things focusing on message-oriented environments [282]. AMQP requires a reliable transport protocol like TCP to exchange messages. Communications are handled by two main components: exchanges and message queues. Exchanges are used to route the messages to appropriate queues. Routing between exchanges and message queues is based on some predefined rules and conditions. Messages can be stored in message queues and then be sent to receivers. Beyond this type of client/server point-to-point communication, AMQP also supports the publish/subscribe communications model.

Data Distribution Service (DDS): Data Distribution Service (DDS) is a publish/-subscribe protocol for real-time M2M communications [17]. In contrast to other publish-subscribe application protocols like MQTT and AMQP, DDS relies on a brokerless architecture and uses multicasting to bring excellent Quality of Service (QoS) and high reliability to its applications. Its broker-less publish-subscribe architecture suits well to the real-time constraints for IoT and M2M communications. DDS supports 23 QoS policies by which a variety of communication criteria like security, urgency, priority, durability, reliability, etc. can be addressed by the developer.

Table 9.5 summarizes and compares the different application protocols described in this section.

Table 9.5: Application protocols in the network of things.

Application Protocol	RESTful	Transport	Publish/ Sub-scribe	Request/ Re-sponse	Security	Header Size (byte)
CoAP	Yes	UDP	Yes	Yes	DTLS	4
MQTT	No	TCP	Yes	No	SSL	2
MQTT-SN	No	UDP	Yes	No	SSL	2
XMPP	No	TCP	Yes	Yes	SSL	-
AMPQ	No	TCP	Yes	No	SSL	8
DDS	No	UDP	Yes	No	DTLS	-
HTTP	Yes	TCP	No	Yes	SSL	-

9.4.1.5 Routing Protocols

Short-range and long-range networks can only be described by star topologies where a central node can maintain an updated view of the topology even in the presence of node mobility. This is not so straightforward in medium-range IEEE 802.15.4 mesh or tree topologies that requires of distributed routing solutions to transmit data through intermediate nodes (multi-hop).

■ Routing Protocol for Low Power and Lossy Networks (RPL): The IETF routing over low-power and lossy links working group standardized a link-independent routing protocol based on IPv6 for resource-constrained nodes called RPL [275]. RPL was created to support minimal routing requirements through building a robust topology over lossy links. This routing protocol supports simple and complex traffic models like multipoint-to-point, point-to-multipoint and point-to-point [25]. A Destination Oriented Directed Acyclic Graph (DODAG) represents the core of RPL that shows a routing diagram of nodes. The DODAG refers to a directed acyclic graph with a single root. Each node in the DODAG is aware of its parents but they have no information about related children. Also, RPL keeps at least one path for each node to the root and preferred parent to pursue a faster path to increase performance. In order to maintain the routing topology and to keep the routing information updated throughout the network (neighbor metrics, parent node, etc.), RPL uses four types of control messages [275].

9.5 Toward a DTN of Things

In this section, we will argue that most of the presented protocols for the network of things have assumed a relatively stable end-to-end connectivity. Even in the presence of mesh topology variations, routing solutions have sought to guarantee a continuous connectivity with all nodes in the network. Despite any routing update effort, node mobility and power constraints might provoke link instabilities and network partitions difficult or impossible to overcome. In these cases, automated store and forward solutions such as DTN can become an appealing approach to connect the network of things.

9.5.1 *Link Instability*

In general, short, medium and long range infrastructure protocols have assumed that the gateway nodes bridging the things with the server side (tag readers, sink nodes, and LPWA gateways) are relatively stable nodes. Therefore, the core of the network of things topology has been assumed to remain fixed and available through time. On the other hand, the edge of the network including the connected objects were allowed to evidence certain mobility and thus, topology variations.

When using short range protocols such as tag readers, edge node mobility will not affect the application as long as the tags are reachable from any tag reader at all times. When a tag is read by a new reader, the tag location is sent and updated on the server, allowing to identify and track the object. More intelligence is required to keep track of the topology in long range protocols where sensing and acting data not only flow from gateways to server but also from server to the devices. Moreover, medium-range protocols like IEEE 802.15.4 designed to operate on mesh or tree multi-hop topologies have to rely on specific routing solutions such as RPL to properly update the topological information among nodes.

Although mobility is a known and studied property in the network of things domain, temporal disconnections or disruptions have received little or no attention. Indeed, **disruptions** are consequence of extreme mobility where nodes goes beyond the reachability area of the closest neighbor. While it has been assumed that small

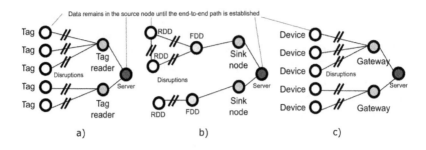

Figure 9.9: Link disruption in the edge of the network of things using a) long, b) medium, and c) long range protocols.

disruptions can happen on edge nodes (the things), more complex scenarios are possible when disruptions occur in the inner part of the network. These require particular attention and are discussed throughout this section.

9.5.1.1 Store and Forward: Single-Hop

In period of disruption between the source and the receiver nodes (typically on the link connecting the device with the gateway as shown in Figure 9.9), existing protocol stacks have sought to discard the data or to temporarily store it on the source node's application layer (either on the device or on the server side). In other words, when the application is informed by the underlying layer that a given node is not reachable, it has to refrain from transmission, decide upon the unexpected situation and eventually retry at a later time. In these cases, a copy of the data can remain in the sender node until a stable path is established with the destination. Since the storing process happens only once along the end-to-end data path, we call this way of handling data as single-hop store and forward. Most application protocols are prepared to use single-hop store and forward on the source node to tolerate this type of unwanted disruption.

9.5.1.2 Store and Forward: Multi-Hop

While single-hop store and forward can be applied at the application layer on the source node, it is not enough to tolerate a more general case of mobility and disconnections involving the gateways or other intermediate node in the system. In particular, there might be scenarios where a continuous end-to-end path might never be established because of a highly disrupted topology.

To illustrate this case let's explore some examples:

■ Consider a mobile gateway that can sporadically establish contact with the things part or the Internet part of the network of things. This would be the case of a NFC or IEEE 801.15.4 enabled smartphone in our pocket that collects data from the connected things as we walk and then delivers it to the server when reaching a WiFi router connected to the Internet. Although gateway mobility can extend the network reachability, it has not been considered in existing network of things solutions. This example involves star topologies.

■ Another store and forward example involving intermediate nodes is the case of road traffic control applications with limited connectivity. Vehicles recording important information on road status may need to store/carry data from other vehicles until they find a suitable opportunity to forward the data (either a road-side equipment or another vehicle). In any of these cases link instability would require intermediate nodes to temporarily store in-transit date. Another similar case is military networks where nodes (e.g., tanks, airplanes, soldiers) may move randomly and are subject to being destroyed. Nodes can be networked using mesh technologies (also rely on peer to peer intermediate nodes) similar to those addressed by IEEE 802.15.4 solutions. The eventual removal or unexpected movement of a node out of range of the neighbors can suddenly partition the network in several simultaneous parts of the topology. These examples involve peer-to-peer mesh topologies.

In these cases and many others, if store and forward is only applied at the application layer on the devices on the edge of the network, data might never have a stable end-to-end path to reach the destination, thus, drastically lowering the performance of the network of things. Instead, store and forward can occur on a multi-hop basis, thus overcoming cases of severe disruption and network partition. This means that intermediate nodes or mobile gateways should be able to receive in-transit data and store it until a suitable link becomes available with a next hop. Since the process of storing data on nodes can now happen several times along the data path, we call this approach multi-hop store and forward.

Multi-hop store and forward allows several disruptions to happen simultaneously in any of the network of topology (Figure 9.10). However, existing Internet-based network of things protocols are not designed to operate in these scenarios. Furthermore, it is worth noticing that network of things applications operating under the store and forward paradigm must also be delay tolerant. In other words, they must not assume the destination will be immediately available to respond.

9.5.2 The DTN Architecture

None of the existing IoT protocols are prepared to execute automated multi-hop store and forward. A different architecture already exists for this type of scenario.

Delay Tolerant networks (DTN) are mobile networks that may never have end-to-end stable path [152]. The characteristics of DTN are different from traditional ad-hoc networks in the sense that in ad-hoc protocols mobility and topology variations are considered, but network connectivity is never disrupted or partitioned. In other words, every node in the topology is reachable at all times. Indeed, DTN architecture is thought to cope with a more general case where topology evolves through time and nodes might be disconnected for arbitrary periods of time.

Researchers have proposed new routing algorithms and protocols to support

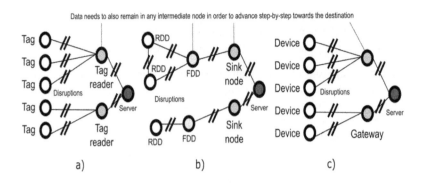

Figure 9.10: Link disruption provoked by gateway or intermediate node mobility using a) long, b) medium, and c) long range protocols.

DTN challenges (mobility, transient connection, long-delay delivery, etc.) [41]. Every DTN routing scheme is based on a multi-hop store and forward pattern by using a dedicated layer known as Bundle Protocol (BP) [244] (Figure 9.11). This layer allows intermediate mobile nodes to store local or in-transit data till it finds an appropriate relay node (to forward the message) in the path towards the destination. The BP provides addressability, a basic traffic flow control known as custody transfer, among other features that can replace layer 3 and 4 (IP and TCP) functionalities in some specific cases. It has also been designed to operate over any existing infrastructure protocol by means of Convergence Layer Adapters (CLAs), including those overviewed for the network of things.

BP has been implemented for space applications in Interplanetary Overlay Network (ION) [52], for research in the implementation of the DTN Research Group in DTN2, for utilization in Android and Linux platforms in IBR-DTN [244], for resource-constrained wireless sensor nodes in μDTN among others.

Enabling delay tolerant communication in the network of things will allow mobile smart objects to better communicate even with the presence of disruption in their connectivity. Specifically, DTN allows intermediate nodes (possibly other networked objects) to act as data-mules to collect data from sensors and carry it until being in proximity of a network infrastructure or a better intermediate node. A detailed survey studying the opportunities and challenges in using DTN strategies in the network of things can be found in [41].

9.6   Experiences with the DTN of Things

In the past few years, significant research efforts have been made on the network of things to widen its application domains and overcome its constraints. One of the most challenging constraints is prolonging network lifetime considering resource limitation while allowing sporadic connectivity in order to deliver data all over the network

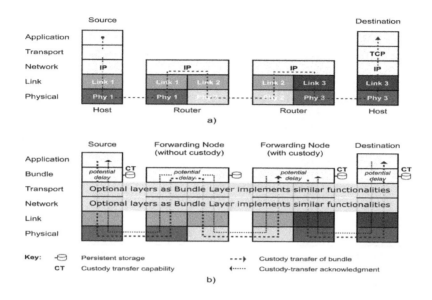

Figure 9.11: Internet data flow in a) and DTN data flow in b).

to the final destination. Also, maintaining a stable connectivity in sparse networks (mostly rural and urban environments) presents significant challenges. In this cases, node's mobility and tolerance to disruptions are a way to overcome resource constraints and network disconnection [41].

Several experiences reported in the first decade of the 2000s have used DTN techniques in the WSN context, without using the standardized BP. One early example of such projects is ZebraNet [157], which aims at tracking wildlife in Kenya. Seal-2-Seal [180] tracks contacts between wild animals, while LUSTER [246] aims at monitoring environmental parameters to be used by ecologists. Also DTNLite was presented in [32] as an implementation of DTN concepts on TinyOS with a proprietary protocol. All those projects use a proprietary delay tolerant communication protocol, whereas also the BP would have been suitable. These early experiences showed, that a need for delay tolerant communication in WSNs exist.

With the publication of the first draft of BP in 2007 [52], successive experiences adopted the DTN terminology with the objective of integrating DTN into the network of things ecosystem traditionally based only on IoT protocols. In this section, we will survey these existing experiments on the DTN of Things. Some of them motivate different application domains and optimization models, while others focused on developing and validating real implementations.

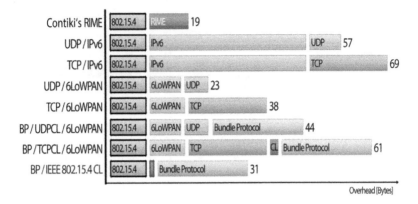

Figure 9.12: IEEE -802.15.4 protocol overhead comparison including BP [217].

9.6.1 Applications and Models

2012 - Delay-Tolerant Data Elevator: Data elevator has been introduced by Pottner et al. [217] to apply the bundle protocol in wireless sensor networks (medium-range infrastructure protocols). Authors demonstrate how the solution can be useful for resource-limited sensors, even for 8-bit based processors. DTN nodes were configured to seamlessly interact with standard BP implementations running on standard PCs. A sensor that was installed on the roof of a 15-story building and opportunistically connects to an elevator which transports bundles carrying measured values to a receiver on the 3rd floor provides a useful use case for smart homes and building automation systems. Authors claim that BP can be used as layers 3 and 4 of the protocol stack, whereas an overlay network is not required. This allows using the BP with a simple convergence layer directly inside IEEE 802.15.4 radio frames, thereby significantly reducing overhead. Figure 9.12 illustrates that the overhead of using BP on top of IEEE 802.15.4 is comparable with existing non-delay tolerant network of things protocols. Indeed, communication overhead of the BP is in the same dimension as existing non-delay-tolerant communication protocols but offers the additional functionality of efficiently handling network disruptions. Authors have validated this concept by implementing a full-featured BP stack for Contiki, which is interoperable with existing BP compliant DTN implementations on Linux.

2013 - Integrated Delay-Tolerant RFID and WSN: Al Turjman et al. have proposed in [26] a combined RFID and sensor platform (i.e., short and medium-range protocols) which uses delay tolerance in case of intermittent connections. Authors explore a network topology where tags are read by several tag readers which in turn are connected via a multi-hop and disruption-prone wireless sensor network to the sink node. They motivate the work by assuming tags provide sensing data to the readers and that courier nodes with high mobility are available to store-carry and forward sensed data. Although results are based on simulation models, authors present optimal localization methods and prove that the DTN scheme can be applied to interconnect very sparse network of things deployments were persistent end-to-end connectivity cannot be guaranteed. In these scenarios, courier nodes can efficiently provide autonomous store carry and forward capabilities by using DTN protocols that automatically select the best courier at any moment.

2014 - Delay-Tolerant Environmental Sensing for Smart Cities: The work in [105] proposes an architecture to interconnect standard based M2M platforms to opportunistic networks in order to communicate with these energy-constrained sensor devices. Authors argue that deploying an on-purpose infrastructure to collect and forward the information from the sensors through large spaces such as a city (smart city application domain) might be costly or unfeasible. Also, in some cases, data monitored from sensors might be not critical, and its transmission to a service platform can be simply delayed or even transferred few times a day without endangering the consistency and sense of the information (trash collection optimization, street furniture maintenance or environmental monitoring). In these cases, providing ways for opportunistic networking would be the alternative. To this end, authors propose an IEEE 802.15.4-based architecture where instead of having a continuously enabled transceiver, a secondary channel in the UHF band is used to opportunistically wake up the devices. The network topology is composed of mobile gateways placed in public buses that go through the city waking up the devices and collecting data. The concept was validated in a pilot deployment in Spain.

2014 - Augmented Reality via Opportunistic Communications: Authors in [294] introduced the discussion on Challenged IoTs. By using Bluetooth Low Energy (BLE) protocol over IEEE 802.11, challenged objects can convey data to mobile users with a smartphone to opportunistically enable augmented reality features. Especially, BLE saves device resources while enabling continuous discovery of smart objects and provides viable throughput performance. Authors base their study on the observation that smart object functionality is largely relevant in the spatial vicinity of an object, e.g., in smart buildings or personal appliances, and hence does not necessarily depend on Internet connectivity. Indeed, this work enables smart objects to define and directly communicate the interface required to interact with their functionality. As a result, users can receive and visualize the respective interfaces and interact with all encountered objects within a single generic app on a smartphone. Authors implemented a prototype in Android and iOS operating systems and evaluated the performance proving the real-life applicability of the concept.

9.6.2 Implementations

2012 - IBR-DTN: a DTN stack for embedded devices: In [200] [201] authors introduced IBR-DTN, an implementation of DTN for embedded systems from smartphones up to PCs or Notebooks. Specifically, IBR-DTN was aimed at turning WLAN (IEEE 802.11) access point into a stand-alone DTN-node for mobile applications. IBR-DTN was successfully tested on various low cost access points e.g. Netgear WGT634U, Linksys WRT54G3G and even on an ultra low budget FON FON2200. Authors validate that IBR-DTN operates very efficiently and consumes few memory resources. Also, the modular software design of "IBRDTN" is centered on the efficient use of resources and interoperability with the DTN2 reference implementation. Indeed, IBR-DTN has been used as a starting point for many other DTN of Things implementation efforts discussed below.

2012 - μDTN: a DTN stack for Wireless Sensor Networks: In [309], authors present μDTN, a DTN implementation for Contiki OS used in the data elevator experiment discussed above. μDTN is compatible to Contiki's network stack which is able to handle disruptions without packet loss. Furthermore, it is interoperable with BP implementations running on Linux. In contrast to all other approaches, μDTN does not build upon traditional transport and network-layer protocols like TCP/IP. It is located directly above the MAC-Layer and is designed around the IEEE 802.15.4 convergence layer. The support of multiple convergence layers has been neglected in favor of higher performance and code efficiency. Also the Bundle Protocol specifications were necessary: Compressed Bundle Header Encoding (CBHE) was implemented to operate with smaller bundles that are easier to process and have less communication overhead. Fragmentation is not supported, but IP-based network discovery was adapted to work on IEEE 802.15.4 frames. On the routing side, a modified version of flooding routing was implemented (the quantity of copies created is bounded). Authors concluded that μDTN can handle network disruptions without losing application data. Interoperability was also evaluated with IBR-DTN.

2015 - CoAP over BP: The work in [36] addresses the implementation of a BP binding for CoAP as a means to enable Delay Tolerant IoT or DTN of Things. Similarly to [275], authors motivate the application on smart cities where sensors and actuators (traffic detectors, instruments for measuring the air pollution, the noise level, the temperature, etc.; and also actuators coupled with valves, gates, signs, etc) are connected to network of things (Figure 9.13). However, a full mesh ensuring the flow of information between the network of sensors, the actuators and the central servers over an entire city at all times is unrealistic due to network partitioning. As a result, the end-to-end connectivity assumption on which CoAP relies may no longer be valid. However, the CoAP protocol itself can operate without immediate response. To tackle the latter, authors developed their own implementation of CoAP, named BoAP,

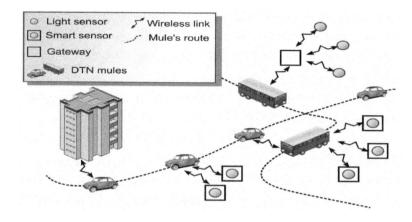

Figure 9.13: DTN mules in a smart city environment [200].

with the objective to firstly include a BP binding. BoAP was implemented as part of the existing DTN implementation in IBR-DTN [200]. Preliminary tests allowed authors to show that the BP can be an effective substitute to UDP as a CoAP binding: BoAP does not largely degrade transmission delays when disconnections are short, and, contrary to CoAP/UDP, it continues to play its role when the connectivity is strongly intermittent.

2015 - DTN Nano: a DTN stack for smart cities: The work in [229] aims to define protocol architectures to respond to the heterogeneity that may arise in a smart context with several devices with several mobility behaviors. Similarly to [275] and [36], the application domain of the delay-tolerant approach towards network of things is smart cities. Indeed, authors propose a test scenario based on fixed sensors and actuators that generates and receives data by means of mobile gateways placed on vehicles or carried by people. In the proposed architecture, IPv6 solutions such as 6LoWPAN are replaced by the Bundle Protocol that actually allows naming all entities in a network. Authors propose nanoDTN, a streamlined implementation based on a reduced version of μDTN. They kept the operating system Contiki using the C language but removed some functionalities such as reporting, custody management, redundancy functions. A single convergence layer adapter was left (IEEE 802.15.4) and flooding routing was the only one allowed together with FREAK. FREAK (Frequency Routing, Encounters And Keenness) is a new routing solution aimed at keeping track of encounter only with the gateways, thus allowing a reduced memory footprint than other probabilistic solutions such as PRoPHET.

2017 - MQTT over BP: The work in [189] focuses on the publish/subscribe messaging transport provided by MQTT protocol. Similarly to CoAP, MQTT does not require a persistent end-to-end connection, but TCP, its underlying protocol does. Authors propose an architecture to increase the robustness of MQTT by integrating a DTN approach. Author highlights that MQTT, does not behave well in mobile scenarios when sensing devices face periods of disconnection followed by a re-connection (possibly on a different sub-network). This issue may be detrimental to the wide adoption of this standard. As a result, they combined MQTT for Sensor Networks (MQTT-SN) with the BP implementation in IBR-DTN [200]. Authors presented experimental results. The set-up included two Raspberry Pi 2 Model B (RPi acting as MQTT-SN gateway) devices and seven Zolertia Re-Mote Sensor Board (motes acting as MQTT-SN clients). The chosen infrastructure protocol was 6LoWPAN. The performance of the DTN-based backbone network was measured under different message sizes and configurations. Bundle processing times and interarrival times were found acceptable for general network of things applications, considering that with this new architecture end-to-end applications do not have to worry about the possible nodes' disconnections periods. However, it is highlighted that scalability can be a limitation in systems with hundreds of nodes.

9.6.3 Persistent Challenges

Despite the presented experiments, several challenges remain to be tackled. Some of them offer interesting research opportunities and even potential business cases in the ever growing network of things market.

Scalability: Most of the existing Internet-based IoT protocols have sought to release the device from making complex calculations. This has been achieved by moving as much functionality to gateways nodes which are always reachable from the devices. Unfortunately, this is not possible in disrupted or partitioned networks where devices can spend significant amounts of time disconnected from the rest of the network. Nodes are thus required to implement routing, addressing, and other functionality so it can operate in a stand alone fashion. Besides demanding more resources, this makes scalability a big issue in DTN of Things. Indeed, the larger the deployment, the more processing power nodes will need to properly perform the required routing or addressing calculations. Designing efficient and scalable solutions to tackle such functionalities is among the most interesting challenges toward enabling large DTN of Things network deployments.

Security: DTN of Things requires data to be stored an unbounded period of time in intermediate devices which might not be trusted. This means that sensible data will be present and potentially accessed by harmful entities in a remote memory storage threatening even the most advanced cryptographic technologies [150]. Secure Bundle Protocol or BPSec was specifically designed and standardized to cope with security in DTNs [268]; however, it has never been tested nor validated in large DTN deployments so it is difficult to assess its robustness. Security will certainly need particular attention and be placed under scrutiny in order to secure future DTN of Things deployments.

Cultural: Store and forward data handling is obviously based on leaving data temporarily stored in intermediate devices. Imagine that these devices can be our own smart phone. This implies that our own device must be able to receive data from a potentially unknown neighbor, store it, carry it for some time, and then transmit it to Internet when available. How would you feel to use your phone memory and energy for this? Are you willing to share your resources for others? While not a problem in controlled and private network of things deployments, there are cultural factors that might limit a full adoption of DTN of Things.

9.7 Summary

In this chapter we have discussed the concept of DTN of Things. Together with Internet of Things (IoT), they can operationalize the present and future network of things. Objects we use on a day to day basis will become more and more connected, making significant changes in our way of life.

In this context, DTN of Things was introduced as means to enable networking over highly disrupted network of things, such as those expected to be deployed in very large areas (building, cities, rural). Furthermore, allowing disruptions to happen will also open the way to temporarily disable devices in medium and close range communications. Indeed, by exploiting a store and forward approach, high node mobility and aggressive power saving features can be embraced in future network of things.

However, store and forward data transmission leads to a list of non-trivial challenges that need to be addressed before the DTN of Things can be operationally applied. These range from security up to cultural factors.

Chapter 10

DTN Congestion Control

Aloizio P. Silva

University of Bristol

Scott Burleigh

JPL-NASA/Caltech

Katia Obraczka

University of California Santa Cruz

CONTENTS

10.1   Introduction

Arbitrarily long delays and frequent connectivity disruptions that set DTNs apart from traditional networks imply that there is no guarantee that an end-to-end path between a pair of nodes exists at a given point in time. Instead, nodes may connect and disconnect from the network over time due to a variety of factors such as node mobility, wireless channel impairments, power-aware mechanisms (e.g., nodes being turned off periodically or when their batteries are low), security concerns, etc. Consequently, in DTNs, the set of links connecting nodes, also known as "contacts," varies over time. This fundamental difference between DTNs and conventional networks results in a major paradigm shift in the design of core networking functions such as routing, forwarding, congestion- and flow control.

The DTN architecture described in [107] uses the so-called *store-carry-and-forward* paradigm, as opposed to the Internet's *store-and-forward*, to deliver messages from source to destination. In *store-carry-and-forward*, nodes store incoming messages and forward them when transmission opportunities arise. Note that in traditional networks, nodes also store messages before forwarding them; however, the time scales at which data is stored locally while waiting to be forwarded are typically orders of magnitude smaller when compared to DTNs. Therefore, DTN nodes need persistent storage capabilities to be able to store data for arbitrarily long periods of time while waiting for an opportunity to either deliver the data to its final destination or forward it to another node.

According to the *store-carry-and-forward* paradigm, when a DTN node "encounters" another DTN node, it decides whether to forward messages it is carrying to the other node. Therefore, the concept of *links* in traditional networks (wired or wireless) is replaced with the notion of *contacts*. In scenarios where these encounters are random, *store-carry-and-forward* is also referred to as *opportunistic forwarding*. On the other hand, when contacts are known a priori (e.g., in deep space communication applications), *store-carry-and-forward* is known as *scheduled forwarding*. Finally, there are scenarios where node encounters follow a probability distribution based on past history; in these cases, *store-carry-and-forward* is based on *probabilistic forwarding* [194]. Note that, since *contact times* are finite and may be arbitrarily short, a node may need to choose which messages to forward based on some priority; a node may also decide whether the new neighbor is a "good" candidate to carry its messages. A node's "fitness" as a relay for a particular message depends on several factors that can be dependent or independent of the message's destination. For instance, the frequency at which the potential relay encounters the message's destination is clearly destination dependent, while the relay's mobility characteristics and capabilities (e.g., storage, energy, etc.) are destination independent [145] [262].

The fact that in DTNs the existence of an end-to-end path between any pair of nodes at all times cannot be guaranteed raises fundamental challenges in end-to-end reliable data delivery. In DTNs, the Internet model of end-to-end reliability (as implemented by TCP) is not applicable. The DTN architecture proposed in [107] replaces TCP's end-to-end reliability with *custody transfer*, which uses hop-by-hop acknowledgments to confirm the correct receipt of messages between two directly connected nodes. Additionally, due to the inability to guarantee end-to-end connectivity at all times, functions based on the TCP/IP model such as congestion and flow control will not always work in DTNs. Instead, hop-by-hop control can be employed.

Controlling congestion is critical to ensure adequate network operation and performance. That is especially the case in networks operating in "challenged"- or "extreme" environments where episodic connectivity is part of the network's normal operation. As such, congestion control is an important research area in computer networking in general and in DTNs in particular since the Internet's end-to-end congestion control model is not adequate for DTNs. Indeed, in the DTN architecture specification [75], it is argued that congestion control is a topic "on which considerable debate still exists". Furthermore, as discussed in [293], the past few years have seen an increase in wireless communication with variable bottleneck rates; datacenter networks with high data rates, short delays, and correlations in offered load; path with excessive buffering (*buffer-bloat*); cellular wireless networks with highly variable, self-inflicted packet delays; links with non-congestive stochastic loss; and networks with large bandwidth-delay products. Under these conditions, the classical congestion control methods embedded in TCP can also perform poorly. Without the ability to adapt its congestion control algorithms to new scenarios, TCP's inflexibility constrains architectural evolution, as dicussed in [292].

As previously pointed out, DTN congestion occurs when storage resources become scarce. A node experiencing such conditions has several options to mitigate the problem as listed below in decreasing order of preference: drop expired messages, move messages to other nodes, cease accepting messages with custody transfer, cease accepting normal messages[1], drop unexpired messages, and drop unexpired messages for which the node has custody.

Given that expired messages are subject to being discarded prior to the onset of congestion, there may be no expired messages to discard. Moving messages somewhere else may involve interaction with routing computations; this is a reasonable approach if storage exists near the congestion point. This amounts to a form of flow control operating at the DTN hop-by-hop layer and can result in backlogs of custody transfer as they accumulate upstream of congested nodes. To cease accepting normal messages, the node essentially disconnects from its neighbors for some period of time. DTN tolerates such disconnections, but doing so frequently may result in upstream congestion.

From an end user perspective, dropping unexpired messages results in lower network reliability. Discarding messages for which a node has taken custody defeats much of DTN's delay tolerance of DTN. DTN attempts to provide a delivery abstraction similar to a trusted mail delivery service; discarding custody messages is clearly antithetic to this goal.

Still, to this date, custody transfer's cost-benefit trade-offs and its impact on DTN congestion behavior remain not well understood. This is not surprising and can be explained by the fact that the DTN architecture is still not widely deployed and does not yet carry significant traffic loads to stress-test the DTN architecture. It is worth noting that a similar situation occurred in the early history of the Internet: recall that the original TCP protocol specification includes no congestion management functionality, and the congestion problem remained unrecognized until 1980, more than 10 years after the first experiments with Internet.

This chapter presents a survey of the state-of-the-art on DTN congestion control. To set the stage for our discussion of existing DTN congestion control mechanisms,

---

[1] Three service classes are defined by DTN architecture in the bundle protocol (bulk, normal, expedited) corresponding to different levels of priority that scheduling algorithms should take into account during routing operation.

we start by discussing network congestion more generally and in the context of the Internet mechanisms.

10.2   What Is Congestion?

Like any real system, networks have limited resources and ideally their utilization should be maximized in order to maximize network service providers' return on investment. As illustrated in Figure 10.1, which shows throughput ("data delivered") as a function of offered load ("data sent"), as the offered load increases, throughput also increases until network capacity is reached. Ideally, if the network keeps operating under those conditions, resource utilization is maximized as well as network throughput. However, if the offered load increases beyond what the network can handle, throughput starts decreasing. And, if these conditions persist, network performance can degrade sharply, and, in the worst-case scenario, can result in "congestion collapse". Congestion collapse was first observed in the early years of the Internet, more precisely in October 1986 and motivated the design and implementation of TCP's congestion control mechanisms.

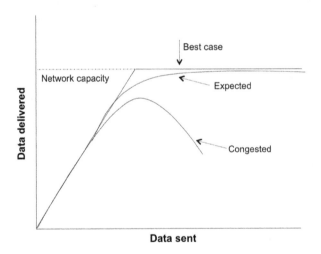

Figure 10.1: Congestion in networks.

Congestion control refers to techniques and mechanisms that can either prevent congestion, before it happens, or remove congestion, after it has happened. In general congestion control mechanisms are based on control theory and can be divided into two categories:

1. **Open-loop** congestion control (prevention). Open-loop solutions attempt to

prevent congestion before it takes place. Generally, open-loop solutions do not utilize runtime feedback from the system and the decisions are made without regard to the current state of the network.

2. **Closed-loop** congestion control (removal). Closed loop solutions use feedback (i.e. measurements of system performance) to mitigate congestion at runtime. It can be divided in three phases when applied to congestion control:

   (a) Phase 1: monitor the system to identify when and where congestion takes place.

   (b) Phase 2: forward the previous information to places where an action can be taken.

   (c) Phase 3: regulate system operation to fix the problem.

The traditional methods based on open loop congestion control are listed as follows.

■ Retransmission Policy : This is the policy by which retransmission of data are effected. If the sender feels that sent data are lost or corrupted, the data must be retransmitted. This transmission may increase the congestion in the network. To prevent congestion, retransmission timers must be designed to prevent congestion and also able to optimize efficiency.

■ Window Policy : The type of window at the sender side may also affect the congestion. Several packets in the Go-back-n window are resent, although some packets may be received successfully at the receiver side. This duplication may increase the congestion in the network and make it worse. Therefore, a selective repeat window should be adopted as it sends the specific packet that may have been lost.

■ Discarding Policy : A good discarding policy in the routers enables the routers to prevent congestion by partially discarding corrupted or less sensitive data while maintaining message quality. For example, in audio file transmission, routers can discard less sensitive packets to prevent congestion and also maintain the quality of the audio file.

■ Acknowledgment Policy : Since acknowledgments (ACKs) are also the part of the load in network, the acknowledgment policy imposed by the receiver may also affect congestion. Several approaches can be used to prevent congestion related to acknowledgment. The receiver should send an acknowledgment for N packets rather than sending an acknowledgment for a single packet. The receiver should send an acknowledgment only if it has to send a packet or a timer expires.

■ Admission Policy : In admission policy a mechanism should be used to prevent congestion. Switches in a flow should first check the resource requirement of a network flow before transmitting it further. If there is a chance of a congestion or there is a congestion in the network, router should refrain from establishing a virtual network connection to prevent further congestion.

In the case of closed loop congestion control the following strategies can be considered:

■ Backpressure : Backpressure is a technique in which a congested node stops receiving packets from an upstream node. This may cause the upstream node or nodes to become congested and reject receiving data from further upstream nodes. Backpressure is a node-to-node congestion control technique that propagates control in the opposite direction of data flow. The backpressure technique can be applied only to a virtual circuit where each node has information of its above upstream node.

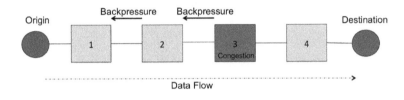

Figure 10.2: Example of closed loop strategy: backpressure.

Figure 10.2 shows one example of backpressure: the 3rd node is congested and stops receiving data; as a result the 2nd node may get congested due to slowing down of the output data flow. Similarly, the 1st node may get congested and inform the source to slow down.

■ Choke Packet Technique : Choke packet technique is applicable to both virtual networks as well as datagram subnets. A choke packet is a packet sent by a node to the source to inform it of congestion. Each router monitors its resources and the utilization at each of its output lines. Whenever resource utilization exceeds the threshold value which is set by the administrator, the router directly sends a choke packet to the source giving it a feedback to reduce the traffic. The intermediate nodes through which the packets have traveled are not warned about congestion. Figure 10.3 shows in high level how choke packet technique operates.

■ Implicit Signaling : In implicit signaling, there is no communication between the congested nodes and the source. The source guesses that there is congestion in a network. For example when sender sends several packets and there is no acknowledgment for a while, one assumption is that there is a congestion.

■ Explicit Signaling : In explicit signaling, if a node experiences congestion it

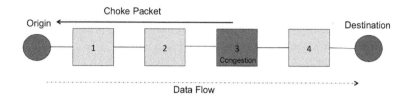

Figure 10.3: Example of closed loop strategy: choke packet technique.

can explicitly send a packet to the source or destination to inform about congestion. The difference between choke packet and explicit signaling is that the signal is included in the packets that carry data rather than creating different packets as in case of choke packet technique. Explicit signaling can occur in either the forward or backward direction.

1. Forward Signaling : a forward signaling signal is sent in the direction of the congestion. The destination is warned about congestion. The receiver in this case adopts policies to prevent further congestion.

2. Backward Signaling : a forward signaling signal is sent in the opposite direction of the congestion. The source is warned about congestion and recognizes that it needs to slow down.

As background, the next section briefly reviews congestion control on the Internet, followed by an overview of DTN congestion control (Section 10.4) and a description of DTN congestion control solutions to-date (Section 10.5).

10.3 Internet Congestion Control

The Internet is a worldwide computer network that uses packet switching to transmit data between source and destination. The Internet architecture, which is based on the TCP/IP protocol suite, specifies two main transport-layer protocols for data transmission end-to-end, i.e., between two communicating processes running on two different hosts connected to the Internet. The first one is UDP (User Datagram Protocol), which provides unreliable data delivery service. In other words, the different pieces of the same application message may arrive out of order, appear duplicated, or go missing without notice. UDP does not provide congestion control or flow control capabilities.

The other Internet transport protocol is TCP (Transmission Control Protocol). TCP provides reliable, ordered delivery of a stream of bytes between two communicating processes. In addition to reliable delivery, TCP also performs flow- and congestion control.

At this point, we should make clear the difference between flow and congestion control. Flow control is "all about the receiver", i.e., it tries to ensure that the sender does not outpace the receiver, sending data faster than the receiver can receive. On the other hand, congestion control is "all about the network" making sure that the sender does not send more data than the network can handle.

Therefore, congestion occurs when resource demands from users/applications exceed the network's *available capacity*. The need for congestion control on the Internet surfaced in 1986 when the Advanced Research and Projects Agency Network (ARPANET), the precursor to the Internet, suffered congestion collapse [151]. Congestion collapse generally occurs at choke points in the network, where the total incoming traffic to a node exceeds the outgoing bandwidth. As described in [297], there are two fundamental approaches to the problem of controlling Internet congestion, namely: capacity provisioning and load control.

1. The **capacity provisioning** approach is based on ensuring that there is enough capacity in the network to meet the offered load.

2. The **load control** approach ensures that the offered load does not exceed the capacity of the network.

The latter approach inspired the development of the first Internet congestion control algorithm [264]. The basic idea behind the algorithm was to detect congestion in the network using *packet loss* as congestion indicator. Upon detecting a packet loss, the source reduces its transmission rate; otherwise, it increases it.

According to the IRTF's Transport Modeling Research Group (TMRG) [114] [115], performance metrics that can be used to evaluate Internet congestion control protocols include:

■ **Convergence speed**: estimates time to reach the equilibrium, i.e., states how much time elapsed between the moment that the congestion was detected and the moment that congestion ceased to exist.

■ **Smoothness**: is defined as the largest reduction in the sending rate in one RTT (Round-Trip Time). In addition, it reflects the magnitude of the oscillations through multiplicative reduction, which is the way TCP reduces its transmission rate.

■ **Responsiveness**: is defined as the number of RTTs of sustained congestion required for the sender to halve the sending rate.

■ **Fairness**: specifies the fair allocation of resources between the flows in a shared bottleneck link.

■ **Throughput**: characterizes the transmission rate of a link or flow typically in bits per second. Most congestion control mechanisms try to maximize throughput, subject to application demand and constraints imposed by the other metrics (network-based metric, flow-based metric and user-based metrics).

■ **Delay**: can be defined as the queue delay over time or in terms of per-packet transfer times.

- **Packet loss rates**: measures the number of packets lost divided by total packets transmitted. Another related metric is the loss event rate, where a loss event consists of one or more lost packets in one round-trip time (RTT).

Some of the metrics discussed above could have different interpretations depending on whether they refer to the network, a flow, or a user. For instance, throughput can be measured as a network-based metric of aggregate link throughput, as a flow-based metric of per connection transfer, and as user-based utility metric.

These metrics were originally proposed for the Internet and in [192], they have been used to provide a categorized description of different congestion control strategies in packet switching networks. While some of them can still be used for DTNs, new metrics are needed. For instance, in DTNs, queueing delays are expected because of the high latencies and intermittent connectivity. Furthermore, paths are lossy, so losses do not necessarily indicate congestion as assumed in TCP.

10.4 Congestion Control in DTN

Initial DTN congestion control focused on buffer management schemes for selectively dropping data in order to improve delivery ratio [247] [248] [61]. Buffer management is important, but it does not address the issue of finite capacity links and discarding data should be a last resort [176] [166].

As previously discussed, congestion control is critical to guarantee satisfactory network performance. According to [266], DTNs must rely on adequate congestion control mechanisms to ensure reliability and stability. Furthermore, in [266], it is argued that most proposed DTN congestion control solutions do not account for Bundles nor the Bundle Protocol; moreover, most DTN congestion control mechanisms have not been fully deployed and tested.

As previously pointed out, the challenges of controlling congestion in DTNs are mainly due to two factors:

1. Episodic connectivity, i.e., end-to-end connectivity between nodes cannot be guaranteed at all times, and

2. Communication latencies can be arbitrarily long caused by high propagation delays and/or intermittent connectivity.

Consequently, traditional congestion control does not apply to DTN environments. A notable example is TCP's congestion control mechanism, which is not suitable for operation over a path characterized by extremely long propagation delays, particularly if the path contains intermittent links. Basically, TCP communication requires that the sender and the receiver negotiate an end-to-end connection that will regulate the flow of data based on the capacity of the receiver and the network. Establishment of a TCP connection typically takes at least one round-trip time (RTT) before any application data can flow. If transmission latency exceeds the duration of the communication opportunity, no data will be transmitted [54] [108]. Furthermore, there is a two-minute timeout implemented in most TCP stacks: if no data is sent or received for two minutes, the connection breaks. However, in some DTNs RTTs can be much longer than two minutes.

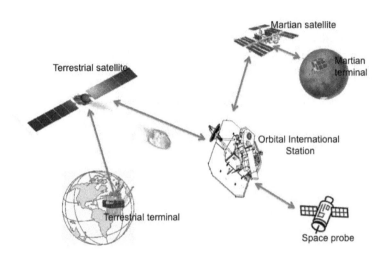

Terrestrial satellite

Martian satellite

Martian terminal

Orbital International Station

Terrestrial terminal

Space probe

Figure 10.4: DTN connection example: deep space communication scenario.

For example, in an interplanetary network supporting Earth-Mars communication (as illustrated in Figure 10.4), the RTT is around 8 minutes when both planets are closest to one another, with a worst-case RTT of approximately 40 minutes. In this scenario, the terrestrial satellite is connected to an Orbital International Station (OIS) in orbit around the Sun, which in turn connects to a Martian satellite and a space probe. Additionally, there is a Martian terminal connected to the Martian Satellite. The link between the OIS and the Martian Satellite is interrupted whenever the planet Mars is between the OIS and the orbiting satellite, as well as whenever the Sun is between Mars and the OIS. Therefore, traffic on the "link" between the Terrestrial and Martian satellites may need to be buffered at the OIS for long and varying periods of time. If the OIS becomes heavily congested, it will significantly hamper communication between the Terrestrial satellite and the space probe.

Alternatives such as UDP or DCCP (Datagram Congestion Control Protocol [164]) are not generally appropriate because they offer limited reliability and, in the case of UDP, no congestion control. Consequently, efficient techniques are thus needed to effectively control congestion in DTNs so that network utilization is maximized.

10.5 DTN Congestion Control Mechanisms

Several DTN congestion control mechanisms have been proposed in the literature. This section presents a brief overview of some these mechanisms focusing on their approach to mitigate congestion.

A congestion control mechanism for interplanetary networks is presented in [47]. This mechanism relies on both storage routing schemes and next hop MADM (Multi-Atributte Decision Making) selection policies. In addition, the implementation of advanced RED (Random Early Detection) and ECN (Explicit Congestion Notification) mechanisms within the delay tolerant network architecture is carried out. Given the challenging peculiarities of interplanetary environments, the proposal mechanism is considered in deep space communication scenario. In this case, the data communication are affected by long propagation delays. The authors consider that congestion events will result in possible buffer overflow, causing thereby bundle discard. They use RED at the bundle protocol and apply to for normal[2] bundle. The proposal mechanism detects congestion when the ratio between the number of queued normal bundles and the difference between buffer capacity and the number of expedited bundles exceeds a certain admittance threshold. In addition, they use also ECN which is implemented at the bundle protocol and applied to expedited bundles. In this case, if the ratio between the number of queued expedited bundles and the difference between buffer capacity and the number of normal bundles exceeds a certain admittance threshold, an ECN flag is set to one. Thus the number of marked bundles is increased accordingly. The mechanism considers the storage-based routing that moves bundles stored in DTN node that are about to experience congestion to other DTN nodes, whose available buffer capacity is large. NS-2 simulator is used by authors for a reference scenario. Non buffer management strategy is explicit presented in spite of the mechanism discards bundles when congestion occurs. The proposal mechanism adopts a closed-loop approach since it combines RED and ECN schemes. As the mechanism drops packets according service classes to solve the congestion problem the packet loss rate could be high and the delivery rate could be lower.

In [273] is presented a local approach to detect and respond to congestion by adjusting the copy limit for new messages based on the current observed network congestion. The authors propose a definition of DTN congestion using specific, observed network metrics. ACKs of generated messages, duplicate ACKs, time-out messages and dropped messages are indicators of network-wide congestion which were identified by the authors. Based on the relationship between duplicates, drops and the delivery rate the authors developed a threshold-based congestion control protocol. The goal of the protocol is to determine the best number of copies for each message based on the global congestion level in order to maximize delivery rate. During congestion detection procedure each node samples the network and creates a congestion view based on the collected metrics. The congestion view value is the ratio of drops and duplicate deliveries, which is compared to the congestion threshold. According to the comparison the copy limit for new messages is lowered or raised

---

[2]Three service classes are defined by bundle protocol (bulk, normal, expedited) corresponding to different levels of priority that scheduling algorithms should take into account during routing operation.

following a back-off algorithm. The authors implemented the proposal congestion control mechanism in the ONE simulator over one scenario represented by a network of 100 nodes with random waypoint mobility. The authors do not specify if the mechanism uses any routing protocol, but they presented a comparative analysis using Spray and Wait routing protocol. In the simulated scenario no discussion of buffer management is presented despite the use of drop count to detect congestion. The proposal mechanism is a closed-loop approach, since based on the value of congestion view the algorithm sets a target copy limit.

The concept of delay and disruption tolerant transport protocol and such protocol LTP-T (Linklider Transmission Protocol - Transport) is defined and presented in [110]. The set of LTP-T extensions has an extension termed Congestion Notification which is a sequence of pairs: the first element of each being on encoding of the LTP node address to which the congestion information pertains, the octets of this element are preceded by a two byte unsigned length field, and then the octets representing the node address itself. The second part of each pair is a 4 octet (network byte order) value specifying the number of seconds from the time of transmission of the segment during which it is believed the node in question will be suffering from storage congestion. To evaluate LTP-T one scenario was modeled, where was constructed a network that emulates a deep-space network consisting of nodes on earth and around Mars. To handle congestion they assume that each LTP implementation has a congestion timer for to-be-forward blocks. The idea is that entire block should be transmitted before this timer expires. If the congestion timer expires, then the node should signal the congested state to upstream nodes. The entire block and all state information associated with it are deleted. It is clear that the LTP-T uses a closed-loop approach when handling congestion. No buffer management strategy is cited by the authors in the context of LTP-T. No discussion related to routing protocol and congestion issues is presented. But, we can note that LTP-T requires a router to select a next-hop destination for each outbound block. The work presented does not use any simulator, but the authors implemented LTP-T using C or C++.

The effect of various environments to buffer occupancy of DTN nodes is analyzed in [172]. A study of buffer management policies in the prospect of how the messages should be deleted and one classification of these policies in the proactive policy and the reactive policy is presented by the authors. The proactive scheme deletes the message depending on the time-to-live (TTL) or feedback acknowledgment as for definitely deleting. On the other hand, the reactive policy deletes some messages due to congestion in nodes' buffer. The paper also identifies the looping problem, which is caused by using popular reactive policy, as a result of degradation in routing performance. Looping problem appears when a node generally drops the carrying messages before they are expired and that the same messages in other nodes during one encounter can be looped back to the node that has just deleted such message. After that, they propose on effective looping control mechanism to prevent such a problem. Finally, they propose a heuristic congestion control policy called CCC (Credit-based Congestion Control). Figure 10.5 shows how CCC works, based on the concept that the message that is obsolete should be deleted first as in drop-oldest. The DTN router along these credit-based policies will drop the message that has the least credit when the buffer becomes full to allocate vacant space for the new coming message. The key point of this policy is refilling and refunding amount of credit when pair nodes encounter at any point of transmission. The congestion detection is performed during nodes' contact, when they exchange the messages depending on applied routing protocol. After, each node refunds the sent message's credit for amount of value named penalty at the sender side and refills the received message's credit for amount of value named reward at the receiver. If the credit exceeds a maximum value congestion exists. The authors use the ONE simulator where buffer management policies and looping control and congestion control are implemented. The scenario modeled by them uses the random waypoint movement for 100 nodes over 1Km x 1Km world size.

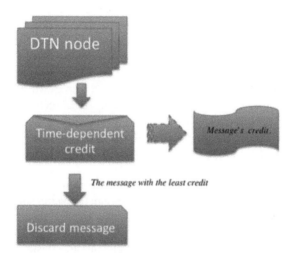

Figure 10.5: Credit-based Congestion Control workflow.

The work presented in [225] introduces a flow control mechanism in IPN (Interplanetary Networks) in order to permit a better operation of the transport layer protocol. This mechanism looks for providing traffic information and local resource availability such as link capacity or buffer space to store bundles through an explicit feedback called ATI-Availability Timeline Information. This approach makes use of flow control in a hop-by-hop manner instead of the classical end-to-end approach typical of the TCP protocol. The authors have implemented their proposal using NS-2 simulator and they have used a simple simulation scenario which is composed of two terrestrial terminals connected to two terrestrial satellites. A satellite, called orbital station, is placed in the middle between Earth and Mars. A martial terminal is connected to a martial satellite. The proposal flow control algorithm works as follows. At the beginning of the bundle k transmission the sender verifies if link is active and if there is bandwidth availability towards next hop receiver for the transmission. If there is not bandwidth availability, k-th bundle is stored in the buffer until bandwidth availability exists. To avoid buffer overflow and guarantee that input traffic does not overcome buffer size the algorithm verifies if the C_i (total traffic received at receiver that can be managed in order to avoid buffer overflow) is smaller than or equal C_r (the input link capacity at receiver). In addition, the proposal algorithm by the authors makes use of a fair bandwidth sharing policy on the basis of the resource availability at the receiver. In this case, the authors consider that the data rate assigned to k-th bundle if it is admitted is the total traffic recieved at receiver divide by the number of active transmissions at receiver plus one.

Taking into account that storage-based congestion at custodians becomes inevitable in scenarios where a custodian is simultaneously not able to offload its stored messages to other nodes, the authors of the paper [248] propose storage routing, a congestion management solution. This solution employs nearby nodes with available storage to store data that would otherwise be lost given uncontrollable data sources. When the congestion occurs, the solution proposed is invoked to determine a set of messages for migration and a set of neighbors as targets for migrated messages. When congestion subsides, the solution is invoked to retrieve message. The algorithm proposed limits the topological scope of node selection by restricting the set of nodes searched alternative custodians to the congested node's k-neighborhood. The k-neighborhood of node v is then defined to be the set of nodes within k hops of v. The node searches in which all the nodes m hops away from the congested node v invoking storage routing are considered for migration prior to any $m + 1$ hops away, up to a maximum of k hops from v. The node selection algorithm returns the node with the lowest migration cost. The migration cost is the sum of the transmission cost and the storage cost it is weighted by the gain constant α, where $0 \leq \alpha \leq 1$, which allows for the transmission or storage cost preference. The proposal solution selects message for migration according to push policy. Three policies are considered by the authors: pushtail, pushhead and pusholdest. The authors presented several simulation experiments over which they study the performance of the proposal solution. And they use two metrics for evaluation: the message completion rate and time weighted network storage. The message completion rate metric was used in evaluating DTN routing schemes, the time weighted storage metric considers the product of the amount of storage used for storing messages with the amount of time they remain stored. The results described by the authors show that the best performance for proposal solution was achieved when it prefers the relatively oldest messages to migrate and when the current amount of storage available for migration in each neighbor is known by the migrating router. According to authors the congestion detection occurs when one message arrives at a node for which sufficient persistent storage is unavailable. They implemented the proposal algorithm using YACSIM simulator as a discrete-event simulator, DTNSIM as a simulator base and GT-ITM as a topology generator. The scenario modeled by the authors includes triggered sensor nodes, where they assume symmetric links with equal bandwidth. In each simulation, they use as buffer management strategy the PushTail policy, but they consider PushHead and PushOldest for comparison.

In [61] it is described that congestion control in DTN as in the Internet is accomplished by inducing flow control at the applications that are the source of the excess traffic and since that flow control is driven by sustained net growth in buffer space occupancy, a natural way to implement DTN congestion control is to propagate buffer utilization stress back through the network to the source bundle nodes. The authors try to accomplish this by declining to take custody of bundles forcing the sending bundle node to retain the bundles and thereby increasing that node's local demand for buffer space, forcing it in turn to refuse custody of bundles. Given this strategy, the authors try to answer the question: how to determine when to decline to take custody of a bundle in order to conserve local buffer space prudently. Then they propose to apply a financial model of buffer space management. The basic notion is that unoccupied buffer space is regarded as analogous to money and routing network traffic is regarded as analogous to the daily financial activities of an investment banker. According to the authors the financial model can be described as following: a router has limitted buffer space, analogous to the fixed amount of capital a banker has to work with. They imagine that the application that owns the sender and receiver of a bundle will pay a conveyance fee to get the bundle delivered. This fee is a function of the bundle size and the transmission priority requested. The banker (router) receives a commission for completing one hop of each bundle's end-to-end route, and the commission considered is some fraction of the bundle's total conveyance fee. Accepting custody for an in bound bundle for forwarding equates to a banker buying a non interest bearing debenture: the bundle is occupied at the cost of certain amount of free buffer space. The rule based congestion control mechanism presented by the authors uses the negative purchasing decisions by refusing to accept bundles for forwarding. The rules considered by the authors are:

■ If the number of bytes occupied by a given bundle added to the current number of bytes that are queued up for transmission at the router is greater than the router's maximum capacity of all bundle buffers (in bytes), refuse the bundle.

■ If the projected usage of buffer space is less than the total buffer capacity, then accept the bundle.

■ If the risk rate of a given bundle exceeds the mean risk rate over the bundle's residual TTL (Time To Live), then the bundle is of above average risk; refuse it. Otherwise, accept the bundle.

When exercise of the second rule requires that a bundle be discarded, the effect is an instance of Drop-last policy because the rules are applied only to newly received bundles. The authors have tested the proposal mechanism in several scenarios. Basically, these scenarios involved two to five computers connected in a line topology. In scenarios with more than two nodes, there was no direct connection between nodes A and E considering the existence of intermediate nodes between A and E, for instance. In these cases, bundles are routed through all routing nodes.

In [143] is proposed a new congestion avoidance mechanism based on path avoidance for DTN. This mechanism tries to optimize the management of node storage and divides the node state into normal state, congestion adjacent state (CAS), and congestion state (CS). Based on these different node states separate strategies are employed by authors to avoid congestion. The authors use simulation through OP-NET [8] to get results. The authors suggest that congestion occurrence at a DTN custody node is likely to be gradual. Congestion detection is performed according to the transitions among the identified node states. The proposed congestion control mechanism adopts a closed-loop approach: when a node comes into CAS or CS state, the node broadcasts information to its neighbors to inform them of this transition. Upon this notification, the neighbor node sets the link connected to the sender node to be "half-hung". This causes the path including the link which was set to be half-hung to be avoided. The scenario cited by the authors is a community-based improved model which is established on a 2000m x 3000m area. This area is divided into six subnets and each has a gateway which provides forwarding of bundles, thus it adopts the drop-least policy as a buffer management strategy. The authors do not take into account the relationship between the mechanism and routing protocol.

Congestion-aware forwarding algorithms within opportunistic networks are the context of the work presented in [131]. The authors propose a distributed congestion control algorithm that adaptively chooses the next hop based on contact history and statistics, as well as storage statistics. They aim to distribute the load away from the storage hotspots in order to spread the traffic around. They show that congestion control is achieved in a fully open loop manner and by just local dissemination of statistics of nodes' availability and connectivity. According to the authors, opportunistic networks require in-network congestion control and should be open-loop as this best addresses the lack of information available to the senders and receivers, together with the delay and cost associated with reporting. They proposed one algorithm termed CAFé (Context Aware Forwarding Algorithm) that is composed of two core components: Congestion manager, which is concerned with the availability metric of network nodes, and Contact manager, which is concerned with the forwarding heuristic used at the nodes. Each node contains two buffers: a forwarding buffer and a sender buffer. The authors believe this is the way to enable a fair distribution between the sending nodes' own data and forwarding the data from others. The congestion detection occurs during each meeting between nodes, where the two nodes exchange their availability information, statistics about their interactions with other nodes, and their locally calculated centralities. Using the information received during this exchange, the contact manager and congestion manager make forwarding decisions. The CAFé evaluation is made through a custom built trace driven opportunistic simulator. For their real connectivity traces they use the INFOCON 2005 dataset, which contains logs for 41 Bluetooth devices. There is a relationship between the proposed congestion control algorithm and the routing protocol that is delineated by forwarding heuristic. The proposed mechanism includes the discarding of packets but does not exclude any buffer management strategy.

The work described in [247] focuses on the problem of handling storage congestion at store-and-forward nodes by migrating stored data to neighbors. The proposed solution includes a set of algorithms to determine which messages should be migrated to which neighbors and when. It also includes an extension to the DTN custody transfer mechanism enabling a "pull" form of custody transfer where a custodian may request custody of a message from another custodian. According to the authors, this approach enables the problem of storage allocation among a relatively proximal group of storage nodes to be decoupled from the overall problem of path selection across a larger network. Using simulation, the authors evaluated the approach and tried to show how migrating custodian in the proposed model can improve message completion rate for some storage constrained DTN networks. The work proposed by the authors detects congestion by checking the availability of DTN custodian buffer. If the buffer is full or the message arriving does not fill on it the DTN custodian becomes congested. When this happens the DTN node must migrate its stored messages to alternative storage locations to avoid loss. In addition, a new approach termed Push-Pull Custody Transfer is constructed. It operates at the DTN layer, so DTN custody transfer mechanisms are available to implement the transfer. Thus a congested custodian must make decisions about when to invoke push or pull operations to satisfy its data routing and congestion goals. This involves a set of algorithms the authors called storage routing (see Figure 10.6). In summary, the algorithms comprising storage routing include:

- Message selection: selects which message to migrate to storage nodes

- Node selection: produces a set of nodes where messages should be migrated to

- Retrieval selection: selects which messages to retrieve and from whom.

The scenario used to study the algorithm includes a dynamic network topology that contains 18 nodes, with a set of connected core nodes having average node degree 4.125. It also includes many degree-1 leaf nodes. They used a publicly-available simulator called DTNSIM. The buffer management strategy used at the algorithms includes Drop-last strategy, in this case data is only supposed to be dropped when no other migration options are available.

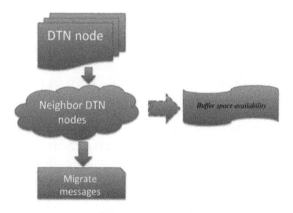

Figure 10.6: Storage Routing for DTN Congestion Control workflow.

To address the challenges of congestion control mechanisms in DTNs a novel congestion control mechanism based on the use of tokens is presented in [85]. In this approach the network nodes must possess a token in order to inject messages into the network. Tokens are initially uniformly distributed but thereafter move randomly through the network. A discrete event simulator is used by the authors to analyze the proposed mechanism. The idea behind this proposal, a token-based congestion control mechanism, originates from the desire to match the amount of traffic entering a network to the total network capacity. The problem with this proposal is that if we have many tokens in the network they can cause congestion. In addition, the authors considered the network capacity as the amount of data a network can deliver to destinations in a bounded time. They do not consider a utility fee in the network, and at the same time they adopt the strategy of discarding messages. The congestion detection at the token-based congestion control mechanism proposed by authors occurs when the amount of traffic entering a network is greater that to the total network capacity. The authors believe that if the network as a whole, or any individual node, is only presented with an amount of traffic that it can forward, situations like congestion collapse can be avoided. The token mechanism is independent of the routing algorithm used and is only triggered if a suitable next hop neighbor is discovered. Thus, the authors seem to adopt a closed-loop approach. The buffer management strategy included in the proposed mechanism is Drop-last, where upon receiving a message the node checks if the queue is full and, if so, the message is dropped. The proposal mechanism is evaluated by the authors using their own discrete event simulator, written in Java. The research scenario is a sensor network where nodes periodically send data among themselves. The network used consists of 60 nodes operating in a 100m x 100m grid.

The usage of buffer space advertisements to avoid congestion in mobile opportunistic DTNs appears in [169]. The authors discuss congestion control in opportunistic networks and present a local congestion control mechanism that proactively acts before message drop induced by congestion occurs. They use two scenarios with two different trace-based mobility models. The number of mobile nodes is 116. In the first mobility trace, the San Francisco taxi cab trace, the area size is 5700 x 6600m. The second mobility trace was obtained in a synthetic fashion from a map of Helsinki City Center for which the dimensions are 4500 x 3400m. Basically, the mechanism allows node's buffer ocuppancy rate be advertised to adjacent nodes. In this case, the node takes local decisions and avoids sending messages to nodes whose buffers are nearly full. The proposed mechanism is evaluated through simulation using NS-2 and ONE simulators.

Radenkovic and Grundy [222] explore a number of social, buffer and delay heuristics to offload the traffic from congested parts of the network and spread it over less congested parts of the networks. They focus on congestion-aware forwarding algorithms that adaptively choose the next hop based on contact history, predictive storage and delay analysis of a node and its ego networks in order to distribute the load away from the storage hotspots and spread the traffic around. The authors define the "ego network" as the set of all or some of the contacts that a node has experienced. A new forwarding heuristic is introduced by authors. This forwarding heuristic includes three parts: a socially-driven part that exploits social relationships, a node-congestion driven part that avoids nodes that have lower availability and higher congestion rates, and an ego-network-driven part that detects and avoids congested parts of the network. The mechanism proposed by authors tries to predict which of the nodes or the ego networks of the two nodes is more likely to get filled up faster. The ONE simulator and RollerNet [277] connectivity traces are used to test the approach.

In [311] the concept of revenue management and employ dynamic programming is applied to develop a congestion management strategy for DTN. For a class of network utility functions, they show that the solution is completely distributed in the nature of DTN where only the local information, such as the available storage space of a node, is required. The authors look to optimize the overall traffic by accepting or forwarding bundle transfer requests under the assumption of minimally cooperative behavior among nodes. A discrete event-driven simulator was developed by the authors to evaluate the congestion management strategy. The problem is that they assumed that each link has infinite bandwidth; this can mask some simulation results pertaining to network congestion. According to the authors, if a node receives a request for custody from its neighbor, it will decide whether to accept or reject the request based on the current available storage space and predefined optimal control strategy. Storage space is assumed by the authors to be a key resource constraint, enabling detection of congestion. The idea behind the optimal control policy is that accepting a request imposes a given benefit and cost if and only if there is sufficient remaining capacity in the relevant resources and the benefit is greater than or equal to the opportunity cost of occupying the storage space. The opportunity cost measures the value of the storage capacity, which is the benefit that may be lost when a higher-benefit request must be rejected as a result of consumption of storage resources by the lower-benefit request. They see the benefit (function) of a request as the movement of a bundle to the next hop. The scenario simulated by the authors consists of static nodes, destinations, and mobile nodes distributed in a $3000m^2$ field. Mobile nodes function as relay nodes and their mobility follows a random waypoint model.

Given the lack of stable end-to-end paths, message replication is commonly used to increase delivery in DTN. However, the limited resources available to the node can be overwhelmed by too much replication, which leads to congestion and ultimately to reduced delivery rates. Taking this into account, the work presented in [274] explores the use of dynamic node-based replication management in intermittently connected networks (ICNs) where individual nodes determine how many messages they can replicate at each encounter (see Figure 10.7). This management is performed locally, independent of the routing protocol used. According to the authors, a routing protocol orders the messages that it wants to transfer at each encounter and the congestion control algorithm determines how many of them are actually sent. To determine such local replication limits, nodes must observe the current level of congestion in the network. The purpose of the work is to find metrics that can track the change in delivery rates due to buffer congestion. To calculate the congestion magnitude, a node monitors the network for a certain time and then computes the ratio of drops to replications counted during the sample period. The proposed congestion control mechanism enables the system to detect congestion, to offer information about congestion level, and to correct the problem, returning to the equilibrium level. To evaluate the proposed mechanism the authors used the ONE simulator in a scenario in which each scenario had 100 nodes in a 4500m x 3400m area. One of the scenarios models the behavior of mobile agents in a disaster. Another simulates a day-time/night-time traffic pattern.

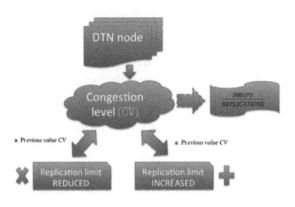

Figure 10.7: Retiring Replicants congestion control workflow.

Because there is no acknowledgment information in Epidemic Routing, redundant copies of packets will be transported pervasively among nodes even if the packets have been received by the destination node. This not only wastes bandwidth and buffer resources, it also leads to congestion at nodes. In order to solve this problem, a congestion control strategy under epidemic routing for nodes in DTN is introduced in [306]; this method is called "average forwarding number based on epidemic routing" (see Figure 10.8). The AFNER strategy is based on two basic assumptions. First, packets are sorted in the buffer in ascending order of fowarding number. Second, the packet has arrived at the destination node when the fowarding number of the packet is bigger than or equal to the average fowarding number. They used NS-2 simulator to do a performance evaluation of the strategy. The scenario used by the authors includes a network of 50 nodes into a 1000m x 1000m area. These nodes move at a speed of 20 m/s. Since the proposal congestion control mechanism operates under epidemic routing, trying to prevent congestion and to check for congestion, we can see it as a closed-loop approach. According to this mechanism the buffer management strategy is to erase all the packets whose forwarding number is bigger that the average forwarding number of all of the packets from sources to the destination. Thus the node receives incoming packets and does not erase the other packets in the buffer until the buffer has been completely occupied. Then the node drops all subsequently arriving packets.

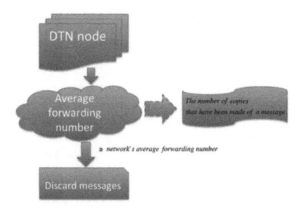

Figure 10.8: Average forwarding number based on epidemic routing workflow.

Smart-DTN-CC is a congestion control mechanism proposed in [93] which employs reinforcement learning [267] that overcomes some of the limitations of existing DTN congestion control schemes. Specifically, the author proposes to use computational intelligence techniques in order to automatically adapt to the dynamics of the underlying network without the need for human intervention, which is particularly important in extreme environments. To this end, Smart-DTN-CC is introduced as a novel DTN congestion control framework which uses computational intelligence to adapt the congestion control effort based on the dynamics and operating conditions of the target environment. Congestion control adaptation happens automatically as the characteristics and conditions of the underlying network change. The proposed approach is based on reinforcement learning [267], a machine learning technique which, given the current *environment*, takes *actions* with the goal of maximizing a cumulative *reward*. Reinforcement learning is based on the idea that if an action is followed by a satisfactory state (e.g., performance improvement), then the tendency to produce that action is strengthened (reinforced). On the other hand, if the state becomes unsatisfactory, then that particular action is penalized. Reinforcement learning is known to be well suited to problems in which the *environment* plays a paramount role but prior knowledge about it cannot be assumed either because it is not available or it is prohibitively expensive to collect. Therefore, the only (practical) way to acquire information about the environment is to learn by interacting with it "online" or in real time. In particular, Smart-DTN-CC works as follows. Let's assume a DTN node with buffer overflow. This node can, for example, broadcast a beacon to inform that it is not able to receive any message. This might prevent other nodes from sending messages, consequently reducing the number of dropped messages. To accomplish this, it would have to run intelligent programs that search the set of possible solutions (actions) and select the best action after analyzing the effects of various combinations. Simultaneously, it could use on-line machine learning programs that suggest which option(s) might be better in the light of past experience. For truly intelligent behavior, the mobile nodes need to do congestion control simply by avoiding buffer overflow and display the ability to adapt intelligently to their changing environments. Smart-DTN-CC can be easily combined with many existing schemes for local congestion control.

Authors in [121] [190] highlight that congestion might arise either by a) excessive local traffic along a route path or b) excessive remote traffic both deriving in an overbooking of contacts or nodes storage. To tackle a), full path-awareness was studied as an extension to CGR routing algorithm showing that it can drastically reduce the overall congestion in some scenarios. The resulting method was coined PA-CGR. To tackle b), a reactive scheme based on feedback messages such as Estimated Capacity Consumption (ECC) can be used [45]. However, it is well known the performance of feedback messages in highly disrupted networks is drastically degraded.

To face the congestion problem in highly disrupted and delayed networks, authors in [121] [190] propose to exploit a common feature in space networking: traffic predictability. In particular, traffic in space missions can be accurately estimated as it is generally product of specific commands destined to the spacecraft payload or platform. Since commands are scheduled from a central decision point on ground (typically a mission control center), the expected traffic behavior of the system can also be known in advance and exploited by this control center for other purposes.

In this context, congestion in DTN can be not only mitigated but completely removed by designing node-specific contact plans in such a way that each node will have a local view of the forthcoming topology with a capacity which is exclusively reserved for that node. Indeed, nodes could still run CGR over these contact plans without any modification. The capacity assigned for each node can be accurately calculated by the control center based on the expected data generation rate which in turn is product of the status and activities scheduled to the network. As a result, and only at a very small overhead of using one contact plan per network node, the communication resources (node's memory and link capacities) will never result overbooked nor in conflict. Evidently, this holds as long as the injected traffic matches the expected one. Authors also discussed the use of capacity safeguard margins to tolerate unexpected traffic variations.

10.5.1   DTN Congestion Control Taxonomy

This section describes a taxonomy to classify DTN congestion control mechanisms [256]. This taxonomy maps the DTN congestion control design space and uses it as backdrop to put existing DTN congestion control techniques in perspective. Eight criteria have been used to underpinning the proposed DTN congestion control taxonomy.

1. Congestion Detection: **How is congestion detected?** In general, congestion occurs when resource demands exceed available capacity. Consequently, congestion detection can consider: *network capacity*, *buffer availability*, and *drop rate*.

- *Network capacity*: mechanisms that use network capacity to gauge congestion try to assess if the traffic arriving at the network is greater than the traffic that could be supported by network.

- *Buffer availability*: some DTN congestion detection strategies check whether the storage capacity at nodes is filling up as new messages are received.

- *Drop rate*: the fact that the drop rate goes beyond a certain threshold is used by certain mechanisms as a way to detect congestion.

2. Open- versus Closed-Loop Control: Another way to classify DTN congestion control mechanisms is based on whether they employ *open loop* or *closed loop* control. In open-loop congestion control, end systems do not rely on feedback from the network; instead they try to "negotiate" their sending rate a priori which can be considered a congestion prevention approach. Closed-loop (or *reactive* control) mechanisms utilize feedback from the network to the end systems. Network feedback usually contains information about current conditions. End systems typically respond to congestion build-up by reducing the traffic load they generate [22]. Notable examples of open-loop congestion control mechanisms include the ones based on buffer management, e.g., dropping messages according to a variety of policies. Some of these techniques use classical drop policies (i.e. drop-random and drop-tail) which were inherited from traditional networks. In Section 10.5.2, we list and discuss some of these classical policies as well as drop policies which have been proposed specifically for DTNs (i.e., E-Drop and Mean-drop).

3. Proactive versus Reactive Control: Congestion control mechanisms can also be classified as *proactive* or *reactive*. Proactive congestion control (also known as congestion avoidance) schemes take a preventive approach and try to prevent congestion from happening in the first place; in reactive congestion control, end systems typically wait for congestion to manifest itself (e.g., router queue build-up, packet loss, etc.) before any action is taken.

   Because DTN systems may exhibit long delays, reactive congestion control may not be sufficient. Proactive congestion control or hybrid approaches combining proactive and reactive control are interesting alternatives.

   We should also point out that, at first glance, proactive and open-loop control may seem to subsume one another. However, a closer look reveals that approaches based on open-loop control can be reactive. For instance, we can have a mechanism that is triggered solely based on the size of the router's queue, and therefore is open-loop. However, it can still be reactive if it starts dropping packets only when the queue is full. In a proactive open-loop approach, packets would start getting dropped sooner in an attempt to avoid letting congestion settle in.

4. Application: DTNs have a wide range of applications from deep space communication to mobile sensor networks on Earth. Depending on the application,

DTNs may exhibit very different characteristics to be able to address the requirements of the driving applications.

5. Contacts: When a communication opportunity appears between two DTN nodes we call this a *contact*. There are different kinds of contacts [138] [161] depending on whether they can be predicted. *Scheduled* contacts are predictable and known in advance. A typical example of a DTN with scheduled contacts is interplanetary networks, in which the mobility of nodes is completely predictable and known a priori. In *probabilistic* DTNs, contacts are probabilistic following a particular distribution that is based on historical data. Finally, contacts in *opportunistic* DTNs are totally random and not known ahead of time.

6. Routing: The taxonomy also considers whether there is any relationship between congestion control and routing. In this case, the congestion control mechanisms can be classified as routing protocol—*independent* or *dependent*. Congestion control approaches that can work with any routing mechanism are said to be routing-protocol independent; on the other hand, congestion control mechanisms that are proposed taking into account specific DTN routing protocols are routing-protocol dependent.

   An interesting aspect of protocol-dependent congestion control is that, often times, the same mechanism that is employed for routing also serves the congestion control mission. A simple example of such "serendipitous" congestion control is controlled epidemic, where the same mechanism used to limit the number of message copies in the network is also performing congestion control.

7. Evaluation Platform: Most experimental evaluations of proposed DTN congestion control mechanisms have employed network simulation platforms. A significant advantage of network simulators is the fact that they make it easy to subject protocols under evaluation to a wide range of network and traffic conditions. They also allow experiments to be reproduced easily.

8. Deployability: In the context of DTN congestion control, *deployability* has been defined as the ability to deploy a protocol in realistic scenarios under real-world conditions. As previously pointed out, most existing DTN congestion control protocols have been implemented and tested using simulation platforms, which do not necessarily expose the mechanisms being evaluated to real-world conditions. In order to explore the deployability of existing congestion control mechanisms, in Section 10.5.2, they are classified using three different levels, namely: low, medium, and high deployability. The criterion enabled to assess the deployability of DTN congestion control mechanisms is whether they rely on global knowledge of the network.

10.5.2 Classification of DTN Congestion Control Mechanisms

In this section, we provide an overview of existing DTN congestion control mechanisms in light of the classification previously presented. The classification of various mechanisms is summarized according to the taxonomy in Table 10.7.

Congestion Detection Congestion detection techniques and the congestion indicators they use are critical to the congestion control effort. They determine how reliably a network is able to detect congestion and how quickly it is able to react to it.

Several DTN congestion control mechanisms use buffer occupancy rate, that is, the availability of buffer space, as an indicator of congestion [47] [61] [78] [81] [110] [131] [132] [143] [169] [172] [222] [247] [248] [287] [288] [303] [311] [184] [31] [93].

As an example, the congestion control technique proposed in [247] checks buffer availability of potential DTN custodians, i.e., nodes that will be assuming responsibility for carrying messages on behalf of other nodes. If the custodian's buffer is full or the new message will not fit, the DTN custodian is considered to be congested.

Network capacity has also been used as congestion indicator in DTNs [85] [306] [56]. Token based congestion control [85] tries to regulate the amount of traffic entering a network based on network capacity, where network capacity is measured by the amount of data a network can deliver to destinations in a given period of time.

Some efforts employ number of dropped packets á la TCP as a way to detect congestion [274] [273]. However, due to frequent topology variations and high error rates, packets may be dropped for reasons other than congestion. For this reason, the congestion detection proposed in [274], partially follows TCP's strategy: the proposed congestion indicator is a function of the volume of message drops and replication.

Open- versus Closed-Loop Control As illustrated in Table 10.7, over 80% of the DTN congestion control mechanisms considered here [225] [248] [61] [143] [131] [132] [85] [247] [303] [81] [274] [169] [222] [47] [273] [110] [172] [184] [31] [56] can be classified as closed loop. For example, the approach described in [143] is based on the observation that congestion in DTN nodes builds up gradually. It then proposes three states for DTN nodes, namely: normal state, congestion adjacent state (CAS), and congestion state (CS). Congestion control is then performed depending on the state of the node. More specifically, the proposed congestion control mechanism adopts a closed loop approach by having nodes broadcasting congestion information to their neighbors once they enter into CAS or CS.

The other mechanisms we examined, i.e., [78] [311] [306] [288] [287], fall into the open-loop control category. For example, in [306] a congestion control strategy called the Average Forwarding Number based on Epidemic Routing (AFNER) is proposed. In AFNER, when a node needs to receive an incoming message and its buffer is full, the node randomly drops a message from those whose forwarding number is larger than the network's average forwarding number. The forwarding number of a message is defined as the number of copies that have been made of a message, and the average forwarding number is the mean forwarding number of all the messages currently in the network. According to the forwarding number queue, the strategy determines the packet forwarding sequence. Open-loop congestion control mechanisms are generally based on dropping messages, while closed-loop approaches use feedback information, i.e., messages from neighbors, to avoid having to drop messages. They typically only drop data as a last resort. Some existing DTN congestion control strategies use drop policies that have been proposed for "traditional" networks. A list of some traditional drop policies along with the DTN congestion control mechanisms that use them is shown in Table 10.1.

Table 10.1: Traditional drop policies

Drop Policy	Description
Drop-random [96] [47] [131] [222]	a message from the queue is selected at random to be dropped.
Drop-head [96] [248]	the first message in the queue, i.e., the head of the queue, is dropped.
Drop-tail [96] [248] [61] [247] [85] [56]	the most recently received message, i.e., the tail of the queue, is removed.
NHop-Drop [306]	any message that has been forwarded over *N* hops is dropped.
Drop-least-Recently-received [96] [303]	the message that has been in the node buffer longest is removed.
Drop-oldest [96][169] [248]	the message that has been in the network longest is dropped.
Drop-youngest [96]	drops the message with the longest remaining life time.

Table 10.2 shows some drop policies (and the mechanisms that use them) which have been proposed specifically for DTNs.

While one could argue that open loop congestion control systems are a better match for DTNs, most existing mechanisms employ a closed-loop approach. However, the SMART-DTN-CC [93] presents a hybrid approach that includes open- and closed-loop scheme.

Proactive versus Reactive Control Congestion control solutions can be classified as proactive (i.e., performs congestion avoidance), reactive (i.e., responds to congestion events), or hybrid (i.e., performs congestion avoidance and reacts to congestion build-up).

From Table 10.7, we observe that most existing DTN congestion control mechanisms (around 52%) adopt a hybrid policy using both proactive and reactive control [47] [61] [78] [131] [132] [143] [172] [225] [274] [303] [306] [184] [31] [93]. An example of a hybrid approach that combines random early

Table 10.2: DTN drop policies

Drop Policy	Description
Drop-largest [227]	the message of the largest size is selected.
Evict Most Forwarded First - (MOFO) [146]	the message forwarded the maximum number of times is selected.
Evict Most Favorably Forwarded First - (MOPR) [146] [306]	each message is related to a forwarding predictability FP. Whenever the message is forwarded the FP value is updated and the message that contains the maximum FP is dropped first.
Evict Shortest Life Time First - (SHLI) [146][169]	the message that contains the smallest TTL is dropped.
Evict Least Probable First - (LEPR) [146]	since the node is less likely to deliver a message for which it has a low P-value (low delivery predictability) and that it has been forwarded at least MF times, drop the message for which the node has the lowest P value.
Global Knowledge Based Drop - (GBD) [166]	based on global knowledge about the state of each message in the network (number of replicas), drop the message with the smallest utility among the one just received and the buffered messages.
History Based Drop - (HBD) [166]	a deployed variant of GBD that uses the new utilities based on estimates of m (the number of nodes, excluding the source, that have seen message since its creation until elapsed time) or n (the number of copies of message in the network after elapsed time). The message with the smallest utility is selected.
Flood Based Drop - (FBD) [166]	accounts only for the global information collected using simple message flooding, that is, without considering message size.
Threshold Drop - (T-Drop) [37]	drops the message from congested buffer only if size of existing queued message(s) falls in Threshold range (T).
Equal Drop - (E-Drop) [228]	drops the stored message if its size is equal to or greater than that of the incoming message; otherwise does not drop.
Message Drop Control - (MDC) [38]	the largest size message will be dropped. This policy controls the message drop through the use of an upper bound.
Mean Drop [226]	drops messages that have size greater than or equal to the mean size of queued messages at the node.
Small-Copies Drop [163]	drops messages with the largest expected number of copies first.
Adaptive Optimal Buffer Management Policies - (AOBMP) [176]	drops messages according to the utility function associated to the message.

detection (RED) and explicit congestion notification (ECN) within the DTN architecture is described in [47].

Note that approximately 31% of mechanisms studied adopt a purely reactive approach [81] [110] [247] [248] [273] [287] [288], while 17% of them are based on a purely proactive policy [85] [169] [222] [311] [56]. The approach proposed in [169] proactively advertises buffer occupancy information to adjacent nodes. Nodes can then avoid forwarding messages to nodes with high buffer occupancy. Following the mechanism described in [247], when a DTN node becomes congested, it tries to migrate stored messages to alternate locations to avoid loss.

Application The challenging peculiarities of interplanetary networking environments inspired the congestion control mechanisms proposed in [47] [110] [225] [56] which target deep space communication scenarios. In the case of [110], to evaluate the Linklider Transmission Protocol - Transport (LTP-T), a scenario that emulates a deep-space network consisting of nodes on Earth and around Mars was modeled. In this context, the proposed congestion control protocol is designed to withstand the noise and delays incurred by communication across astronomical distances.

Terrestrial applications are considered in [248] [61] [143] [131] [132] [85] [169] [222] [274] [184]. In [274] one of the scenarios discussed models the behavior of mobile agents in disaster relief operations.

Dynamic scenarios with random generation of nodes following a statistic model are used in [78] [81] [273] [172] [287] [247] [311] [306] [288] [303] [31]. Most of them are based on using mobile terrestrial communication applications to study the proposed congestion control mechanisms. In the case of [78] the authors use a simple scenario where nodes and network parameters are generated randomly and a mobility model is set.

It is interesting to note that SMART-DTN-CC [93] is the only one mechanism able to operate in different DTN scenarios. It adjusts its operation automatically as a function of the dynamics of the underlying network. It employs reinforcement learning, a machine learning technique known to be well suited to problems in which the *environment*, in this case the network, plays a crucial role; however, no prior knowledge of the target environment can be assumed, i.e., the only way to acquire information about the environment is to interact with it through continuous online learning.

Contacts We observe from Table 10.7 that around 17% of DTN congestion control mechanisms [274] [288] [248] [78] assume in their scenarios both predicted and opportunistic contacts. Among the remaining congestion control mechanisms [47] [110] [225] [56], another 13% assume scheduled contacts. In particular, the hop-by-hop based mechanism proposed in [225] experiments with an interplanetary network scenario which employs scheduled contact between planets.

Approximately 39% of the mechanisms we studied, namely [81] [287] [172] [169] [303] [306] [311] [184] [31] assume opportunistic contacts while around

30%, namely [61] [131] [132] [143] [222] [85] [247] [273] use predicted contacts.

One example of opportunistic contact can be seen in [81] where a congestion control routing algorithm for security and defense based on social psychology and game theory is presented. In this case, DTN nodes are assumed to be randomly distributed and to perform random routing. Random routing results in randomness of each node's encounters.

A distributed congestion control algorithm that adaptively chooses the next-hop based on contact history and statistics is described in [131]. It has a component called contact manager that executes a forwarding heuristic taking into account predicted contacts.

Observe that [93] does not specify any type of contact. Since it was designed to operate in dynamic environments it can easily take opportunity of opportunistic, scheduled and predicted contacts to mitigate DTN congestion.

Routing Most of the DTN congestion control mechanisms, e.g., [47] [110] [131] [132] [143] [222] [248] [274] [273] [287] [288] [303] [306] [184] [31], were proposed taking into account specific DTN routing protocols. We have classified them as routing protocol dependent. For example, the AFNER approach described in [306] proposes a congestion control strategy for epidemic routing. On the other hand, there are mechanisms such as [61] [78] [81] [85] [169] [172] [225] [247] [248] [274] [311] [56] [93] that are independent of the routing protocol. This means that congestion control does not act or rely on the underlying routing mechanism and thus can be used with any routing infrastructure. A notable example is the Credit-Based Congestion Control mechanism [172] which uses the age of a message as a heuristics to decide when the message will be discarded when congestion builds up in nodes.

We should point out that for DTNs that require interoperability through the Bundle Layer [245], congestion control mechanisms should be independent of the routing protocol. Intuitively, congestion control mechanisms that work independently from the underlying routing protocol are more general and applicable to a wide array of scenarios.

Evaluation Platform Approximately 17% of the mechanisms covered in this survey were evaluated using custom simulators [61] [131, 132] [85] [311]. For example, a discrete event-driven simulator was developed in [311] to test a congestion management strategy that uses the concept of revenue management and employs dynamic programming.

Another 17% of the techniques surveyed were evaluated using the *ns-2* network simulator [14], a simulation platform widely used in network research [225] [169] [47] [306].

Almost 44% of the approaches we researched [81] [287] [288] [274] [222] [273] [303] [172] [184] [31] [93] use the ONE simulator [160]. ONE was designed specifically to evaluate DTN protocols and has become popular within the DTN research community. For instance, a local approach to detect and respond to congestion by adjusting the copy limit for new messages [273] has been implemented and evaluated using the ONE simulator.

The remaining approaches employ other simulator platforms such as OP-NET [143], YACSIM [248], GT-ITM [248], Weka [78], Netem [110] and ION [56] [154].

We should also highlight the Interplanetary Overlay Network (ION) [154] implementation of the DTN architecture. ION accomplishes congestion control by computing *congestion forecasts* based on published contact plans. These forecasts are presented to the mission teams so that they can take corrective action, revising contact plans before the forecast congestion occurs. The transmission rates in the contact plans are enforced automatically by built-in rate control mechanisms in the ION bundle protocol agent. In the case of rate control failure, causing reception rate to exceed what was asserted in the contact plan, the receiving bundle protocol agent drops data according to a drop-tail policy to avoid congestion. The insertion of new bundles into the network can also lead to congestion. To avoid this, ION implements an admission control mechanism that may either function in a drop-tail manner or simply block the application until insertion of the new bundle no longer threatens to congest the node.

The general trend that can be observed is that most proposed DTN congestion control mechanisms have been evaluated experimentally using simulation platforms. Employing more realistic experimental environments including real world scenarios and testbeds should become a priority in DTN protocol research.

Deployability DTN congestion control mechanisms that rely on global network knowledge are classified as "low" deployability. A notable example is the protocol proposed in [306] which uses the network's average forwarding number, i.e., the average forwarding number considering all messages currently in the network, to decide which message(s) to drop in case nodes get congested (see Table 10.3). The low deployability of this approach is due to the fact that, in order to compute the average forwarding number, nodes need global knowledge of the network, which is hard to acquire, especially in DTNs.

Around 70% of the surveyed DTN congestion control mechanisms ([248] [143] [131] [132] [85] [247] [303] [81] [169] [222] [47] [110] [225] [274]

Table 10.3: DTN congestion control mechanisms with low deployability

Mechanism	Description
Average forwarding number based on epidemic routing (AFNER) [306]	Nodes drop messages whose forwarding number is larger than the network's average forwarding number (a message's forwarding number is the number of copies of that message floating in the network).

[273] [288] [287] [31]) can be classified as "medium" deployability. For example, the approach presented in [273] tries to respond to congestion by adjusting the maximum number of message copies based on the current level of network congestion. Nodes estimate global congestion levels using the ratio between drops and duplicate deliveries obtained during node encounters. From Table 10.4, which shows a brief description of medium-deployability

Table 10.4: DTN congestion control mechanisms with medium deployability

Mechanism	Description
Hop-by-hop Local Flow Control [225]	Nodes use hop-by-hop flow control where the sender verifies if the link is active and if there is resource availability towards the next-hop receiver.
Storage Routing [248]	Under congestion, nodes use a migration algorithm to transfer messages to other, less congested nodes.
Congestion Avoidance Based on Path Avoidance [143]	This scheme manages node storage and defines three states: normal, congestion adjacent, and congested. Nodes broadcast their current state to their neighbors who avoid forwarding messages to congested nodes.
Context Aware Forwarding Algorithm (CAFÉ) [131] [132]	Nodes adaptively choose a message's next hop based on contact history and statistics.
Token Based Congestion Control [85]	In this scheme all nodes must have a token in order to inject messages into the network. Tokens are initially uniformly distributed and after some time move randomly throughout the network.
Push-Pull Custody Transfer [247]	This approach uses buffer space availability information from neighbors to mitigate congestion. It includes a set of algorithms to determine which messages should be migrated to which neighbors and when.
Incentive Multi-Path Routing with Alternative Storage (IMRASFC) [303]	This scheme uses an incentive mechanism to stimulate mal-behaving nodes to store and forward messages and also try to select alternative neighbors nodes with available storage.
Congestion Control Routing Algorithm for Security Defense based on Social Psychology and Game Theory (CRSG) [81]	This approach uses social psychology and game theory to balance network storage resource allocation. It does this by obtaining node buffer utilization during node encounters.
Node-Based Replication Management (RRCC) [274]	This scheme detects and responds to congestion by adjusting the message copy limit. It uses local measurements, e.g., the ratio of the number of dropped messages to the number of message replicas measured by a node.
Congestion Avoidance Based on Buffer Space Advertisement [169]	Nodes advertise their buffer occupancy to adjacent nodes; neighboring nodes then use this information to decide their next hop when forwarding messages.
Congestion Aware Forwarding [222]	Congestion avoidance strategy which utilizes heuristics to infer shortest paths to destinations from social information (e.g., connectivity); it uses buffer occupancy and communication latency information to avoid areas of the network that are congested.
Combined Congestion Control for IPN [47]	This mechanism combines ECN (Explicit Congestion Notification) and RED (Random Early Detection) with storage-based routing strategies and makes use of neighbor buffer occupancy to mitigate congestion.
Threshold-Based Congestion Control [273]	This scheme tries to respond to congestion by adjusting the copy limit for messages based on the current observed network congestion level. The congestion level is estimated based on information collected during node encounters.
Licklider Transmission Protocol Transport (LTP-T) [110]	LTP-T maintains a congestion timer for forwarded blocks. The idea is that an entire block has to be transmitted before the timer expires. If the congestion timer expires at an intermediate node, this node should signal the presence of congestion to upstream nodes using LTP-T's congestion notification.
Simulated Annealing and Regional Movement (SARM) [288]	SARM adopts a message deleting strategy based on regional characteristics of node movement and message delivery. As nodes move, the algorithm records the cumulative number of encounters with other nodes. When congestion happens, the cumulative number of the node as the transferred node and the destination node for all messages in these two nodes is calculated and then the message is selected to be deleted.
Following Routing (FR) [287]	FR assumes all nodes are mobile; if a node *A* tries to relay a message to node *B* but *B* is not able to receive the message (e.g., because its buffer is full), *A* tries to follow *B*'s trajectory hoping to encounter a suitable next hop or the destination node.
Dynamic Congestion Control Based Routing (DCCR) [184]	DCCR is based on replication quotas. Each message is associated with an initial quota. During its time-to-live, a message can have its quota value updated according to a quota allocation function. This function reduces or increases the message's replication quota. Moreover, each node maintains buffer occupancy and contact probability with other nodes. In this case each node can compute the local measurement of congestion level information in order to update replication quotas.

Table 10.5: DTN congestion control mechanisms with high deployability

Mechanism	Description
Autonomous Congestion Control [61]	This mechanism adopts a financial model and compares the receipt and forwarding of messages to risk investment. When a new message arrives, the node decides whether to receive it or not according to a risk value of receiving and storing the message. The risk value is determined by local metrics, such as the node's own buffer space and the input data rate.
Distributive Congestion Control for Different Types of Traffic [78]	Congestion control is accomplished by distributing traffic according to different priority levels. Messages with higher priority are ensured minimum bounded delay whereas those with lower priority are discarded at higher congestion levels.
Congestion Management Based on Revenue Management and Dynamic Programming [311]	Nodes accept custody of messages (or bundles) if and only if there is sufficient remaining resource capacity and the resulting benefit of acting as relays is greater than or equal to the cost associated to the use of local resources.
Credit-Based Congestion Control [172]	According to this strategy, some credit is associated to each message; when two nodes encounter each other, the amount of credit for each message is updated. When buffers become full, messages that have the least credit are dropped.
Message Admission Control Based on Rate Estimation (MACRE) [31]	This congestion control scheme decides whether to admit a message according to the relationship between a node's input rate and output rate.
Interplanetary Overlay Network (ION) [154] [56]	ION's congestion control is anticipatory and is performed based on the "contact plan" between the nodes. The maximum projected occupancy of a node is based on the computation of the congestion forecast for the node. Thus, congestion control is performed essentially manually.
SMART-DTN-CC [93]	SMART-DTN-CC adjusts its operation automatically as a function of the dynamics of the underlying network. It employs reinforcement learning [133] [267], a machine learning technique known to be well suited to problems in which the *environment*, in this case the network, plays a crucial role; however, no prior knowledge of the target environment can be assumed, i.e., the only way to acquire information about the environment is to interact with it through continuous online learning. A Smart-DTN-CC node gets input from the environment (e.g., its buffer occupancy, set of neighbors, etc), based on that information chooses an *action* to take from a set of possible actions, and then measures the resulting *reward*. Smart-DTN-CC's overall goal is to maximize the reward which means minimizing congestion.

mechanisms, we observe that such mechanisms rely on local neighborhood information to perform congestion control.

The remainder of the investigated mechanisms, namely [61] [172] [78] [311] [184] [56] (see Table 10.5) can be included in the high-deployability category since they try to perform congestion control by using only local information, i.e., information about the node itself. For example, the scheme presented in [311] proposes a congestion management technique that decides whether to accept custody of a bundle if and only if the benefit of accepting custody is greater than or equal to the cost of the resources used to store the bundle.

Table 10.6: Flow and Congestion Control Mechanisms

Mechanism	Congestion Detection	Open- or Closed-Loop Control	Proactive or Reactive Control	Application	Contacts	Routing	Evaluation Platform	Deployability
Hop-by-hop local flow control [225]	Network capacity	Closed-loop	Hybrid	Interplanetary Communication	Scheduled	Independent	NS-2	medium
Storage routing for DTN congestion control [248]	Buffer availability	Closed-loop	Reactive	Environmental Monitoring	Predicted and Scheduled	Dependent	YACSIM and GT-ITM	medium
Autonomous congestion control [61]	Buffer availability	Closed-loop	Hybrid	Data Dissemination	Predicted	Independent	–	high
Congestion avoidance based on path avoidance [143]	Buffer availability	Closed-loop	Hybrid	Community-based Terrestrial Communication	Predicted	Dependent	OPNET	medium
Context aware forwarding algorithm (CAFE) [131][132]	Buffer availability	Closed-loop	Hybrid	Community-based Terrestrial Communication	Predicted	Dependent	own trace driven opportunistic simulator	medium
Token based congestion control [85]	Network capacity	Closed-loop	Proactive	Environmental Monitoring	Predicted	Independent	own discrete event simulator	medium
Push-Pull custody transfer [247]	Buffer availability	Closed-loop	Reactive	Mobile Terrestrial Communication	Predicted	Independent	DTNSim	medium
Incentive Multi-paths Routing with Alternative Storage (IMRASFO) [303]	Buffer availability	Closed-loop	Hybrid	Mobile Terrestrial Communication	Opportunistic	Dependent	ONE	medium
Distributive congestion control for different types of traffic [78]	Buffer availability	Open-loop	Hybrid	Mobile Terrestrial Communication	Predicted and Opportunistic	Independent	Weka undeployed	high
Congestion Control Routing Algorithm for Security Defense based on Social Psychology and Game Theory (CRSG) [81]	Buffer availability	Closed-loop	Reactive	Community-based Terrestrial Communication	Opportunistic	Independent	ONE	medium
Node-based replication management algorithm (RRCC) [274]	Drop rate	Closed-loop	Hybrid	Mobile Terrestrial Communication	Predicted and Opportunistic	dependent	ONE	medium
Buffer space advertisement to avoid congestion [169]	Buffer availability	Closed-loop	Proactive	Mobile Terrestrial Communication	Opportunistic	Independent	NS-2 and ONE	medium
Congestion aware forwarding algorithm [222]	Buffer availability	Closed-loop	Proactive	Data Dissemination	Predicted	Dependent	ONE	medium

Table 10.7: Flow and Congestion Control Mechanisms

Mechanism	Congestion Detection	Open- or Closed-Loop Control	Proactive or Reactive Control	Application	Contacts	Routing	Evaluation Platform	Deployability
Combined congestion control for IPN [47]	Buffer availability	Closed-loop	Hybrid	Interplanetary Communication	Scheduled	Dependent	NS-2	medium
Threshold-based congestion control protocol [273]	Drop rate	Closed-loop	Reactive	Mobile Terrestrial Communication	Predicted	Dependent	ONE	medium
Licklider transmission protocol transport (LTP-T) [110]	Buffer availability	Closed-loop	Reactive	Interplanetary Communication	Scheduled	Dependent	network emulation with Netem [136]	medium
Congestion management strategy based on revenue management and dynamic programming [311]	Buffer availability	Open-loop	Proactive	Mobile Terrestrial Communication	Opportunistic	Independent	own discrete event-driven simulator	high
Credit-based congestion control [172]	Buffer availability	Closed-loop	Hybrid	Mobile Terrestrial Communication	Opportunistic	Independent	ONE	high
Average forwarding number based on epidemic routing (AFNER) [306]	Network capacity	Open-loop	Hybrid	Mobile Terrestrial Communication	Opportunistic	Dependent	NS-2	low
Simulated annealing and regional movement (SARM) [288]	Buffer availability	Open-loop	Reactive	Data dissemination	Predicted and Opportunistic	Dependent	ONE	medium
Following routing (FR) [287]	Buffer availability	Open-loop	Reactive	Mobile Terrestre Communication	Opportunistic	Dependent	ONE	medium
Dynamic congestion control based routing (DCCR) [184]	Buffer availability	Closed-loop	Hybrid	Mobile Terrestre Communication	Opportunistic	Dependent	ONE	medium
Message Admission Control based on Rate Estimation (MACRE) [31]	Buffer availability	Closed-loop	Hybrid	Data dissemination	Opportunistic	Dependent	ONE	high
Interplanetary Overlay Network (ION) [154][56]	Network capacity	Closed-loop	Proactive	Interplanetary Communication	Scheduled	Independent	ION framework	high
SMART-DTN-CC [93]	Buffer availability	Hybrid	Hybrid	Interplanetary and Terrestrial	Predicted, Opportunistic and Scheduled	Independent	ONE	high

10.6 Summary

In delay tolerant networks, the high mobility of nodes and their dramatically changing topologies lead to intermittent connectivity. The store-carry-forward principle is used by most DTN routing protocols to forward messages. With limited storage space, excessive copies of messages can easily lead to buffer overflow, especially when the bandwidth is also limited and the message sizes differ. In this situation, the question of how to allocate network resources becomes important.

This chapter presented a definition of DTN congestion control and described some existing DTN congestion control mechanisms. A survey of the existing DTN congestion control mechanisms suggests that the design of a more general congestion control mechanism for DTN depends on three important factors.

First, it needs to operate independently of the routing protocol so that it may be suitable for different DTN scenarios.

Second it must embrace a hybrid approach, being able to prevent congestion from taking place (proactive control) and also to promptly react to congestion when it happens (reactive control). Since messages can be buffered for long periods of time before being acknowledged, buffer saturation becomes highly common resulting in excessive latencies and messages losses. This is aggravated by the fact that, in this context, reactive congestion control would be severely impacted by delayed control decisions. To solve this problem a proactive strategy is extremely important.

Finally, a congestion control mechanism must consider both open- and closed-loop strategies, where an open-loop scheme is fundamentally important because it takes into account policies that carefully look for minimizing congestion in the first place, since a closed-loop scheme is not always suitable for DTN. Any closed-loop scheme uses feedback information, and in the context of DTNs such feedback may not be immediate and may take some time before nodes are informed. This surely has a negative impact on network performance. Thus the use of closed-loop feedback control exclusively to maintain consistent state at the source and destination cannot work.

Congestion control is a fundamental issue in DTNs and it is expected to receive much attention from the DTN research community when DTN starts to be deployed in large scale.

Chapter 11

Licklider Transmission Protocol (LTP)

Nicholas Ansell

Independent Researcher

CONTENTS

11.1   Chapter Preface

This chapter first provides an introduction to LTP in section 11.2. The position of LTP within a typical protocol hierarchy is covered as well as protocol encapsulation within an example hierarchy. A description of how payload data is prepared for transmission is then provided. Finally, section 11.2 concludes with two typical transmission sequences and a description of two key LTP concepts.

The sections of the chapter following the introduction provide further detail in a number of different areas. Section 11.3 provides a detailed look at the structure and content of LTP segments. The Internal procedures used by LTP are found in section 11.4. Sender and receiver state transition diagrams are provided in section 11.5. Security is covered in section 11.6, including the security mechanisms currently built into the LTP protocol. Section 11.7 provides a practical view of LTP by analyzing network traffic between nodes in two different scenarios. Finally, section 11.8 concludes the chapter with a summary and predictions for the future.

11.2   Introduction

Many Delay and Disruption Tolerant Networking applications can successfully employ traditional networking protocols to achieve reliable communication. Such networks will more than likely seek to take advantage of the ability of DTNs to achieve reliable communication despite disruptions in the network path. One implementation pattern using this approach is DTN data mules, where a data mule transfers data between two or more otherwise disconnected endpoints. Data is transferred from one endpoint to the mule over a short-distance, high-speed network link. The connection is then broken while the mule travels to another endpoint. Once the destination endpoint is reached, a connection is established between the mule and the destination endpoint and the data is transferred. In this scenario the Transmission Control Protocol over Internet Protocol (TCP/IP) could be used over an IEEE 802.11 wireless connection for example, as the data mule and endpoints are likely to be in close proximity to each other during data transmission.

However, there are circumstances where quotidian protocols such as TCP/IP are not suitable. For example, in deep-space where networks spanning huge distances have latency which is measured in minutes, rather than milliseconds. Another characteristic found in deep-space networks which negatively affects TCP/IP is network asymmetry. This asymmetry refers to the return path from a spacecraft having faster data-rates than the outbound path from ground stations, normally due to equipment or power limitations on the spacecraft. Finally, TCP/IP is also found to be unreliable in networks with high-loss, specifically networks with Bit Error Rates (BER) of 10^{-6} and greater. A BER of 10^{-6} refers to one detected bit error per 1 million bits of transmitted data.

The Licklider Transmission Protocol (LTP) was developed by the Delay Tolerant Networking Research Group (DTNRG) to provide reliable network communication over links with high latency and frequent interruptions [55]. The protocol also accommodates the aforementioned network asymmetry and high-loss scenarios much better than TCP/IP. This chapter provides the reader with a comprehensive understanding of LTP, including the methods LTP employs to achieve dependable communication over unreliable networks.

LTP was named after the Internet pioneer Joseph Carl Robert Licklider, an American psychologist and computer scientist, who is known for formulating some of the earliest global networking ideas. These pioneering thoughts led C.R. Licklider to be credited as one of the key thinkers who set the stage for ARPANET, the predecessor of the Internet.

11.2.1 Protocol Overview

There are a number of features factored into the design of LTP which optimize transmission of data over long-haul space network links. These features are summarized below and explained in more detail throughout the chapter.

- Transmission of all segments within an LTP block without waiting for acknowledgment of previous segments. Due to the long round-trip times in deep space, long gaps between data transmissions would result if LTP waited for a positive acknowledgment before sending further segments. In order to make the best use of available bandwidth and transmission equipment timeslots, LTP transmits an entire block of data and waits for acknowledgment. This behavior can be overridden, by introducing intermediate acknowledgment checkpoints if required.

- Multiple parallel sessions are used between LTP nodes to avoid underutilization of the bandwidth available on a network link.

■ Use of Automatic Repeat reQuests (ARQs) to retransmit data if not acknowledged by the receiving node.

■ LTP is "stateful". This means that until all data transmit has been acknowledged, an LTP node must maintain a record of all data sent and also retain the data for retransmission if required.

■ Network asymmetry between nodes is accommodated using configurable, unidirectional connections. This allows for the most optimal use of both the outbound and inbound network paths. The ability to fine-tune each connection also allows bandwidth utilization to be controlled if some bandwidth must be reserved on a link. This fine-tuning is achieved by manipulating configuration parameters such as segment size, aggregation time/size limit, number of import/export sessions and one way light time between nodes.

■ No handshake is used to establish a session before commencement of transmission. Unlike TCP, which conducts a 3-way handshake before transmitting data, LTP foregoes this to make the best use of the available timeslot. This is particularly beneficial over links with long round-trip times. For example, the one-way-light-time (data travelling at the speed of light) between Earth and Mars at their closest point is approximately 4 minutes. Many minutes would pass before any payload transmission was started if a 3-way handshake was needed.

■ LTP supports transmission of both "red-part" data (which must be acknowledged) and "green-part" data (which is sent on a best efforts basis).

11.2.2   LTP Placement within a Protocol Hierarchy

In space networks, LTP is typically deployed over radio frequency (RF) data links using the Consultative Committee for Space Data Systems (CCSDS) standards [18]. LTP can also be operated over the User Datagram Protocol (UDP). The ability to use LTP over UDP was originally implemented for testing and development purposes, but more recently this has been used by NASA for communication with the International Space Station. Implementations of LTP using UDP should use port 1113, which is the ltp-deepspace port assigned by the Internet Assigned Numbers Authority (IANA).

The placement of LTP in relation to other network layers and protocols within a DTN architecture, used over UDP or CCSDS, is shown in Figure 11.1. First an application requests the bundle protocol agent (BPA) to send the payload data to a destination node. Then the BPA pushes the payload into one or more bundles, based on the size of the file and the configured bundle size. The BPA passes the bundles to the LTP convergence layer (LTPCL), which uses the LTP protocol to send the data to the destination node via the underlying network layer.

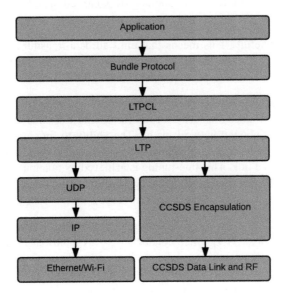

Figure 11.1: LTP Protocol Hierarchy

11.2.3 Protocol Encapsulation

During transmission of data, encapsulation occurs as the operation transfers from higher-level to lower-level protocols. A simplified encapsulation illustration is shown in Figure 11.2. In this example, an LTP segment containing one bundle of payload data is sent over UDP on an Ethernet network. As can be observed each layer places an overhead on the received data before passing down to the next layer. The overhead, consisting of a header and also sometimes a footer, includes source and destination identifiers plus required configuration or session data.

In the provided example, the payload data is passed to the bundle protocol agent, which places the payload into a single bundle. Each bundle overhead contains source and destination endpoint identifiers (eid) plus required configuration or session data. The bundle is pushed into an LTP block which is then placed into a single LTP segment. The LTP overhead contains source and destination LTP engine ids and required configuration or session data. The LTP segment is placed into a UDP datagram, with the UDP overhead containing source and destination ports, plus configuration and session data. The UDP datagram is placed into an IP packet with an overhead containing the source and destination IP addresses, plus configuration and session data. Finally, the IP packet is placed into an Ethernet frame, with the Ethernet overhead

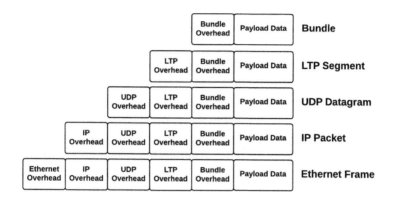

Figure 11.2: LTP Protocol Encapsulation

containing source and destination hardware addresses plus configuration and session data.

11.2.4 Bundles, Blocks and Segments

Payload data undergoes a number of transformations in the bundle and LTP layers shown in Figure 11.2, before being passed to the transport protocol. These transformations are shown in Figure 11.3. After receiving the payload data, the bundle protocol agent pushes the data into one or more bundles (based on the size of the data and configured bundle size). The bundles are then passed to LTP via the LTP convergence layer. LTP pushes the bundle data into LTP blocks. This may result in bundle fragmentation (i.e. a single bundle spanning more than one block) or aggregation (i.e. multiple bundles pushed into a single block). The likelihood of bundle aggregation or fragmentation is governed by a number of factors including the size of the bundles and LTP configuration parameters such as aggregation time limit or aggregation size limit. The LTP blocks are then broken down into segments, which are queued for transmission to the destination node. The size of an LTP segment can be configured on a node, but the maximum segment size is ultimately governed by the underlying maximum transmission unit (MTU) of the network layer.

There are a number of different types of LTP segments, including report segments (which acknowledge the receipt of data) and cancellation segments (which are used to cancel LTP sessions). Further explanation of each segment type, including actions taken upon receipt of each segment is found throughout this chapter.

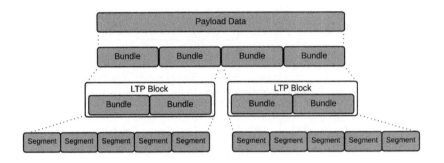

Figure 11.3: Bundles, Blocks and Segments

11.2.5 *LTP Engines and Sessions*

When configuring and using an LTP network, one must be familiar with the concept of LTP engines and sessions, a graphical overview of which is provided in Figure 11.4. An engine id is a unique identifier that must be assigned to each node in an LTP network and is analogous to an IP address in a TCP/IP network. Like TCP/IP a single node can also be assigned more than one engine id if running multiple instances of LTP. When LTP is used underneath the bundle protocol, the LTPCL is responsible for translating between DTN endpoint ids and LTP engine ids.

In order to send an LTP block containing multiple segments, the sender node must establish a session with the receiver node. Unlike TCP, this is not done using a multi-directional handshake. A sender node simply generates a session id then adds that id to the header of all segments related to that LTP block. The receiver node, when sending any segments to the sender node (for example a report segment) adds this session id to the header of each segment. This method allows for multiple sessions to be established simultaneously, thus increasing the likelihood of using available bandwidth effectively. An LTP session is closed once the LTP block has been successfully transmitted. A session can also be cancelled by either node for various reasons. Further information on session cancellation is provided later in the chapter.

11.2.6 *Typical LTP Sequence*

Figures 11.5 and 11.6 show typical examples of a single block of red part data (data that must be acknowledged) being sent from one LTP node to another.

In Figure 11.5 a single block of red part data is sent from the sender to the receiver with no transmission errors experienced. First, the sender sends the data segments, the last of which has the checkpoint (CP), end of red part (EORP) and end of block

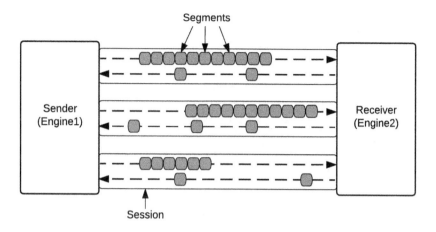

Figure 11.4: LTP Engines and Sessions

(EOB) markers. The receiver, upon receiving the CP, compiles a report segment (RS) to acknowledge the successful receipt of the red part data. The sender, upon receiving the RS, sends a report acknowledgment (RA). Having confirmed that all red part data within the block was both sent and acknowledged, both nodes close the session.

Figure 11.6 shows the same block of red part data sent from the sender to the receiver. However, in this scenario data is lost during transmission. The sequence is almost identical to Figure 11.5, but in this case the first RS does not acknowledge all red part data within the block. The sender establishes which data is missing by analyzing the RS and resends the required red part data, including the required CP, EOB and EORP markers in the last segment. The receipt of the CP triggers the receiver to send a corresponding RS. The RS is acknowledged by the sender by sending an RA. Following the successful transmission and acknowledgment of all red part data in the block, both the sender and receiver close the session.

11.2.7   Link State Cues and Transmission Queues

LTP is dependent on link state cues in order to operate in a time and power efficient manner. Link state cues are internal triggers used by LTP to communicate between different tasks. For example there are link state queues used to indicate the status of a path to a node. This may be a link state cue indicating that radio frequency communication has been established, for example following antenna alignment or frequency calibration. This same link state queue may also be an event triggered by prior knowledge of contact to a node. For example using contact plans, which can be generated

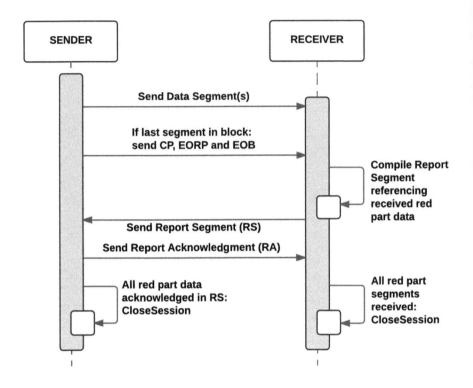

Figure 11.5: LTP Sequence - No Errors

ahead of time (perhaps using a combination of ephemeris data for planetary orbits and known schedules of orbiting satellites). Link state queues may also indicate other events such as the sending of a CP, to trigger the starting of a CP timer. Moreover, link state cues are used to signal that a path to a node is no longer available, in which case segments are generated but their transmission is deferred into one of two queues.

The two transmission queues used by LTP are named internal operations queue and application data queue. Both queues operate on a first-in, first-out basis i.e. the oldest segment in the queue is always dequeued for transmission first. The internal operations queue takes priority over the application data queue and therefore must be empty before any segments are sent from the application data queue. If however, the link status to a node is active and both queues are empty, a segment is in concept sent immediately to the destination node.

Segments such as report segments (RS) or report acknowledgments (RA) are placed in the internal operations queue. This in theory gives a higher priority to acknowl-

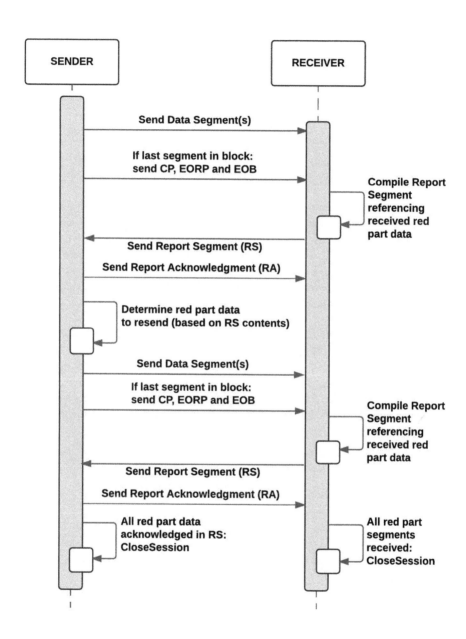

Figure 11.6: LTP Sequence - With Errors

edgment segments and accelerates any required retransmission within a given transmission window. In concept, it can therefore be assumed that this method prioritizes

the completion of prior blocks over transmission of new blocks. This in turn, should reduce the size of storage required at each node, as the smallest possible number of open blocks need to be maintained within a node's state.

11.3   Segment Structure

This section details the structure of LTP segments and the encoding used within them. A detailed description of the segment header and content is provided, along with examples of various different segment types.

LTP uses self-delimiting numeric values (SDNV) to provide extensibility and scalability, which is based on the abstract syntax notation one (ASN1) scheme. Data encoded as an SDNV uses the most significant bit (MSB) to indicate whether or not an octet of data is the last octet of the SDNV. This feature allows for data of variable length to be transmitted, providing it is terminated by an octet with an MSB of 0. Correspondingly, an octet with an MSB of 1 indicates it is the first or middle octet within a series of octets using the SDNV scheme.

An example of the letters A (1010), B (1011) and C (1100) encoded as a two octet SDNV (10010101 00111100) is provided in Figure 11.7. Octet 1 has the MSB set to 1, indicating this is the first or middle octet in the SDNV. Octet 2 has the MSB set to 0, indicating it is the last octet of the SDNV. Note that each octet must total eight bits, or seven bits plus the MSB. If zero padding is required to achieve a total of 8 bits, this must be applied at the start of the octet, after the MSB as shown in octet 1.

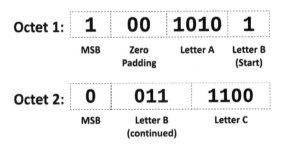

Figure 11.7: SDNV Encoding of 0xABC. Source: Adapted from [224]

Another example, demonstrating the numbers 1 (0001), 2 (0010), 3 (0011) and 4

(0100) encoded as a two octet SDNV (10100100 00110100) is shown in Figure 11.8. Like the 0xABC example, octet 1 has the MSB set to 1 and Octet 2 has the MSB set to 0. Note in this example the leading zeros are trimmed from the start of the binary representation of the number 1.

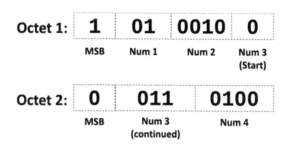

Figure 11.8: SDNV Encoding of 0x1234. Source: Adapted from [224]

The transmission of data by LTP uses three segment types. Additionally there is a forth type of LTP segment used to manage the transmission session. Data segments are sent by the sender to the receiver and contain the payload data passed by the Bundle Protocol Agent to LTP. RS are sent by the receiver to the sender to identify to the sender which data was successfully received. RA segments are sent by the sender to the receiver, to signal the successful receipt of a report segment. Report acknowledgment segments contain the serial number of the report segment to ensure correct identification of the report segment being acknowledged.

In addition to the aforementioned three segment types, LTP also uses two session management segments: cancellation segments (Cx) and cancellation acknowledgment segments (CAx). Session management segments can be sent by either the sender or receiver and are used to terminate an LTP session. A cancellation segment contains a code describing the reason for the cancellation, which can be used for troubleshooting LTP transmission problems. A cancellation acknowledgment segment however has no content and simply signals the successful receipt of a cancellation segment.

Figure 11.9 shows a high-level overview of the structure of an LTP segment, which consists simply of a header, a content section and a trailer section. The trailer section is reserved for any extensions used for the transmission.

Header	Segment Content	Trailer Extensions

Figure 11.9: LTP Segment Overview

11.3.1 Segment Header

Figure 11.10 shows a detailed representation of an LTP header. The first octet - the control byte, is broken down into two four bit sections. The first four bits indicate the protocol version ("0000" at the time of writing). The last four bits are used to denote the type of segment using segment type flags. The session id portion of the segment contains two SDNVs. The first SDNV identifies the sender using its engine id. The second SDNV is the session id, which is a random number generated by the sending node to uniquely identify the session. This is important as a single LTP transmission will often use more than one session to make the best use of available bandwidth and equipment timeslots.

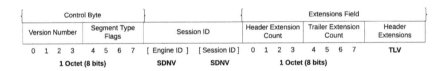

Figure 11.10: LTP Segment Header. Source: Adapted from [224]

11.3.1.1 Segment Type Flags

The segment type flags present within the last 4 bits of the control byte indicate the segment types contained within each segment. Each of the flags are named in order as CTRL, EXC, flag 1 and flag 0. The sixteen different combinations of flags, their unique code and a description of each is shown in Table 11.1. Rules can be noticed which are applied to the flags in some segment types. For example, all segment types with the CTRL bit set to zero are data segments. Another example is segments with both the CTRL and EXC flags set to one are used for session cancellation. A further description of each segment type, including the actions taken by nodes upon receipt of each can be found later in this chapter.

Table 11.1: Segment Type Flags. Source: Data from [224]

CTRL	EXC	Flag 1	Flag 0	Code	Description
0	0	0	1	1	Red data, CP, NOT EORP or EOB
0	0	1	0	2	Red data, CP, EORP, NOT EOB
0	0	1	1	3	Red data, CP, EORP, EOB
0	1	0	0	4	Green data, NOT EOB
0	1	0	1	5	Green data, undefined
0	1	1	0	6	Green data, undefined
0	1	1	1	7	Green data, EOB
1	0	0	0	8	Report Segment (RS)
1	0	0	1	9	Report-acknowledgment segment (RA)
1	0	1	0	10	Control segment, undefined
1	0	1	1	11	Control segment, undefined
1	1	0	0	12	Cancel segment from block sender (CS)
1	1	0	1	13	Cancel-acknowledgment segment to block sender (CAS)
1	1	1	0	14	Cancel segment from block receiver (CR)
1	1	1	1	15	Cancel-acknowledgment segment to block receiver (CAR)

11.3.1.2 Extensions Field

The extensions field within an LTP segment header is broken into two sections. The first section is one octet and broken into two four bit sections. The first four bits denote the number of extensions present in the header extensions portion of the segment. The second four bits denote the number of extensions present in the trailer portion of the segment. If there are no extensions present in the header or trailer sections, all four bits in the corresponding part of the octet must be set to zero.

When extensions are present, they are defined in a type, length, value (TLV) structure. The type section contains one octet identifying the type of extension, using the extension tags shown in Table 11.2. All LTP extensions are maintained by the Internet Assigned Numbers Authority (IANA). The length parameter is an SDNV indicating the length of the data within the value section. The value section contains the data of the length specified in the length section.

One example of an extension used in the header extension portion of a segment is the LTP cookie extension (0x01), which is used to make denial of service attacks more difficult.

Table 11.2: Extension Tags. Source: Data from [224]

Extension Tag	Description
0x00	LTP Authentication Extension
0x01	LTP Cookie Extension
0x02 to 0xAF	Unassigned
0xB0 to 0xBF	Reserved
0xC0 to 0xFF	Private / Experimental Use

11.3.2   Segment Content

The content section varies depending on whether it is a data segment, a report segment, report acknowledgment segment or session management segment.

11.3.2.1   Data Segment

A data segment is sent by the sending node and primarily used for client service data, but can also contain a checkpoint or report serial number as shown in Figure 11.11. The start of a data segment contains an SDNV to identify the client service receiving

Client Service ID	Offset	Length	Checkpoint Serial No.	Report Serial No.	Client Service Data
SDNV	SDNV	SDNV	SDNV	SDNV	Array of octets

Figure 11.11: Data Segment. Source: Adapted from [224]

the data. The client service id is required to allow multiple client service applications to exist on a single LTP node and is analogous to a TCP port number.

The offset SDNV is used when the client service data is fragmented across multiple segments. The offset value is simply the number of bytes from the original client service data, which precede the data within this client service data chunk. If all data is contained within a single segment, or it is the first segment in the block, the offset will be zero.

The length SDNV denotes the number of octets which are present in the client service data section.

A checkpoint marks the end of a section of data which must be verified by the receiving node. Each checkpoint has a unique serial number, sent as an SDNV which is generated by the sending node. The first checkpoint serial number is normally a random number for security reasons; each subsequent checkpoint serial number is simply incremented by one. If a checkpoint is transmitted and not acknowledged within a calculated time by the receiving node, it is resent using the original checkpoint serial number.

Checkpoints are sometimes queued for transmission following the retransmission of data, as identified in a report segment. In this case, the report serial number that triggered the retransmission is contained within the report serial number section.

The client service data section contains the client service data (or payload) to be sent, starting at the provided offset if used.

11.3.2.2   Report Segment (RS)

Report Serial No.	Checkpoint Serial No.	Upper bound	Lower bound	Claim Count	Reception Claims
SDNV	SDNV	SDNV	SDNV	SDNV	Offset and Length SDNV(s)

Figure 11.12: Report Segment. Source: Adapted from [224]

A report segment is sent by a receiving node to indicate to the sending node which red data was successfully received. An overview of the report segment structure is shown in Figure 11.12.

The start of a report segment contains an SDNV denoting a unique serial number generated by the receiving node. A report serial number is normally a randomly generated number for security reasons and, like a checkpoint serial number, must be incremented by one with each subsequent report serial number. If a report segment is resent, the same report serial number as the original report segment is used.

The next SDNV contains the serial number of the checkpoint which the report segment is responding to. The third and fourth SDNVs denote which part of an LTP block the report segment relates to using upper and lower bound identifiers. The upper bound SDNV denotes the size of the LTP block prefix for which the reception claims are confirming receipt. The lower bound SDNV denotes the size of the LTP block prefix that the reception claims do not relate to. Or in other words the scope of an LTP block which the reception claims refer to, is between the lower bound SDNV value and the upper bound SDNV.

The final two sections of a report segment contain the number of claims (claim count SDNV) and one or more reception claims. Each reception claim contains an offset SDNV and length SDNV. The offset SDNV, in combination with the length and lower bound SDNV, is used to identify the data within an LTP block being acknowledged.

A report segment example is shown in Figure 11.13. In this example, the report segment is sent in response to a checkpoint with a serial number of 3554. The report segment contains a single claim to acknowledge the successful receipt of data from byte zero to byte 2048 of an LTP block.

Report Serial No.	Checkpoint Serial No.	Upper bound	Lower bound	Claim Count	Reception Claims
1221	3554	2048	0	1	Offset:0 Length 2048

Figure 11.13: Report Segment Example 1

A second example is shown in Figure 11.14. In this case, the report segment is sent in response to checkpoint 3555. This segment contains two claims for data within an LTP block from bytes 2048 (the lower bound SDNV) to 6144 (the upper bound SDNV). The first reception claim is for bytes 2048 to 4096. This can be calculated by first determining the start point within the block for the reception claim. The start point is simply the sum of the lower bound value and the offset (2048 + 0 = 2048). The end point within a block for the reception claim in this case is determined by the sum of the length of all consecutive claims, plus the offset provided by the lower bound value (2048 + 2048 + 2048 = 6144).

Report Serial No.	Checkpoint Serial No.	Upper bound	Lower bound	Claim Count	Reception Claims
1222	3555	6144	2048	2	Offset:0 Length 2048 Offset : 2048 Length 2048

Figure 11.14: Report Segment Example 2

A third example is shown in Figure 11.15, which provides an alternative scenario to the example in Figure 11.14. In this case, the segment contains one reception claim for bytes 2048 to 4096 and a second reception claim for bytes 5048 to 6144. This report segment therefore indicates that bytes 2049 to 2999 were lost and require retransmission.

Report Serial No.	Checkpoint Serial No.	Upper bound	Lower bound	Claim Count	Reception Claims
1222	3555	6144	2048	2	Offset:0 Length 2048 Offset : 3000 Length 1096

Figure 11.15: Report Segment Example 3

11.3.2.3   *Report Acknowledgment Segment (RA)*

A report acknowledgment segment is sent by the sending node to the receiving node to acknowledge receipt of a report segment. This segment contains a single SDNV which represents the serial number of the report segment being acknowledged.

11.3.2.4   *Session Management Segments (CS, CR, CAS, CAR)*

Session management segments are sent by either the sending or receiving node and are used to cancel a particular LTP session. As can be observed in Table 11.3, cancellation segments from the sending node use different segment type codes to cancellation segments sent from the receiving node. This feature is implemented to allow loopback testing to be conducted.

Cancellation segments (CS or CR) contain a single byte code to indicate the reason for the cancellation, as shown in Table 11.3. Upon receipt of a cancellation segment, a node will respond with a cancellation acknowledgment segment (CAS or CAR) which has no content.

Several scenarios may trigger a session cancellation; one commonly observed code is RLEXC (0x02). This code is seen during times of high transmission loss and can indicate that a checkpoint has been resent the maximum number of times, therefore the sending node has chosen to cancel the session and start a new one. Another possible cause for a RLEXC reason-code in high-loss conditions is when a node is not able to acknowledge all received data before reaching the maximum allowed number of report segments.

Table 11.3: Session Cancellation Codes. Source: Data from [224]

Code	Mnemonic	Description
0	USR_CNCLD	The client service (application requesting the LTP transmission), cancelled the session
1	UNREACH	The client service is unreachable
2	RLEXC	Retransmission limit exceeded
3	MISCOLOURED	Offset calculations determined that red part or green part data has overlapped
4	SYS_CNCLD	System error resulting in a cancellation
5	RXMTCYCEXC	Exceeded the retransmission cycles limit
06-FF	Reserved	Reserved for future use

11.3.3 Segment Trailer

In addition to the header extension portion of the segment, extensions are also placed into the trailer section using the same TLV format as the header extensions. One example using the trailer section is the LTP authentication extension, details of which are found in section 11.6 of this chapter.

11.4 Internal Procedures

This section describes the procedures used by LTP while responding to various events within an LTP session. In order to avoid any unnecessary processing, before any internal procedures are triggered, the procedure in Figure 11.16 is followed.

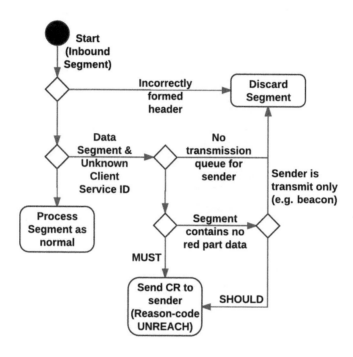

Figure 11.16: Internal Procedures Initial Flow. Source: Adapted from [224]

A summary of all internal procedures is shown in Table 11.4. In this table, the event that triggers each procedure is summarized in the trigger column. The action summary column provides an overview of the most probable actions. Potential deviations from the actions within the table and a further explanation of each procedure can be found in the subsections following Table 11.4.

Table 11.4: Internal Procedures. Source: Adapted from [224]

Trigger	Procedure	Action Summary
Link State Cue "Start of transmission to node"	Start Transmission	Dequeue and deliver segments to the specified node
Link State Cue "Cease of transmission to node"	Stop Transmission	Stop dequeuing and delivery of segments to the specified node
Link State Cue "CP dequeued for transmission"	Start Checkpoint Timer	Calculate expected arrival time of RS, start CP timer
Link State Cue "RS dequeued for transmission"	Start Report Segment Timer	Calculate expected arrival time of RA, start RS timer
Link State Cue "Cx dequeued for transmission"	Start Cancel Timer	Calculate expected arrival time of CAx, start Cx timer
CP Timer Expiry	Retransmit Checkpoint	Resend corresponding CP or if retry limit exceeded send Cx, call "Cancel Session" procedure and trigger "Transmission-Session Cancellation" notice to client service.
RS Timer Expiry or receive redundant CP	Retransmit Report Segment	Resend corresponding RS or if retry limit exceeded send Cx, call "Cancel Session" procedure and trigger "Transmission-Session Cancellation" notice to client service.
Cx Timer Expiry	Retransmit Cancellation Segment	Resend corresponding Cx or if retry limit exceeded call "Close session" procedure
Receive RS	Retransmit Data	Send RA and cancel CP timer, resend missing data if missing data identified in RS or no action if RS redundant or contains unknown serial.

Continuation of Table 11.4

Trigger	Procedure	Action Summary
Receive CP with or after the corresponding EORP	Signify Red-Part Reception	Trigger "Red-Part Reception" notice to client service
Receive Green Data Segment	Signify Green-Part Segment Arrival	Trigger "Green-Part Segment Arrival" notice to client service
Receive CP / Asynchronously triggered by specific implementation circumstance	Send Reception Report	Send RS or if retry limit exceeded send CR, call "Cancel Session" procedure and trigger "Reception-Session Cancellation" to client service
Link State Cue "Cease of transmission to node"	Suspend Timers	Suspend all timers (CP, RS or Cx) with an expected arrival time $>=$ current time
Link State Cue "Start of transmission to node"	Resume Timers	Recalculate expected arrival time of all RS, RA or CAx identified in suspended CP, RS or Cx timers, then resume the suspended timers
Receive RA	Stop Report Segment Timer	Delete corresponding RS timer identified in the RA. If no other corresponding RS timers are running, call the "Close Session" procedure.
Receive Cx	Acknowledge Cancellation	Send CAx, call "Cancel Session" procedure, trigger "Reception-Session Cancellation" notice to client service, then call "Close Session" procedure
Receive CAx	Stop Cancel Timer	Delete corresponding Cx timer identified in the CAx and call "Close Session" procedure
EOB Sent and RS(s) received acknowledging all red data	Signify Transmission Completion	Trigger "Transmission-Session Completion" notice to client, then call "Close Session" procedure.

Continuation of Table 11.4

Trigger	Procedure	Action Summary
Call by other internal procedure	Cancel Session	Delete corresponding segments from outbound queue and cancel all corresponding timers
Call by other internal procedure	Close Session	Delete all corresponding timers and session state record
Receive Mis-colored/Mis-ordered segment	Handle Mis-colored Segment	Discard segment, call "Cancel Session" procedure, trigger "Reception-Session Cancellation" notice to client service
Irrecoverable System Error	Handle System Error Condition	Call "Cancel Session" procedure, send Cx and trigger notice to client "Transmission-Session Cancellation" (if sender) or "Reception-Session Cancellation" (if receiver)

11.4.1 Start Transmission

Following the arrival of a link state cue indicating a path to a node is available, segments are dequeued and delivered to the remote LTP engine specified in the link state cue. Often, this link state cue is generated based on a known event. For example, an LTP administrator may configure a contact window based on prior knowledge of orbital mechanics, which would affect direct communication between two nodes. In this case, when the start of a configured contact window is reached, the "start transmission to node" link state cue is generated.

11.4.2 Stop Transmission

Following arrival of a link state cue indicating a path to a node is unavailable, all transmission is stopped to the node specified in the link state cue. Often, this link state cue is generated based on a known event. For example, an LTP administrator may configure a contact window based on prior knowledge of orbital mechanics, which would affect direct communication between two nodes. In this case, when the end of a configured contact window is reached, the "cease transmission to node" link state cue is generated.

11.4.3   Start Checkpoint Timer

After a checkpoint (CP) is dequeued for transmission, a report segment (RS) is expected from the receiver node to acknowledge the transmitted data. This procedure computes the expected arrival time of the RS and starts a countdown timer (CP timer). The CP serial number is used to uniquely identify each respective CP timer. If however, it is known that a remote node has ceased transmission, it is impossible to compute the expected arrival time of the RS and the countdown timer is suspended.

11.4.4   Start Report Segment Timer

This procedure is similar in function to the start checkpoint timer procedure. After an RS has been dequeued for transmission, a report acknowledgment (RA) is expected from the sender node, to acknowledge the receipt of the report segment. This procedure computes the expected arrival time of the RA and starts the RS timer. Like the CP timer procedure, the RS serial number is used to uniquely identify each RS timer. Also like the start checkpoint timer procedure, an RS timer will be suspended if it is known the remote node has ceased transmission.

11.4.5   Start Cancel Timer

This procedure is triggered following the de-queuing for transmission of a cancellation segment "Cx" (CS or CR). Like the CP and RS timer operations, this procedure first calculates the expected arrival time of the required acknowledgment segment. In this procedure the corresponding cancellation acknowledgment for the transmitted cancellation segment is a CAx (CAS or CAR). In a similar fashion to the CP and RS timer functions, the CS timer will be suspended if it is known the report node has ceased transmission.

11.4.6   Retransmit Checkpoint

When a countdown timer for a CP expires, (i.e. the node has not received an RS within the calculated time period) this procedure is triggered. The usual action taken is to simply resend a new copy of the CP identified in the CP timer. However, if the retry limit is exceeded the node transmits a CS segment containing the RLEXC reason-code. The node then calls the "cancel session" procedure and triggers the "transmission-session cancellation" notice to the client service.

11.4.7   Retransmit Report Segment

When an RS timer expires, (i.e. the node has not received an RA within the calculated time period) this procedure is triggered. The usual action taken is to simply resend a new copy of the RS identified in the RS timer. However, if the retry limit is exceeded the node transmits a CR segment containing the RLEXC reason-code. The node then calls the "cancel session" procedure and triggers a "transmission-session cancellation" notice to the client service.

11.4.8 Retransmit Cancellation Segment

When a CS timer expires (i.e. the node has not received a CAS or CAR within the calculated time period) this procedure is triggered. The usual action taken is to simply resend a new copy of the Cx. However, if the retry limit is exceeded the node calls the "close session" procedure to end the session.

11.4.9 Retransmit Data

When a node receives an RS this procedure is called. The first action taken is to send an RA containing the serial number of the received RS. At this point, there are two conditions which cause no further action to be taken:

■ The RS has already been processed (i.e. a redundant RS resent by the receiver following an RS timer expiry)

■ The RS serial number is unknown (this may relate to a session which has already been closed)

If none of the above conditions are true, the node deletes the corresponding CP timer. Next, any data acknowledged in the RS is cleared from the retransmission buffer. If the RS indicates some data was lost during transmission, the required data segments are queued for re-transmission to the receiver node. The last data segment contains a new CP using a serial number equal to the CP identified in the RS incremented by one. The CP also includes the report serial number for the received RS, which indicates to the sender node this is a CP related to data which has been retransmitted.

If however, a retry limit has been reached for this block (for example the CP retry limit) no data is resent. Instead, the node calls the "cancel session" procedure and sends a CS containing reason-code RLEXC. Finally, the node triggers the "transmission-session cancellation" notice to the client service.

11.4.10 Signify Red-Part Reception

This procedure is triggered when a node receives a data segment containing a CP and EORP marker, or when a CP is received after the corresponding EORP. In many cases, the segment containing the CP will also contain the EORP. This procedure simply triggers a "red-part reception" notice to the client service.

11.4.11   Signify Green-Part Reception

When a node receives a data segment containing a green-part, this procedure simply triggers a "green-part segment arrival" notice to the client service.

11.4.12   Send Reception Report

Upon receiving a CP, which was sent immediately after all red part data segments or asynchronously, this procedure is called. If the retry limit has been reached for the sending of report segments, this procedure sends a CR containing reason-code RLEXC, calls the "cancel session" procedure and triggers a "reception-session cancellation" notice to the client service.

If the retry limit has not been reached, one (or multiple if required) report segments (RS) are prepared and sent. Report segments will contain upper and lower references indicating which data within the block is being acknowledged. Report segments also contain the serial number of the CP it is acknowledging. If however, the CP is determined to be out of order, no RS is sent.

11.4.13   Suspend Timers

This procedure is triggered by a link state cue indicating cessation of transmission to a node, in the same way the "stop transmission" procedure is triggered. First this procedure calculates the expected arrival of any pending RS, RA or CAx segments. The expected arrival time is a combination of known latency and node processing time. For example, the One-Way Light Time (OWLT) between nodes plus any configured processing time or delay. The node then suspends all CP, RS and Cx timers with an expected arrival time of the current time or later.

11.4.14   Resume Timers

This procedure is triggered by a link state cue indicating the start of transmission to a node, in the same way the "start transmission" procedure is triggered. First, this procedure calculates the new expected arrival time of any previously suspended CP, RS and Cx timers using the same method as the "suspend timers" procedure. Once the expected arrival time is calculated, each previously suspended timer is resumed.

11.4.15   Stop Report Segment Timer

When a node receives an RA, the corresponding RS timer identified by the RS serial number in the RA segment is deleted. If there are no other RS timers running for this transmission session (identified by the session id in the segment header), all required RA segments have been received, therefore the "close session" procedure is called.

11.4.16   Acknowledge Cancellation

When a node receives a cancellation segment "Cx" (i.e. CS or CR), this procedure is called. In some cases, a Cx may refer to a session with a node to which there is no transmission queue, for example when a receiver is a read-only node. In this case no further action is taken. A cancellation acknowledgment segment (CAx) is then sent, however the type depends on which node is receiving the cancellation segment. If the Cx is a cancellation segment from the sender (CS), a CAS is sent. Alternatively, if the Cx is a cancellation segment from the receiver (CR) a CAR is sent.

In some cases, the received Cx has been previously received and processed, this may occur if the Cx was resent due to a CAx which was lost in transmission. In which case no further action is taken. However, if the Cx has not been previously processed, the "cancel session" procedure is called, followed by a "reception-session cancellation" notice to the client service. Finally, the node calls the "close session" procedure.

11.4.17   Stop Cancel Timer

When a node receives a cancellation acknowledgment (CAR or CAS), the corresponding CS timer is deleted and the "close session" procedure is called for the session identified in the segment header.

11.4.18   Signify Transmission Completion

This procedure is triggered following the successful transmission and acknowledgment of all red-data within a block. The end of the transmission is indicated by the transmission of an EOB marker and the receipt of one or more report segments referencing all red-data within the transmitted block indicates the successful acknowledgment of all red-data. Once triggered this procedure sends a "transmission-session completion" notice to the client, then calls the "close session" procedure for the session identified in the segment header.

11.4.19   Cancel Session

This procedure is triggered by other internal procedures, which also provide the reason for the session cancellation in the form of a reason-code. When triggered, this procedure deletes all segments from the outbound queue related to the identified session. All timers associated with the session are also deleted. As LTP is a stateful protocol, a sender node must maintain a copy of all sent data until it is acknowledged, in case retransmission is needed. Therefore, if this procedure is triggered on a sender node, all data within the retransmission buffer for this session is also deleted.

11.4.20 Close Session

Like the "cancel session" procedure, this is triggered by other internal procedures. When triggered, all timers associated with the session are deleted as is the associated session state record. Therefore, after this procedure is called the node cannot process any more segments related to the identified session.

11.4.21 Handle Mis-Colored Segment

During normal LTP operation, all red-data is sent before green-data within a block. However, if within a single session, a node receives green-data before the end of the red-data, or red-data after green data, this procedure is called. First, the node discards the received segment then should send a cancellation segment with reason-code MISCOLORED. Finally, a "reception-session cancellation" notice is sent to the client service. If the node receives a mis-colored segment from a node to which there is no outbound transmission queue, no cancellation segment is sent.

11.4.22 Handle System Error Condition

There may be occasions when an unrecoverable system error occurs, where the node continues to run but is unable to continue a session satisfactorily. For example, a memory or paging error that prevents a node from maintaining its state sufficiently. In this case, the node calls the "cancel session" procedure and sends a cancellation segment with reason-code SYS_CNCLD. If the node is the sender, a CS is sent and a "transmission-session cancellation" notice is sent to the client service. If the node is the receiver, a CR is sent and a "reception-session cancellation" notice is sent to the client.

11.4.23 Client Service Notifications

During many internal procedures described previously, the client service running on the sender or receiver node is sent notifications. For example, when a session is started or when all data within a block is received. These notices allow the client service to take any necessary action, in order to manage higher level functions on the LTP node. The trigger, name and the contents of each notice is shown in Table 11.5.

Table 11.5: Client Notices. Source: Adapted from [224]

Trigger	Notice	Notice Summary
Sender: "Start Transmission" procedure. Receiver: Triggered by receiving the first valid data segment with a new session id in the segment header	Session Start	Notify client service that the session has started. Notification contents: Session id
Triggered by "Signify Green-Part Segment Arrival" procedure	Green-Part Segment Arrival	Notify client service that a green part segment has arrived. Notification contents: Session id, source engine id, client service data, data length, data offset, indication if last byte in block
Triggered by "Signify Red-Part Reception" procedure	Red-Part Reception	Notify client service that a red part segment has arrived. Notification contents: Session id, engine id, client service data, data length, data offset, indication if last byte in block
Triggered by "Signify Transmission Completion" procedure	Transmission-Session Completion	Notify client service that all data within the block has been successfully transmitted AND acknowledged. Notification contents: Session id
Triggered by "Retransmit Checkpoint", "Retransmit Report Segment" or "Handle System Error Condition" procedure	Transmission-Session Cancellation	Notify (sender) client service that the session was cancelled. Notification contents: Session id, cancellation reason-code.
Triggered by "Acknowledge Cancellation", "Send Reception Report" or "Handle Mis-colored Segment" procedure	Reception-Session Cancellation	Notify (receiver) client service that the session was cancelled. Notification contents: Session id, cancellation reason-code.
EOB sent	Initial-Transmission Completion	Notify client service that all data within the block has been transmitted BUT NOT YET acknowledged. Notification contents: Session id.

11.5 State Transition Diagrams

This section describes how LTP nodes transition between various states while sending or receiving segments.

11.5.1 LTP Sender

Figure 11.17 shows the main state transition diagram for the LTP sender. Note three operations (CP, RX and CX) are shown separately. The CP operation triggered after

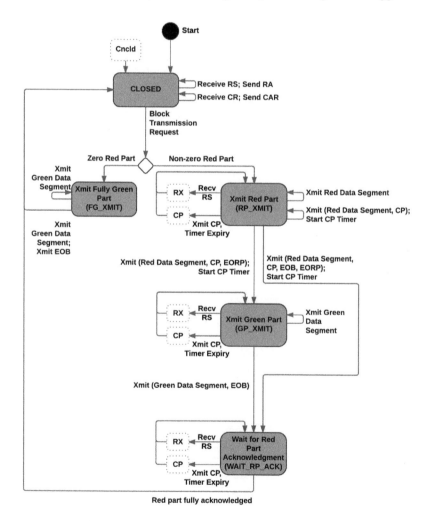

Figure 11.17: LTP Sender Main State Transition Diagram. Source: Adapted from [224]

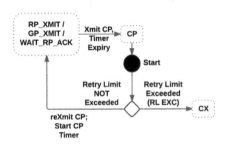

Figure 11.18: LTP Sender CP Operation. Source: Adapted from [224]

transmission of a checkpoint (CP) or following a timer expiry is shown in Figure 11.18. Upon receiving a report segment (RS), the RX operation is triggered, shown in Figure 11.19. Finally the CX operation which handles session cancellation is shown in Figure 11.20.

An LTP sender node, as shown in Figure 11.17 starts in the CLOSED state, as either an initial state or having returned to the state following a session cancellation, block

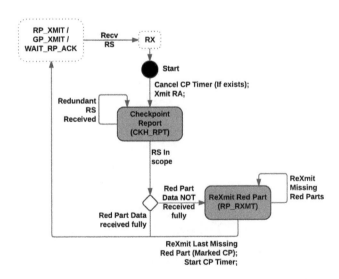

Figure 11.19: LTP Sender RX Operation. Source: Adapted from [224]

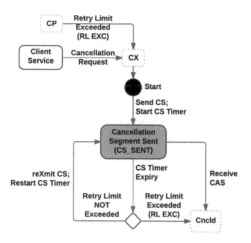

Figure 11.20: LTP Sender CX Operation. Source: Adapted from [224]

transmission attempt or full acknowledgment for transmitted red part data. From this state, there are three possible transitions:

1. Receiving an RS triggers transmission of a corresponding report acknowledgment segment (RA) (using the "retransmit data" procedure) and a return to the CLOSED state (using the "close session" procedure).

2. Receiving a cancellation segment from the receiver (CR) triggers a cancellation acknowledgment segment to the receiver (CAR) (using the "acknowledge cancellation" procedure) and a return to the CLOSED state (using the "close session" procedure).

3. Upon receipt of a block transmission request, the action taken depends on whether there is data requiring acknowledgment (non-zero red part) or not (zero red part i.e. all green data).

11.5.1.1   CP Operation

If a checkpoint timer expires while in either the RP_XMIT, GP_XMIT or WAIT_RP_ACK state, this triggers the "CP" operation shown in Figure 11.18. The CP operation first checks if the retry limit has been exceeded (RLEXC). If the retry limit has not been exceeded, the sender retransmits the checkpoint (using the "retransmit checkpoint" procedure) using the previously used CP serial, then starts a new corresponding CP timer (using the "start checkpoint timer" procedure). If however, the retry limit has been exceeded, the CX operation is triggered.

11.5.1.2 RX Operation

If the sender receives an RS while in the RP_XMIT, GP_XMIT or WAIT_RP_ACK state, the "RX" operation shown in Figure 11.19 is triggered. This operation (using the "retransmit data" procedure) first cancels the corresponding CP timer, sends an RA to the receiver and moves into the CHK_RPT state. In some cases the received RS may be identified as redundant upon checking the RS serial number. In this case no action is taken. If the RS is not redundant the segment is checked to see if all red part data within the block is acknowledged. If all red part data is acknowledged, the node returns to the state from which it came, either the RP_XMIT, GP_XMIT or WAIT_RP_ACK state.

If the RS does not acknowledge all red part data, the sender node moves to the RP_RXMIT state. In this state, the node compiles the missing red part data into segments and retransmits the missing red part data. The last segment contains the CP and a corresponding CP timer is started (using the "start checkpoint timer" procedure). The node then returns to the state that triggered the RX operation, either the RP_XMIT, GP_XMIT or WAIT_RP_ACK state.

11.5.1.3 CX Operation

The CX operation is shown in Figure 11.20. If a CP timer expires within the CP operation and the retry limit has been exceeded, the CX operation is triggered. This causes the node to send a cancellation segment (CS), start a CS timer (using the "start cancel timer" procedure) and move to the CS_SENT state. In this case the reason-code within the CS in this case is RLEXC.

Additionally, the CX operation may be triggered asynchronously if the sender receives a cancellation request from the local client service. This also causes the node call the "acknowledge cancellation" procedure, then follow the "cancel session" procedure to send a CS, start a CS timer (using the "start cancel timer" procedure) and move to the CS_SENT state. In this case the reason-code within the CS is USR_CNCLD.

While in the CS_SENT state, if the sender node receives a cancellation acknowledgment (CAS), the node returns to the CLOSED state after calling the "stop cancel timer" procedure (Cncld pointer in the diagram).

If a node does not receive a CAS before the CS timer expires and the retry limit is not exceeded, a cancellation segment to the receiver (CR) is retransmitted (using the "retransmit cancellation segment" procedure) and a CS timer restarted (using the "start cancel timer" procedure). If however, the CS timer expires and the retry limit is exceeded, the node returns to the CLOSED state using the "close session" procedure (Cncld operation in the diagram).

11.5.1.4 Transmission of 100 Percent Green Data (Zero Red Part)

Upon receiving a block transmission request containing zero red parts (100 percent green data), the sender moves into the FG_XMIT state. The block data is broken into multiple segments based on the size required by the underlying protocol layer and the last segment contains an end of block (EOB) marker. After all segments are sent to the receiver, the sender returns to the CLOSED state.

11.5.1.5 Transmission of Non-Zero Red Parts

A block transmission request containing non-zero red parts (100 percent red data, or a mix of both green and red data) causes the sender node to move to the RP_XMIT state. The red data is broken into multiple segments based on the underlying protocol layer (for example the MTU size). An end of red part (EORP) marker and checkpoint (CP) is also added to the last segment. The transmission of any CP is followed by starting a CP timer (using the "start checkpoint timer" procedure). Optionally a node, may also choose to send an asynchronous CP from this state, before reaching the end of the red part data. This method potentially achieves faster acknowledgment and retransmission of data. A CP, if marking a resubmission of missing data indicated by an RS, will also contain the RS serial number. Otherwise, a report serial number of zero will be included.

If there is green data to send, the sender node moves to the GP_XMIT state. In this state the sender breaks the green part data into segments based on the underlying protocol layer and transmits all green part segments. The last segment contains the end of block (EOB) marker; the node then moves into the WAIT_RP_ACK state.

If there is no green data to send, the last segment also contains the end of block (EOB) marker and the sender moves to the WAIT_RP_ACK state.

A sender node stays in the WAIT_RP_ACK state, executing RX or CP operations when triggered until all red part data is fully acknowledged. Full red part acknowledgment causes the sender node to call the "signify transmission completion" procedure, followed by the "close session" procedure and then move into the CLOSED state.

11.5.2 LTP Receiver

The main state transition diagram for a receiving node is shown in Figure 11.21. Note that like the state transition diagrams for the sending node, there are operations which are shown in separate diagrams. First, the operations carried out when a RS timer expires (RX) are shown in Figure 11.22. Additionally, the "CX" operation carried out upon receiving a mis-colored segment (using the "handle mis-colored segment" procedure), or a segment identifying an unknown service id is shown in Figure 11.23.

The LTP receiver diagram in Figure 11.21 shows the LTP receiver starting in the

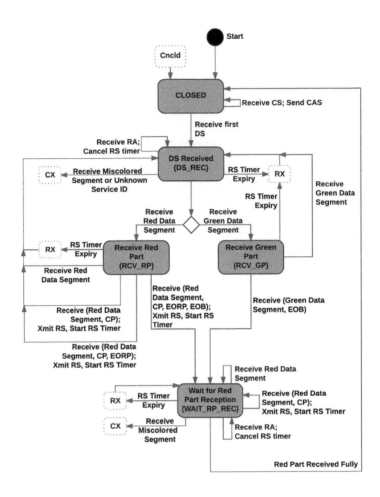

Figure 11.21: LTP Receiver Main State Transition Diagram. Source: Adapted from [224]

CLOSED state. This is either an initial state, following a session cancellation, or a return to the state following the receipt of all red parts within a block. From this state, there are two possible transitions:

1. Receiving a cancellation segment from the sender (CS) triggers a cancellation acknowledgment segment to the sender (CAS) and a return to the CLOSED state (using the "acknowledge cancellation" procedure). While in the CLOSED state, a node has no active session. Therefore receiving a CS in this state is likely to be a retransmission due to a CAS lost in transmission, or premature CS timer expiry.

2. Receipt of the first data segment causes the receiver node to move to the "data segment received" (DS_REC) state.

While in the DS_REC state, a receiver node has four possible transitions:

1. Receiving a report acknowledgment (RA) causes the node to cancel the corresponding RS timer (using the "stop report segment timer" procedure).

2. The expiration of an RS timer causes the receiver node to execute the RX operation.

3. Receiving a mis-colored segment (for example a green part when a red part is expected), or receiving a segment referring to an unknown service id causes the node to execute the CX operation using the "handle mis-colored segment" procedure.

4. Otherwise the operations required to process either red part or green part data are followed.

11.5.2.1 RX Operation

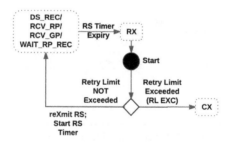

Figure 11.22: LTP Sender RX Operation. Source: Adapted from [224]

The RX operation is triggered when an RS timer expires. A receiver node can trigger this operation either from the "data segment received" (DS_REC), "receive red part" (RCV_RP), "receive green part" (RCV_GP) or "waiting for red part reception" (WAIT_RP_REC) states.

First the node checks if the retry limit has been exceeded. If the retry limit has not been exceeded, the node retransmits the RS (using the "retransmit report segment" procedure) including the same previously used report serial and starts a new RS timer. If however, the retry limit has been exceeded, the CX operation is triggered.

11.5.2.2 CX Operation

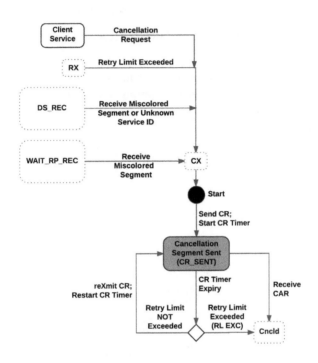

Figure 11.23: LTP Sender CX Operation. Source: Adapted from [224]

A sender node can trigger the CX operation from the RX operation after reaching the RS retry limit. The reason-code within the CS in this case is RLEXC. The CX operation is also triggered from the "data segment received" (DS_REC) state following the receipt of a mis-colored segment, or a segment referring to an unknown client service. The reason-code for a mis-colored segment should be MISCOLORED. For an unknown client service the reason-code should be UNREACH. Finally, the CX operation is also triggered from the "wait for red part reception" (WAIT_RP_REC) state if a mis-colored segment is received. In this latter case the MISCOLORED reason-code is again used.

Like the sender node, a receiver node may also receive an asynchronous cancellation request from the client service. When the session cancellation request originates from the client service, the reason-code within the cancellation segment should be USR_CNCLD.

The CX operation first sends a cancellation segment (CR) to the sender node, starts the CR timer (using the "start cancel timer" procedure) then moves into the "cancellation segment sent" (CR_SENT) state. While in the CR_SENT state, a receiver node may receive a cancellation acknowledgment (CAR), in which case the receiver node calls the "acknowledge cancellation" procedure, stops the CR timer (using the "stop cancellation timer" procedure), then returns to the CLOSED state by following the "close session" procedure (Cncld pointer in the diagram). If the node does not receive a CAR within the defined timer period the CR timer expires, the node then checks if the retry limit has been exceeded. If the retry limit has not been exceeded the CR is retransmit (using the "retransmit cancellation segment" procedure) and CR timer restarted (using the "start cancellation timer" procedure). Otherwise, if the retry has been exceeded the node returns to the CLOSED state calling the "close session" procedure (Cncld pointer in the diagram).

11.5.2.3 Receiving a Green Data Segment

Following the receipt of a green data segment, the node transitions to the "receive green part" state (RCV_GP). If the received segment is a regular green data segment, the node returns to the DS_REC state. The received segment may also contain an EOB marker, in which case the node moves to the "wait for red part reception" (WAIT_RP_REC) state. The WAIT_RP_REC state is described further in the next section.

11.5.2.4 Receiving a Red Data Segment

Following the receipt of red data segment, the node transitions to the "receive red part" (RCV_RP) state. If the segment is a regular red data segment, the node returns to the DS_REC state. If the segment contains a CP or EORP, an RS is sent containing the CP serial number (using the "send reception report" procedure), an RS timer started (using the "start report segment timer" procedure) then the node returns to the DS_REC state to wait for further segments from the block. If the red data segment also contains the EOB marker an RS is sent (containing the CP serial number) and a corresponding RS timer is started (using the "start report segment timer" procedure). However, in this case the node transitions to the "wait for red part reception" (WAIT_RP_REC) state. It should also be noted, a receiver can send an asynchronous RS at any point. This functionality exists to allow a node to accelerate retransmission of data, for the same reason a sender node can send intermediate checkpoints.

Once transitioned to the WAIT_RP_REC state, the node waits for all red parts to be received and report segments acknowledged. This may include retransmitted segments. While in this state if the node receives an RA the corresponding RS timer is cancelled (using the "stop report segment timer" procedure). If a normal red data segment is received, the node returns to the WAIT_RP_REC state. If that red data segment also contains a CP, the node sends an RS (using the "send reception report" procedure), or resends an RS (using the "retransmit report segment" procedure) if the CP is a retransmission. Once all red parts have been fully received, the node calls the "signify transmission completion" procedure and returns to the CLOSED state after calling the "close session" procedure.

11.6   Security

This section provides details of the various security features implemented in LTP. It provides an overview of potential attack vectors, as well as the LTP features that could be implemented in an attempt to prevent such attacks.

A DTN could potentially be used for transferring sensitive or high priority data and therefore could be subject to similar threats as a traditional network. Thus justifying development and implementation of mechanisms to ensure confidentiality, authenticity and integrity. For example, a malicious party could gain access to sensitive information by eavesdropping and redirecting data during transmission (similar to a TCP sequence prediction attack). Traffic could also be sent from an untrusted node with the intention of disrupting a specific transmission (similar to a TCP veto attack). A malicious attacker could also attempt to disrupt an entire service (e.g. a denial of service attack).

In some cases, data received by LTP from a bundle protocol agent, may already have security features applied. For example, using the bundle security protocol [268] to ensure confidentiality, integrity and authenticity of each bundle. Additionally, LTP is normally deployed over a private point-to-point network link, thus reducing the possibility of eavesdropping. However, there remain a number of attack vectors at the LTP level, for which counter-measures are required. This section first details the main security features available for LTP, following this potential attack vectors are presented along with possible prevention techniques.

11.6.1   LTP Security Features

There are a number of features within the specification of LTP that provide a level of protection from a security perspective. LTP silently drops malformed or suspect packets (more information can be found in Figure 11.16). This reduces the amount of information a malicious party could learn from an existing network by sending packets to an LTP node. Silently dropping malformed packets provides a level of protection against attacks similar to a TCP tear-drop denial of service attack, which sends overlapping or oversized packets to a node. LTP also randomly generates numbers used for checkpoint and report serial numbers. This randomness makes attacks similar to a TCP sequence prediction attack much more difficult. In addition to these aforementioned security features, LTP also has two security extensions which provide more comprehensive protection. These extensions are named the "LTP Authentication Extension" and "LTP Cookies Extension".

11.6.1.1 LTP Authentication Extension

The LTP authentication extension uses cryptography to verify the integrity of an LTP segment. It can also provide a degree of authenticity confidence as attackers are not expected to easily gain access to the pre-shared keys used to encrypt and verify the data. This extension provides a level of protection against an attacker injecting a malicious segment into an LTP stream, or attempting a denial of service attack.

When using this extension, a node uses pre-shared keys to generate an "authval", which is either a message authentication code (MAC) or digital signature depending on the cipher suite used. The entire LTP segment (excluding the trailer section) is used in conjunction with a pre-shared key to produce the authval, which is placed into the segment trailer. To verify the integrity of a segment, a receiving node repeats the same procedure to produce an authval, then compares the newly generated authval with the authval provided in the segment trailer. If the newly generated authval matches the authval in the segment trailer, the integrity of the segment is considered correct. If the segment integrity was verified using a trusted pre-shared key, known to be provided only to a trusted node, a receiving node can also consider the segment authentic.

Due to the nature of many DTNs, it may not be possible to implement a public key infrastructure, or implement a key exchange similar to the Internet Key Exchange (IKE). Therefore, LTP is currently limited to using pre-shared keys. The risk of this is partially mitigated by an option to pre-load nodes with multiple keys, which can be selected at random by the sending node. However, this method of pre-sharing keys does not easily support key revocation.

To identify use of this extension to a receiving node, a sending node places the extension type "0x00" into the extensions field of the header of the first segment. The extension can also be placed in subsequent segments to avoid the risk of lost segments. The sending node also identifies the cipher suite and key used to generate the authval in the segment header.
LTP supports up to 256 cipher suites, using values maintained by the IANA. The cipher suites available and respective values are shown in Table 11.6 and described further in the next paragraphs.

HMAC-SHA1-80 cipher suite

When using this cipher suite, a key is employed to generate a message authentication

Table 11.6: Cipher Suite Values. Source: Data from [109]

Cipher suite	Value
HMAC-SHA1-80	0
RSA-SHA256	1
Unassigned	2-127
Reserved	128-191
Private/Experimental	192-254
NULL	255

code (MAC) which results in a fixed 10 octet value. The entire message authentication code (authval) is placed in the trailer section of the LTP segment.

RSA-SHA256 cipher suite

Use of this cipher suite generates a digital signature of the LTP segment, which is of variable length; for example 128 octets for a 1024 bit signature. The generated digital signature (authval) is placed into the trailer section of the LTP segment.

NULL cipher suite

Use of the NULL cipher suite produces a MAC like the HMA-SHA1-80 cipher suite. Though the NULL cipher suite uses the same hard-coded key (c37b7e64 92584340 bed12207 80894115 5068f738) every time. Therefore, use of this cipher suite serves to verify only the integrity of a segment and not the authenticity, as a malicious node could produce an authval using the same hard-coded key.

LTP Authentication Extension Example

Figure 11.24 shows an LTP segment example using the LTP authentication extension and HMAC-SHA1-80 cipher suite. The control byte contains the protocol version,

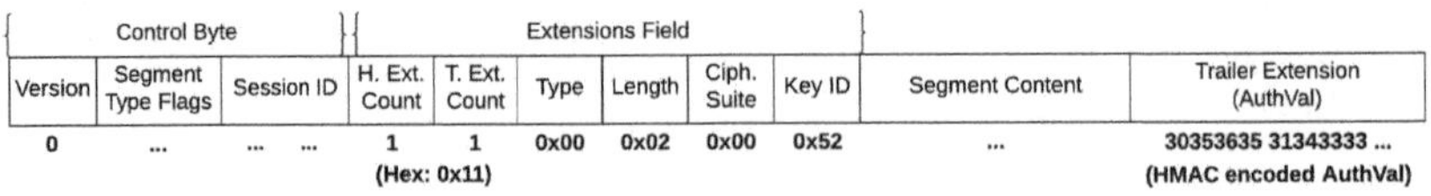

Control Byte			Extensions Field							
Version	Segment Type Flags	Session ID	H. Ext. Count	T. Ext. Count	Type	Length	Ciph. Suite	Key ID	Segment Content	Trailer Extension (AuthVal)
0	...	... ...	1 (Hex: 0x11)	1	0x00	0x02	0x00	0x52	...	30353635 31343333 ... (HMAC encoded AuthVal)

Figure 11.24: Example LTP Segment using HMAC-SHA1-80. Source: Adapted from [109]

segment type flags and session id (engine id and session id). The extensions field has a value of 1 for both the header and trailer extension count fields, indicating that there is one extension in both the header and trailer sections of the segment. The extension type is 0x00 indicating the segment uses the LTP authentication extension. The length field has a value of 0x02 indicating two remaining fields within the extensions field, the cipher suite and key id. The cipher suite has a value of 0x00 indicting the use of HMAC-SHA1-80 and finally the key id in this case is 0x52, indicating which key was passed to the cipher function. The next section of the segment is the content section, which is followed by the trailer extension section. As the HMAC-SHA1-80 cipher suite is used, the authval in the trailer extension section contains the output of the HMAC-SHA1-80 cipher function.

11.6.1.2 LTP Cookies Extension

This extension uses cookies to make a denial of service attack more difficult. To use this extension, a node uses the extension type "0x01" and places a cookie into the LTP segment header. The segment recipient then validates this coookie before processing the segment. A cookie is simply a random number and can be sent by either the sending or receiving node at any time during a transmission, after which all segments must contain the cookie to be considered valid.

Cookies, can also be extended at any time by either the sending or receiving node. After a cookie has been extended, all LTP segments must contain the extended cookie value. After a segment containing a cookie is sent, all future segments received without a good cookie are silently discarded.

Cookies are sent in plain text (i.e. unencrypted), thus an attacker could read a cookie while in transit. This cookie could be replayed to inject malicious segments into the stream. An obtained cookie could also be extended and injected into the stream, therefore causing the receiving node to silently reject all future segments. In order to make such an attack more difficult, the cookies extension could be used in combination with the LTP authentication extension. This would cause any segments injected into the stream by an attacker to be dropped, as the attacker in theory would not have any valid keys to build a valid authval.

11.6.2 Potential Attack Vectors

The following sections describe potential attacks and methods that can be used in an attempt to prevent them.

11.6.2.1 Eavesdropping

There are no mechanisms within LTP to maintain the confidentiality of payload or session data sent between nodes. To protect payload data, features within the bundle security protocol specification must be used, or the payload data encrypted before passing to the bundle protocol agent. Still, the bundle security protocol does not protect the LTP specific elements contained within an LTP segment. This potentially allows a malicious agent to gather information about a current session by sniffing data during transmission. This session data, for example engine or session ids, could possibly be used in other forms of attack.

11.6.2.2 Corrupting Data

When the bundle security protocol is used, the integrity of payload data can be maintained. Additionally, the LTP authentication extension allows a receiving node to verify the integrity of segments as they are received. Thus, if a segment was in any way modified during transmission, the receiving node would detect this and silently drop the segment. The LTP authentication extension also provides a mechanism to verify the authenticity of a segment using one or multiple pre-shared keys. Therefore, if the LTP authentication extension is used, an attacker would need to know a pre-shared key for the segment to be processed by the receiving node. The LTP cookies extension also makes the corrupting of data more difficult but not impossible, as the cookies are sent in plain text and can therefore be replayed. Using LTP cookies in combination with the LTP authentication extension is a more robust solution.

11.6.2.3 Disrupting a Transmission or Service

An attacker could potentially disrupt transmission in a number of ways. For example, a cancellation segment could be sent to prematurely cancel a session. Alternatively, malicious report segments could be sent causing unnecessary retransmission. The LTP authentication extension can be used to defend against such attacks, as this causes the receiving node to verify the authenticity of each segment before processing.

LTP service disruption could potentially occur if a denial of service attack was used. For example, an LTP node could be bombarded with LTP segments. As shown in Figure 11.16, an LTP node discards segments which are received from a node with no transmission queue, which in this context is an unknown node. This prevents undue segment processing, which should reduce the impact of a rudimentary denial of service attack to simply bombard a node with junk segments. Yet, in order to determine whether or not to process a segment, an LTP node first processes the segment header. Therefore, although small, some degree of available resources are used by the node upon receiving each segment. A resource constrained node, if receiving a rate of segments faster than can be processed could be taken out of service, working in a similar way to a UDP flood attack. A denial of service attempt could potentially be made worse if a valid session id, engine id and client service id was obtained by eavesdropping beforehand, in conjunction with the spoofing of a source address. In this case, the receiving node would incur more resource overhead per segment while attempting to process the segment.

11.7 Practical LTP Analysis

This section builds on the theory from previous sections to provide analysis of an LTP implementation. The implementation used is the Interplanetary Overlay Network (ION) developed by Scott Burleigh [52] [60].

11.7.1 Test-Bed Architecture

The environment used, shown in Figure 11.25, consists two nodes (TX and RX) communicating directly via IP over Ethernet. Traffic is captured at node RX using Wireshark. Loss is artificially generated in one test using netem - a utility available in most GNU Linux implementations. The nodes are configured to use only one LTP session over UDP on port 1113. The maximum LTP segment size is configured as 1400 Bytes to comfortably align with the underlying MTU of 1500 Bytes.

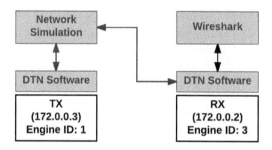

Figure 11.25: Test-bed Architecture

11.7.2 TX Node Configuration File

```
## begin ionadmin
#        Engine id 1
1 1 ''
s

#        Define contact plan between nodes
a contact +1 +3600 1 1 100000000
a contact +1 +3600 1 3 100000000
a contact +1 +3600 3 1 100000000
```

```
#          Define 1sec OWLT between  nodes
a range +1 +3600  1  1  1
a range +1 +3600  1  3  1
a range +1 +3600  3  1  1

m production  1000000
m consumption  1000000
## end ionadmin

## begin ltpadmin

1  256  50000

#a span  peer_engine_nbr
#          max_export_sessions  1
#          max_import_sessions  1
#          max_segment_size  1400
#          aggregation_size_limit  1400
#          aggregation_time_limit  1
#          'LSO_command'
#          [queuing_latency]

# send  to  rx
a span  3  1  1  1400  1400  1  'udplso  172.0.0.2:1113'
s  'udplsi  0.0.0.0:1113'

## end ltpadmin

## begin bpadmin
1
#          Use  the  ipn  eid  naming  scheme
a scheme ipn  'ipnfw'  'ipnadminep'

#          Define  a  single  endpoint  queue
a endpoint ipn:1.1  q

#          Define  ltp  as  the  protocol  used
a protocol ltp  1400  100

#          Listen
a induct ltp  1  ltpcli

#          Send  to  yourself
a outduct ltp  1  ltpclo
```

```
#           Send  to  rx
a  outduct  ltp  3  ltpclo

s
## end bpadmin

## begin ipnadmin

#           Send  to  yourself
a  plan  1  ltp/1

#           Send  to  node  rx
a  plan  3  ltp/3

## end ipnadmin

## begin ionsecadmin
#           Enable  bundle  security  to  avoid  error  messages
#           in  ion.log
1
## end ionsecadmin
```

11.7.3   RX Node Configuration File

```
## begin ionadmin
#           Engine  id  3
1  3  ''
s

#           Define  contact  plan  between  nodes
a  contact  +1  +3600  1  3  100000000
a  contact  +1  +3600  3  1  100000000
a  contact  +1  +3600  3  3  100000000

#           Define  1sec  OWLT  between  nodes
a  range  +1  +3600  1  3  1
a  range  +1  +3600  3  1  1
a  range  +1  +3600  3  3  1

m  production  1000000
m  consumption  1000000
## end ionadmin

## begin ltpadmin

1  256  50000
```

```
#a  span  peer_engine_nbr
#        max_export_sessions  1
#        max_import_sessions  1
#        max_segment_size  1400
#        aggregation_size_limit  1400
#        aggregation_time_limit  1
#        'LSO_command'
#        [queuing_latency]

#        Create  a  span  to  tx
a  span  1  1  1  1400  1400  1  'udplso  172.0.0.3:1113'

#        Start  listening  for  incoming  LTP  traffic
s  'udplsi  0.0.0.0:1113'

## end ltpadmin

## begin bpadmin
1
#        Use  the  ipn  eid  naming  scheme
a  scheme  ipn  'ipnfw'  'ipnadminep'
#        Create  a  single  endpoint  queue
a  endpoint  ipn:3.1  q
#        Define  ltp  as  the  protocol  used
a  protocol  ltp  1400  100

#        Listen
a  induct  ltp  3  ltpcli

#        Send  to  yourself
a  outduct  ltp  3  ltpclo

#        Send  to  tx
a  outduct  ltp  1  ltpclo

s
## end bpadmin

## begin ipnadmin

#        Send  to  yourself
a  plan  1  ltp/1

#        Send  to  rx
```

```
a  plan  3  ltp /3

## end  ipnadmin

## begin  ionsecadmin
#          Enable  bundle  security  to  avoid  error  messages
#          in  ion.log
1
## end  ionsecadmin
```

11.7.4   *Interplanetary Overlay Network Tests*

Two tests are conducted, with 10 x 64KB bundles sent from TX to RX in each test. Each 64KB bundle is a single file named "payload-64k". All data is sent as red data which requires acknowledgment. The first test has no network simulation applied. The second test has a simulated loss of 10 percent to allow the behavior of ION-LTP under abnormally high-loss conditions to be observed. Two utilities which are provided with the ION package are used, bpsendfile and bprecvfile to send and receive bundles respectively. One DTN endpoint is created on each node, ipn:1.1 (TX) and ipn:3.1 (RX).

The bpsendfile command, executed 10 times at 0.5 second intervals on tx:

```
bpsendfile  ipn:1.1  ipn:3.1  payload –64k
```

The bprecvfile command, executed on node RX:

```
bprecvfile  ipn:3.1
```

11.7.4.1   *Send 10 Bundles with No Loss*

Figure 11.26 shows the Wireshark representation of the LTP segment. Wireshark does not dissect all of the LTP header, therefore the raw data from the LTP header is also shown (as bits) in the highlighted section at the bottom of the figure. The segment consists of a header followed by the data segment within the content section. By analyzing the header the following can be observed:

- Byte 1 contains version zero (0000), followed by segment type zero (0000), indicating this is a data segment containing red data with no checkpoint, no end of red part and no end of block markers

- Byte 2 indicates the engine id of the sending node is one (00000001)

- Byte 3 indicates the session id is one (00000001)

- Byte 4 indicates there are no header or trailer extensions (0000 0000)

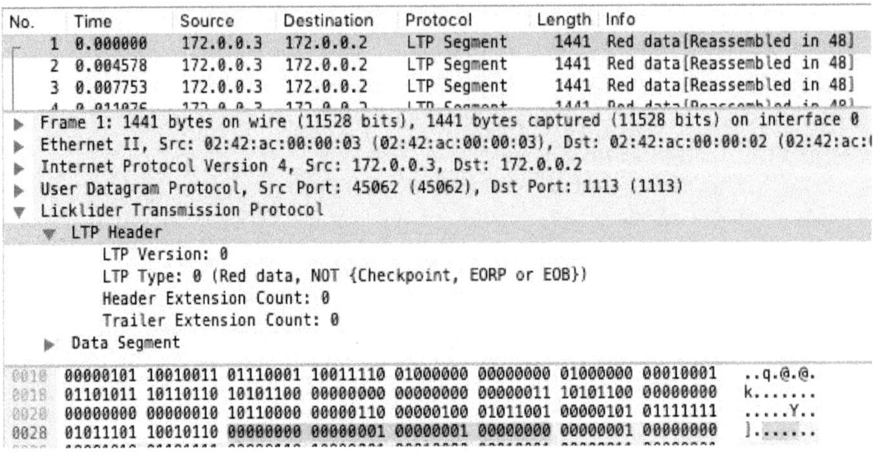

No.	Time	Source	Destination	Protocol	Length	Info
1	0.000000	172.0.0.3	172.0.0.2	LTP Segment	1441	Red data[Reassembled in 48]
2	0.004578	172.0.0.3	172.0.0.2	LTP Segment	1441	Red data[Reassembled in 48]
3	0.007753	172.0.0.3	172.0.0.2	LTP Segment	1441	Red data[Reassembled in 48]
4	0.011076	172.0.0.3	172.0.0.2	LTP Segment	1441	Red data[Reassembled in 48]

▶ Frame 1: 1441 bytes on wire (11528 bits), 1441 bytes captured (11528 bits) on interface 0
▶ Ethernet II, Src: 02:42:ac:00:00:03 (02:42:ac:00:00:03), Dst: 02:42:ac:00:00:02 (02:42:ac:
▶ Internet Protocol Version 4, Src: 172.0.0.3, Dst: 172.0.0.2
▶ User Datagram Protocol, Src Port: 45062 (45062), Dst Port: 1113 (1113)
▼ Licklider Transmission Protocol
  ▼ LTP Header
       LTP Version: 0
       LTP Type: 0 (Red data, NOT {Checkpoint, EORP or EOB})
       Header Extension Count: 0
       Trailer Extension Count: 0
  ▶ Data Segment

```
0010  00000101 10010011 01110001 10011110 01000000 00000000 01000000 00010001   ..q.@.@.
0018  01101011 10110110 10101100 00000000 00000000 00000011 10101100 00000000   k.......
0020  00000000 00000010 10110000 00000110 00000100 01011001 00000101 01111111   .....Y..
0028  01011101 10010110 00000000 00000001 00000001 00000000 00000001 00000000   ].......
```

Figure 11.26: ION Data Segment 1

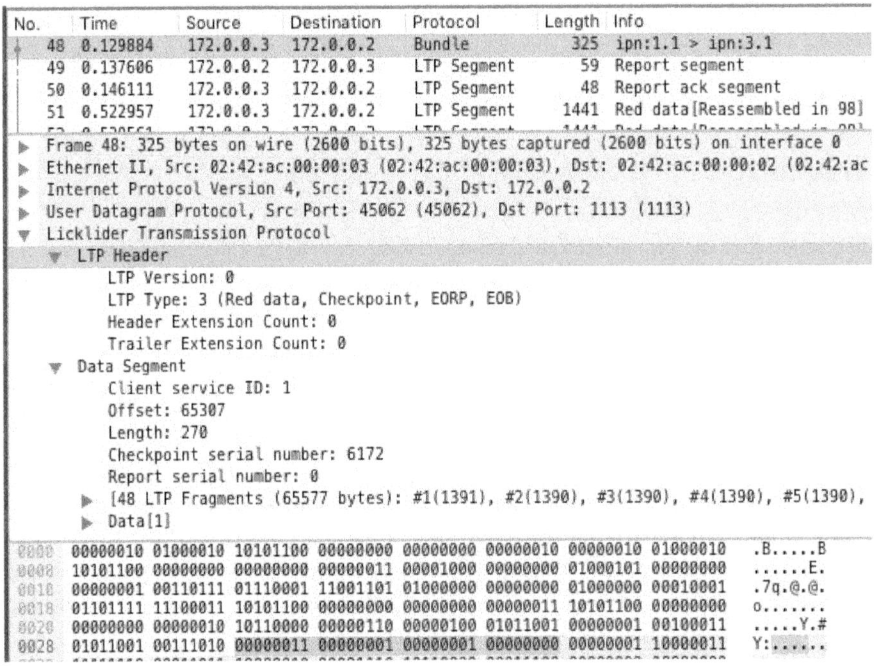

No.	Time	Source	Destination	Protocol	Length	Info
48	0.129884	172.0.0.3	172.0.0.2	Bundle	325	ipn:1.1 > ipn:3.1
49	0.137606	172.0.0.2	172.0.0.3	LTP Segment	59	Report segment
50	0.146111	172.0.0.3	172.0.0.2	LTP Segment	48	Report ack segment
51	0.522957	172.0.0.3	172.0.0.2	LTP Segment	1441	Red data[Reassembled in 98]
52	0.530561	172.0.0.3	172.0.0.2	LTP Segment	1441	Red data[Reassembled in 98]

▶ Frame 48: 325 bytes on wire (2600 bits), 325 bytes captured (2600 bits) on interface 0
▶ Ethernet II, Src: 02:42:ac:00:00:03 (02:42:ac:00:00:03), Dst: 02:42:ac:00:00:02 (02:42:ac
▶ Internet Protocol Version 4, Src: 172.0.0.3, Dst: 172.0.0.2
▶ User Datagram Protocol, Src Port: 45062 (45062), Dst Port: 1113 (1113)
▼ Licklider Transmission Protocol
  ▼ LTP Header
       LTP Version: 0
       LTP Type: 3 (Red data, Checkpoint, EORP, EOB)
       Header Extension Count: 0
       Trailer Extension Count: 0
  ▼ Data Segment
       Client service ID: 1
       Offset: 65307
       Length: 270
       Checkpoint serial number: 6172
       Report serial number: 0
     ▶ [48 LTP Fragments (65577 bytes): #1(1391), #2(1390), #3(1390), #4(1390), #5(1390),
     ▶ Data[1]

```
0000  00000010 01000010 10101100 00000000 00000000 00000010 00000010 01000010   .B.....B
0008  10101100 00000000 00000011 00001000 00000000 01000101 00000000   ......E.
0010  00000001 00110111 01110001 11001101 01000000 00000000 01000000 00010001   .7q.@.@.
0018  01101111 11100011 10101100 00000000 00000000 00000011 10101100 00000000   o.......
0020  00000000 00000010 10110000 00000110 00000100 01011001 00000000 00100011   .....Y.#
0028  01011001 00111010 00000011 00000001 00000001 00000000 00000001 10000011   Y:......
```

Figure 11.27: ION Data Segment 2

The transmission continues with several more regular data segments, like those shown in Figure 11.26. The last data segment in the LTP block is shown in 11.27. By analyzing this segment, it can be observed that bytes 2,3 and 4 are identical to the previous data segment. However the last four bits of Byte 1 indicate this segment is type 3 (0011). This segment therefore contains red data, plus a checkpoint, end of red part and end of block marker. The content portion of this segment also provides additional information:

- An Offset of 65537 and length of 270 indicates the length of the payload data, as well as the position of the data within the overall LTP block

- The checkpoint has a serial of 6172

- The report serial number is zero, indicating that this is the initial transmission and not a retransmission as a result of processing a report segment

```
No.    Time        Source      Destination  Protocol      Length  Info
  47  0.127366    172.0.0.3   172.0.0.2    LTP Segment    1442   Red data[Reassembled in 48]
  48  0.129884    172.0.0.3   172.0.0.2    Bundle          325   ipn:1.1 > ipn:3.1
  49  0.137606    172.0.0.2   172.0.0.3    LTP Segment      59   Report segment
  50  0.146111    172.0.0.2   172.0.0.3    LTP Segment      48   Report ack segment
▶ Frame 49: 59 bytes on wire (472 bits), 59 bytes captured (472 bits) on interface 0
▶ Ethernet II, Src: 02:42:ac:00:00:02 (02:42:ac:00:00:02), Dst: 02:42:ac:00:00:03 (02:42:ac:
▶ Internet Protocol Version 4, Src: 172.0.0.2, Dst: 172.0.0.3
▶ User Datagram Protocol, Src Port: 34737 (34737), Dst Port: 1113 (1113)
▼ Licklider Transmission Protocol
  ▼ LTP Header
        LTP Version: 0
        LTP Type: 8 (Report segment)
        Header Extension Count: 0
        Trailer Extension Count: 0
  ▼ Report Segment
        Report serial number: 1445
        Checkpoint serial number: 6172
        Upper bound: 65577
        Lower bound: 0
        Reception claim count: 1
      ▼ Reception claims
            Offset[0] : 0
            Length[0] : 65577
0000  00000010 01000010 10101100 00000000 00000000 00000011 00000010 01000010   .B.....B
0008  10101100 00000000 00000000 00000010 00001000 00000000 01000101 00000000   ......E.
0010  00000000 00101101 00001100 01100101 01000000 00000000 01000000 00010001   .-.e@.@.
0018  11010110 01010101 10101100 00000000 00000000 00000010 10101100 00000000   .U......
0020  00000000 00000011 10000111 10110001 00000100 01011001 00000000 00011001   .....Y..
0028  01011000 00110000 00001000 00000001 00000001 00000000 10001011 00100101   X0.....%
0030  10110000 00011100 10000100 10000000 00101001 00000000 00000001 00000000   ...▯
```

Figure 11.28: ION Report Segment

In response to the checkpoint, the receiving node sends a report segment which is shown in Figure 11.28. In this segment it can be observed that this is a report segment identified by segment type 8 (1000). The report segment has a serial number of

1445 and also contains the serial number of the checkpoint it was sent in response to (6172). The report segment has one total claim which acknowledges data from byte 0 to byte 65577 within the LTP block.

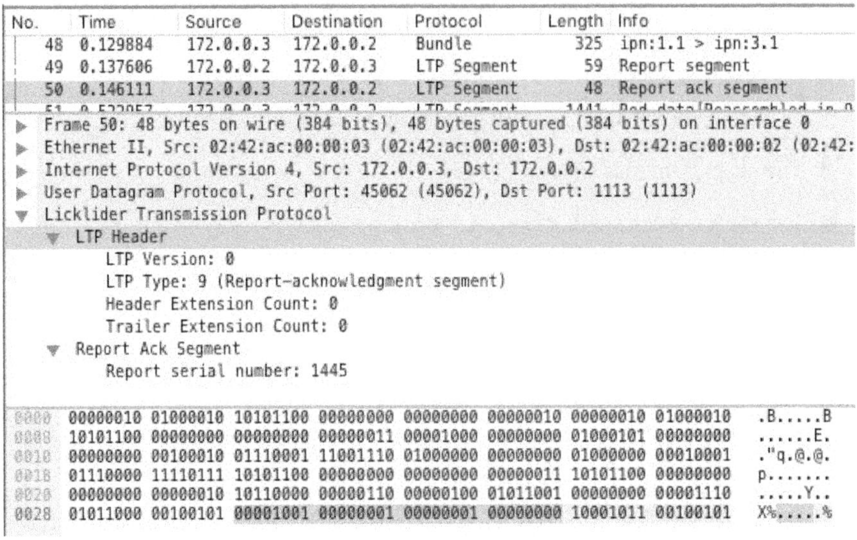

No.	Time	Source	Destination	Protocol	Length	Info
48	0.129884	172.0.0.3	172.0.0.2	Bundle	325	ipn:1.1 > ipn:3.1
49	0.137606	172.0.0.2	172.0.0.3	LTP Segment	59	Report segment
50	0.146111	172.0.0.3	172.0.0.2	LTP Segment	48	Report ack segment
51	0.522057	172.0.0.3	172.0.0.2	LTP Segment	1441	Red data [Reassembled in 0

▶ Frame 50: 48 bytes on wire (384 bits), 48 bytes captured (384 bits) on interface 0
▶ Ethernet II, Src: 02:42:ac:00:00:03 (02:42:ac:00:00:03), Dst: 02:42:ac:00:00:02 (02:42:
▶ Internet Protocol Version 4, Src: 172.0.0.3, Dst: 172.0.0.2
▶ User Datagram Protocol, Src Port: 45062 (45062), Dst Port: 1113 (1113)
▼ Licklider Transmission Protocol
  ▼ LTP Header
       LTP Version: 0
       LTP Type: 9 (Report-acknowledgment segment)
       Header Extension Count: 0
       Trailer Extension Count: 0
  ▼ Report Ack Segment
       Report serial number: 1445

```
0000  00000010 01000010 10101100 00000000 00000000 00000010 00000010 01000010  .B.....B
0008  10101100 00000000 00000000 00000011 00001000 00000000 01000101 00000000  ......E.
0010  00000000 00100010 01110001 11001110 01000000 00000000 01000000 00010001  ."q.@.@.
0018  01110000 11110111 10101100 00000000 00000000 00000011 10101100 00000000  p.......
0020  00000000 00000010 10110000 00000110 00000100 01011001 00000000 00001110  .....Y..
0028  01011000 00100101 00001001 00000001 00000001 00000000 10001011 00100101  X%,....%
```

Figure 11.29: ION Report Acknowledgment

Figure 11.29 shows the report acknowledgment sent by the sending node in response to the report segment. In this figure, it can be observed that the segment is a report acknowledgment segment of type 9 (1001) and that the report acknowledgment is acknowledging a report segment with serial number 1445.

Figure 11.30 shows the next data segment sent after the report acknowledgment. It can be observed that this is a regular data segment, like Figure 11.26. However, note the last 4 bits of the 3rd byte within the header (0010 or 2 in decimal). This indicates that this segment is sent within a new session with id 2. As the nodes are configured to use only one parallel session, this indicates that the node closed the previous session and has begun to send the next LTP block within a new session. Thus, it can be determined that the report segment acknowledged all the data sent in the block.

11.7.4.2   Send 10 Bundles with Simulated Loss

In this test, it was observed that the transmission of data from tx to rx was conducted in the same way as the first test. Regular data segments were sent, followed by a final segment containing a checkpoint, end of red part and end of block marker. However, in this test data loss can be observed by analyzing the report segment. The report segment shown in Figure 11.31, unlike the previously shown report segment, has

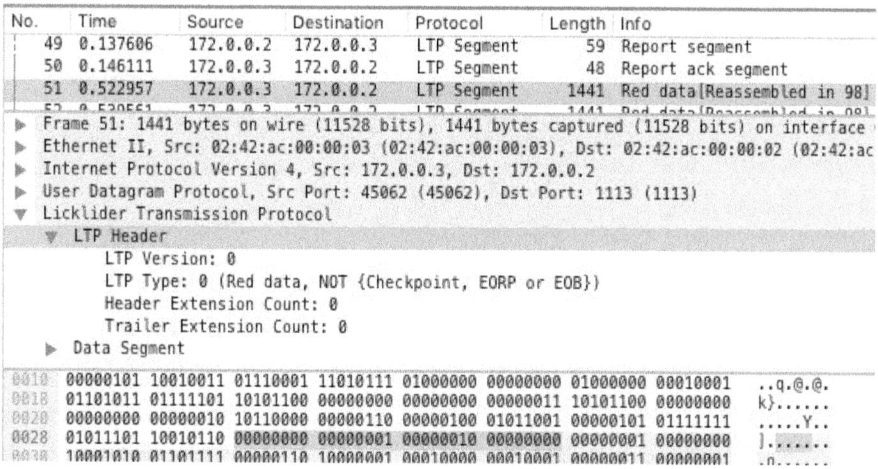

No.	Time	Source	Destination	Protocol	Length	Info
49	0.137606	172.0.0.2	172.0.0.3	LTP Segment	59	Report segment
50	0.146111	172.0.0.3	172.0.0.2	LTP Segment	48	Report ack segment
51	0.522957	172.0.0.3	172.0.0.2	LTP Segment	1441	Red data[Reassembled in 98]
52	0.530551	172.0.0.3	172.0.0.2	LTP Segment	1441	Red data[Reassembled in 98]

▶ Frame 51: 1441 bytes on wire (11528 bits), 1441 bytes captured (11528 bits) on interface
▶ Ethernet II, Src: 02:42:ac:00:00:03 (02:42:ac:00:00:03), Dst: 02:42:ac:00:00:02 (02:42:ac
▶ Internet Protocol Version 4, Src: 172.0.0.3, Dst: 172.0.0.2
▶ User Datagram Protocol, Src Port: 45062 (45062), Dst Port: 1113 (1113)
▼ Licklider Transmission Protocol
  ▼ LTP Header
      LTP Version: 0
      LTP Type: 0 (Red data, NOT {Checkpoint, EORP or EOB})
      Header Extension Count: 0
      Trailer Extension Count: 0
  ▶ Data Segment

```
0010  00000101 10010011 01110001 11010111 01000000 00000000 01000000 00010001   ..q.@.@.
0018  01101011 01111101 10101100 00000000 00000000 00000011 10101100 00000000   k}......
0020  00000000 00000010 10110000 00000110 00000100 01011001 00000101 01111111   .....Y..
0028  01011101 10010110 00000000 00000001 00000000 00000000 00000001 00000000   ].......
0030  10001010 01101111 00000110 10000001 00010000 00010001 00000011 00000001   .n.....
```

Figure 11.30: ION Data Segment 3

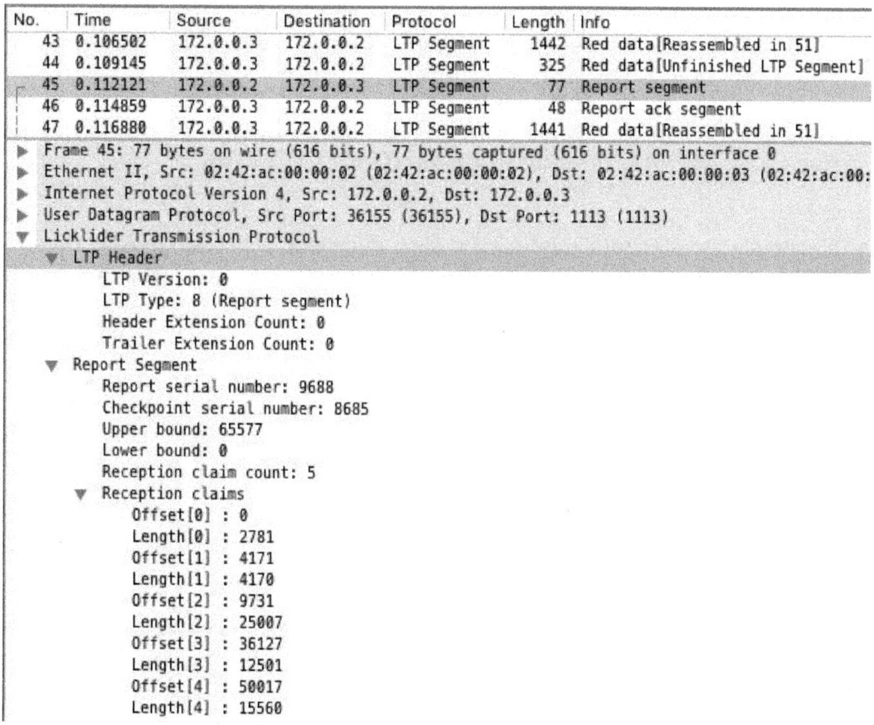

No.	Time	Source	Destination	Protocol	Length	Info
43	0.106502	172.0.0.3	172.0.0.2	LTP Segment	1442	Red data[Reassembled in 51]
44	0.109145	172.0.0.3	172.0.0.2	LTP Segment	325	Red data[Unfinished LTP Segment]
45	0.112121	172.0.0.2	172.0.0.3	LTP Segment	77	Report segment
46	0.114859	172.0.0.3	172.0.0.2	LTP Segment	48	Report ack segment
47	0.116880	172.0.0.3	172.0.0.2	LTP Segment	1441	Red data[Reassembled in 51]

▶ Frame 45: 77 bytes on wire (616 bits), 77 bytes captured (616 bits) on interface 0
▶ Ethernet II, Src: 02:42:ac:00:00:02 (02:42:ac:00:00:02), Dst: 02:42:ac:00:00:03 (02:42:ac:00:
▶ Internet Protocol Version 4, Src: 172.0.0.2, Dst: 172.0.0.3
▶ User Datagram Protocol, Src Port: 36155 (36155), Dst Port: 1113 (1113)
▼ Licklider Transmission Protocol
  ▼ LTP Header
      LTP Version: 0
      LTP Type: 8 (Report segment)
      Header Extension Count: 0
      Trailer Extension Count: 0
  ▼ Report Segment
      Report serial number: 9688
      Checkpoint serial number: 8685
      Upper bound: 65577
      Lower bound: 0
      Reception claim count: 5
    ▼ Reception claims
        Offset[0] : 0
        Length[0] : 2781
        Offset[1] : 4171
        Length[1] : 4170
        Offset[2] : 9731
        Length[2] : 25007
        Offset[3] : 36127
        Length[3] : 12501
        Offset[4] : 50017
        Length[4] : 15560

Figure 11.31: ION Report Segment with errors

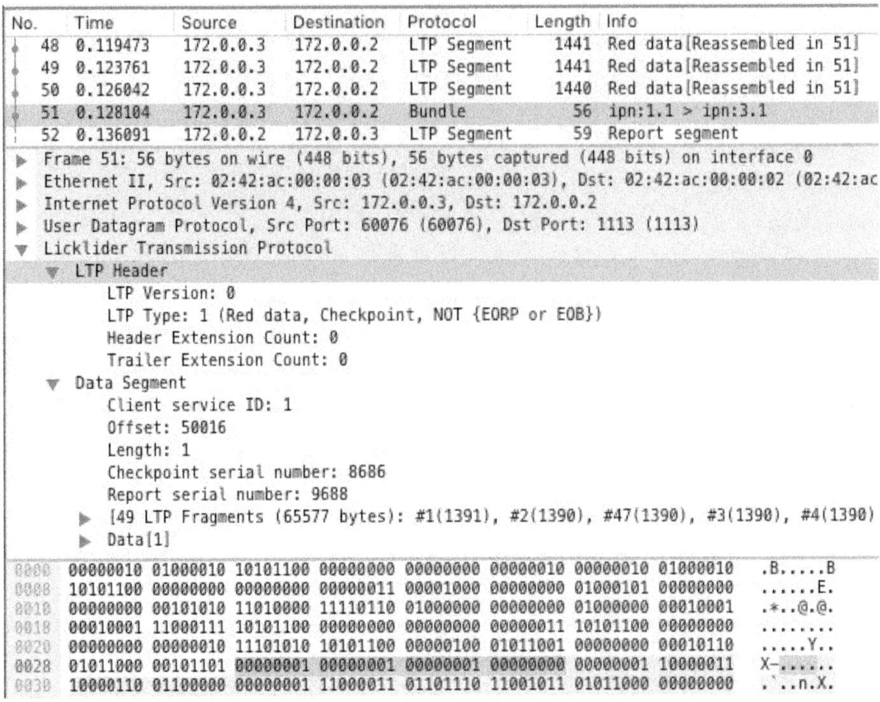

No.	Time	Source	Destination	Protocol	Length	Info
48	0.119473	172.0.0.3	172.0.0.2	LTP Segment	1441	Red data[Reassembled in 51]
49	0.123761	172.0.0.3	172.0.0.2	LTP Segment	1441	Red data[Reassembled in 51]
50	0.126042	172.0.0.3	172.0.0.2	LTP Segment	1440	Red data[Reassembled in 51]
51	0.128104	172.0.0.3	172.0.0.2	Bundle	56	ipn:1.1 > ipn:3.1
52	0.136091	172.0.0.2	172.0.0.3	LTP Segment	59	Report segment

```
▶ Frame 51: 56 bytes on wire (448 bits), 56 bytes captured (448 bits) on interface 0
▶ Ethernet II, Src: 02:42:ac:00:00:03 (02:42:ac:00:00:03), Dst: 02:42:ac:00:00:02 (02:42:ac
▶ Internet Protocol Version 4, Src: 172.0.0.3, Dst: 172.0.0.2
▶ User Datagram Protocol, Src Port: 60076 (60076), Dst Port: 1113 (1113)
▼ Licklider Transmission Protocol
    ▼ LTP Header
          LTP Version: 0
          LTP Type: 1 (Red data, Checkpoint, NOT {EORP or EOB})
          Header Extension Count: 0
          Trailer Extension Count: 0
    ▼ Data Segment
          Client service ID: 1
          Offset: 50016
          Length: 1
          Checkpoint serial number: 8686
          Report serial number: 9688
       ▶ [49 LTP Fragments (65577 bytes): #1(1391), #2(1390), #47(1390), #3(1390), #4(1390)
       ▶ Data[1]
```

```
0000   00000010 01000010 10101100 00000000 00000000 00000010 00000010 01000010   .B.....B
0008   10101100 00000000 00000000 00000011 00001000 00000000 01000101 00000000   ......E.
0010   00000000 00101010 11010000 11110110 01000000 00000000 01000000 00010001   .*..@.@.
0018   00010001 11000111 10101100 00000000 00000000 00000011 10101100 00000000   ........
0020   00000000 00000010 11101010 10101100 00000100 01011001 00000000 00010110   .....Y..
0028   01011000 00101101 00000001 00000001 00000001 00000000 00000001 10000011   X-......
0030   10000110 01100000 00000001 11000011 01101110 11001011 01011000 00000000   .'..n.X.
```

Figure 11.32: ION Data Segment with errors (CP, EORP, EOB)

five claims with different lengths and offsets. Reception claim one indicates that all data from byte 0 to 2781 was received successfully. However, the offset of reception claim two is 4171, indicating that data was lost between bytes 2782 and 4171.

The next point to note is found by analyzing the data segment containing the second checkpoint, shown in Figure 11.32. It can be observed that this segment does not have a value of zero for the report serial number. Instead, the report serial number is 9688, which indicates this checkpoint is related to data resent in response to a report segment with serial 9688.

This transmission was conducted with a very high simulated loss (10 percent), additionally the ION software was configured to only accommodate low levels of loss. This configuration mismatch allows session cancellation to be observed, as the software will reach the maximum number of retries before a block is successfully sent (More detail can be found in the "CP operation" subsection within section 11.5). Figure 11.33 shows one such cancellation segment. By analyzing this segment, it can be observed this is a cancellation segment sent from the sender (1100 in binary or c in hex). It can also be observed that the cancellation code is 2 (Retransmission limit exceeded).

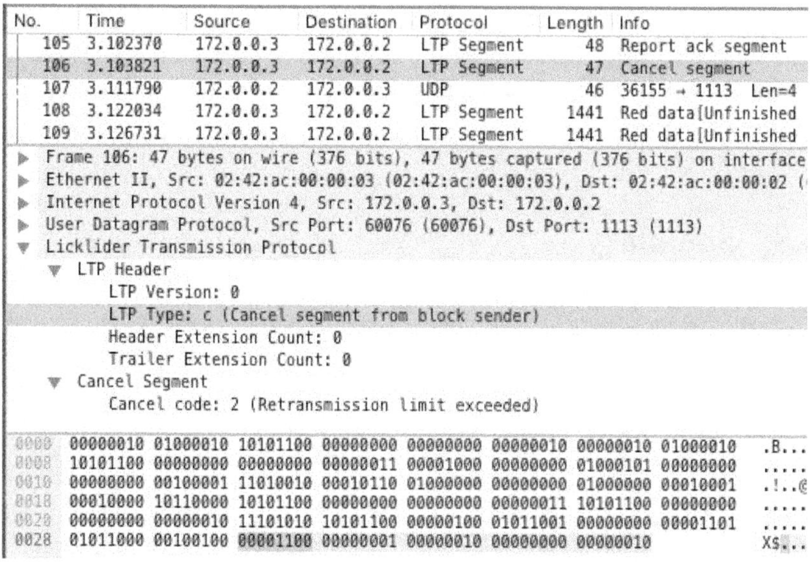

No.	Time	Source	Destination	Protocol	Length	Info
105	3.102370	172.0.0.3	172.0.0.2	LTP Segment	48	Report ack segment
106	3.103821	172.0.0.3	172.0.0.2	LTP Segment	47	Cancel segment
107	3.111790	172.0.0.2	172.0.0.3	UDP	46	36155 → 1113 Len=4
108	3.122034	172.0.0.3	172.0.0.2	LTP Segment	1441	Red data[Unfinished
109	3.126731	172.0.0.3	172.0.0.2	LTP Segment	1441	Red data[Unfinished

```
▶ Frame 106: 47 bytes on wire (376 bits), 47 bytes captured (376 bits) on interface
▶ Ethernet II, Src: 02:42:ac:00:00:03 (02:42:ac:00:00:03), Dst: 02:42:ac:00:00:02 (
▶ Internet Protocol Version 4, Src: 172.0.0.3, Dst: 172.0.0.2
▶ User Datagram Protocol, Src Port: 60076 (60076), Dst Port: 1113 (1113)
▼ Licklider Transmission Protocol
   ▼ LTP Header
      LTP Version: 0
      LTP Type: c (Cancel segment from block sender)
      Header Extension Count: 0
      Trailer Extension Count: 0
   ▼ Cancel Segment
      Cancel code: 2 (Retransmission limit exceeded)
```

```
0000  00000010 01000010 10101100 00000000 00000000 00000010 00000010 01000010   .B...
0008  10101100 00000000 00000000 00000011 00001000 00000000 01000101 00000000   .....
0010  00000000 00100001 11010010 00010110 01000000 00000000 01000000 00010001   .!..@
0018  00010000 10110000 10101100 00000000 00000000 00000011 10101100 00000000   .....
0020  00000000 00000010 11101010 10101100 00000100 01011001 00000000 00001101   .....
0028  01011000 00100100 00001100 00000001 00000010 00000000 00000010           XS...
```

Figure 11.33: ION Cancellation Segment

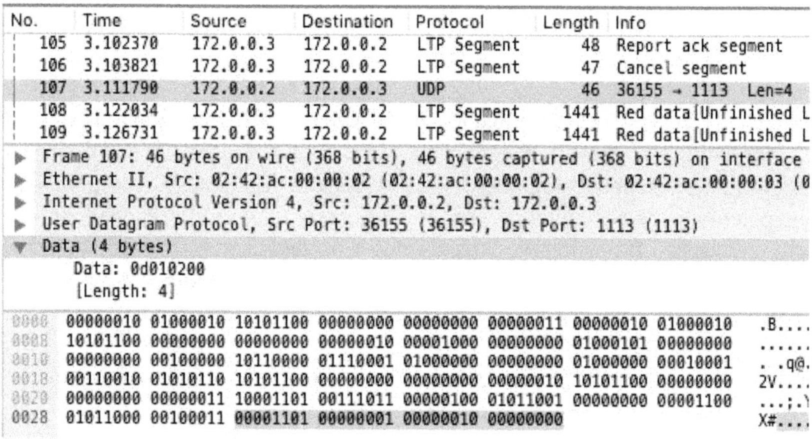

No.	Time	Source	Destination	Protocol	Length	Info
105	3.102370	172.0.0.3	172.0.0.2	LTP Segment	48	Report ack segment
106	3.103821	172.0.0.3	172.0.0.2	LTP Segment	47	Cancel segment
107	3.111790	172.0.0.2	172.0.0.3	UDP	46	36155 → 1113 Len=4
108	3.122034	172.0.0.3	172.0.0.2	LTP Segment	1441	Red data[Unfinished L
109	3.126731	172.0.0.3	172.0.0.2	LTP Segment	1441	Red data[Unfinished L

```
▶ Frame 107: 46 bytes on wire (368 bits), 46 bytes captured (368 bits) on interface
▶ Ethernet II, Src: 02:42:ac:00:00:02 (02:42:ac:00:00:02), Dst: 02:42:ac:00:00:03 (0
▶ Internet Protocol Version 4, Src: 172.0.0.2, Dst: 172.0.0.3
▶ User Datagram Protocol, Src Port: 36155 (36155), Dst Port: 1113 (1113)
▼ Data (4 bytes)
      Data: 0d010200
      [Length: 4]
```

```
0000  00000010 01000010 10101100 00000000 00000000 00000011 00000010 01000010   .B....
0008  10101100 00000000 00000000 00000010 00001000 00000000 01000101 00000000   ......
0010  00000000 00100000 10110000 01110001 01000000 00000000 01000000 00010001   . .q@.
0018  00110010 01010110 10101100 00000000 00000000 00000010 10101100 00000000   2V....
0020  00000000 00000011 10001101 00111011 00000100 01011001 00000000 00001100   ...;.\
0028  01011000 00100011 00001101 00000001 00000010 00000000                     X#....
```

Figure 11.34: ION Cancellation Acknowledgment to Sender

In response to the cancellation segment, the receiving node sends a cancellation acknowledgment segment which is shown in Figure 11.34. Wireshark does not dissect the segment correctly, instead showing a UDP segment containing 4 bytes. However, by showing the raw data as bits, the segment can be manually dissected. Referring to the "segment type flags" subsection within section 2, it can be observed that the flags "1101" within the last four bits of Byte 1 identify this is a cancellation acknowledgment segment to the sender. Thus indicating to the sending node that the receiver has cancelled the session.

11.8 Summary

It is clear that traditional protocols such as TCP/IP are unsuitable for use in some scenarios including deep space communication. Although TCP/IP could perhaps be modified to accommodate extremely high latency, it is certainly not going to provide efficient transmission of data in deep space.

LTP as described in detail throughout the chapter, provides a robust method to achieve both optimal and reliable communication in severely challenging network conditions. Unlike TCP/IP however, LTP is not yet as user-friendly. Reliable TCP/IP communication can be achieved using a few simple configuration parameters such as IP address and port. Yet, as can be observed in section 11.7 within the node configuration files, LTP requires many more parameters to get started. LTP may also necessitate a considerable amount of time and effort to optimize transmission for each particular situation. For most operators though, the amount of effort could be considered commensurate with the difficulty or unreliability of communication in such challenging conditions by any other means.

Although current LTP use is limited to a niche area (space communication), this niche area is growing in popularity. It is now common to see new space-bound experiments all the time, for example using micro-satellites named "CubeSats". Additionally, LTP could potentially be beneficial in other scenarios providing similarly challenging network conditions, such as remote sensor fields. Consequently, it is possible that use of LTP could grow considerably over the coming years and develop into a widely used standard for both space and terrestrial communication.

Chapter 12

Delay-/Disruption-Tolerant Networking Performance Evaluation with DTNperf_3.

Carlo Caini

University of Bologna, Italy

CONTENTS

Delay-/Disruption-Tolerant Networking (DTN) aims to enable communications in the so-called "challenged networks", where long delays, disruption, link intermittency and other challenges prevent the use of the ordinary Internet architecture. Because of these challenges, and the heterogeneity of possible applications scenarios, ranging from InterPlanetary Networks to the Internet of Things, the evaluation of DTN performance is a complex task. To cope with this complexity, it is necessary to develop suitable tools, powerful and flexible at the same time. This is the aim of DTNperf, now at its third release, called DTNperf_3. In this chapter its characteristics and use are presented in detail, but also in a modular way to allow different layers of reading (i.e. technical details are present, but can easily be skipped by the reader more interested in DTNperf_3 use). DTNperf_3 supports DTN2, ION and IBR-DTN bundle protocol implementations, is released as free software and is included in ION.

12.1   Introduction

Performance assessment in DTN (Delay-/Disruption-Tolerant Networking) environments [54][195][75] is much more complex than in ordinary TCP/IP networks because of the possible presence of many impairments such as long delays, disruption, intermittent links, network partitioning, etc. The DTN architecture, as described by RFC 4838 [75] is based on the introduction of a new layer, the Bundle layer, and of its related protocol, the Bundle Protocol (BP) [244][16]. Although all BP implementations, such as DTN2 [244], ION [16] and IBR-DTN [5], offer a few basic tools (to send or receive single bundles, transfer a file, "ping" another node, etc.), by means of which to evaluate performance, they have two limits. First, they are quite elementary and require scripts to carry out non trivial experiments; second, although they perform similar task, their syntax and features differ, depending on the implementation, which makes it more difficult to carry out experiments in mixed environments. As a curiosity, we note here that basic tools in DTN2 and IBR-DTN have the same name (dtnsend, dtnrecv, etc.) and thus can be distinguished only by their locations when both implementations are present, unless suitably renamed. Although this is not a big theoretical problem, to enter the same commands with different syntaxes, is clearly prone to errors for users that must concurrently deal with different implementations, for example to carry out interoperability tests.

DTNperf_3 aims to overcome these limitations. First, its application scope is very large, ranging from simple to very complex experiments. To cover all the range, it makes an intensive use of default values, forcing the user to add syntax elements, as options, only when really needed; moreover, it consists of just one application. DTNperf_3 can be used in three modes, client, server and monitor, to send and receive bundles, or to collect status reports [244], respectively. The first two modes can replace all basic tools for sending and receiving bundles, while the third is unique, as at present there are no tools dedicated to the monitoring, or collection, of status reports. To have just an application instead of three simplifies the use, as it naturally leads to a better consistency of the user interface (the three modes share many options). A second key aspect, is that DTNperf_3 has been explicitly designed to support multiple BP implementations. To this end, it is based on the use of the Abstraction Layer (AL), a library that creates a unique interface to the APIs (Application Programming Interface) of different implementations. At present, AL supports DTN2, ION and IBR-DTN, i.e. all the major open source BP implementations.

The multiple implementation support proved particularly useful in interoperability tests recently carried out by NASA and JAXA to validate the BP CCSDS specifications [16]. This is a typical example of DTNperf advanced use; another example, is the analysis of CGR routing [33][44]. In these cases and in many others, having all status reports collected in one CSV (Comma Separated Value) file that can be easily imported and elaborated into a spreadsheet, allows for a fast and accurate analysis of the experiments, by eliminating the need to manually inspect multiple logs, or to carry out low level analysis with Wireshark, when unnecessary (i.e. when the user is interested only in studying performance at bundle layer, such as routing, priority enforcement, custody transfer, fragmentation and reassembly, delivery).

The current version of DTNperf is the result of a long evolution. Its first version, which dates back to 2005, was developed for DTN2 only; then it was greatly improved in 2008, with the development of the second major release, called DTNperf_2 [33]. This version was used for many years by the authors to evaluate DTN performance in space communications [68][70] and in other fields as well. The third release, DTNperf_3, was issued in 2013, and was completely rewritten from scratch [69]; it added the external monitor and the support of multiple BP implementations (originally DTN2 and ION, from 3.5.0 also IBR-DTN). DTNperf_3 is released as free software and is included in ION.

The aim of this chapter is to give a comprehensive and detailed presentation of the current version of DTNperf_3 (3.5.0), thus updating and extending [69]. The structure is modular, consisting of three logical parts. The first contains a general presentation of DTNperf_3 (Section 12.2); the second part, the detailed description of all its components (client, server and monitor, in Sections 12.3, 12.4 and 12.5, respectively; AL in Section 12.6). The third is a sort of user guide to make the most of the application, embracing both normal and advanced use (Sections 12.7 and 12.8). Conclusions are drawn at the end of the chapter.

12.2　DTNperf_3 General Description

DTNperf_3 is a DTN application running on top of BP. Its main characteristics are described below.

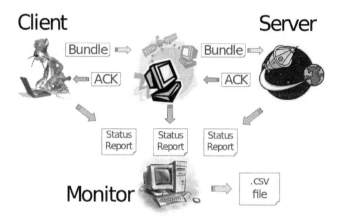

Figure 12.1: DTNperf_3 operating modes. The client sends bundles to the server and receive acknowledgments from it; the monitor collects status reports issued by all DTN nodes and saves them into a .csv file.

12.2.1 Operating Modes

DTNperf_3 has three operating modes: client, server and monitor. The client generates and sends bundles; the server receives and (optionally) acknowledges them; the monitor collects status reports and some control bundles (Figure 12.1). Client, server and monitor correspond to the BP addresses "from", "to" and "report to" [75][244].

12.2.2 Transmission Modes

The client can send bundles in three modes: Time, Data and File (T|D|F). In Time-mode, it generates a series of bundles with a dummy payload of desired dimension and passes these bundles to the BP daemon for transmission, until the pre-set transmission time elapses. In the Data-mode the process is the same but the generation terminates when a pre-set amount of data is reached. In both cases dummy payloads are not stored by the DTNperf server. Finally, in File-mode the content of a file is transferred, instead of dummy data and if the file size is larger than the bundle payload, the file is split into multiple bundles. As a consequence, in this case the DTNperf server has to buffer the incoming payloads until the entire file is received (or a time threshold is reached). If all bundles have arrived in time the file is saved, otherwise discarded.

12.2.3 Congestion Control Policies

Independently of the Tx mode, DTNperf can use two alternative congestion control policies: window-based and rate-based (W|R). They look like the mechanisms at the basis of TCP and UDP, but are not actually the same, as explained below.

12.2.3.1 Window-Based Congestion Control

If the window-based congestion control is selected, bundles sent by the DTNperf client must be acknowledged by the server. To allow multiple bundles in flight, DT-Nperf_3 uses a "congestion window" W, which represents the maximum number of bundles sent but not acknowledged yet. The mechanism is similar to the TCP congestion window, but the dimension of W is fixed (being set by the user) and in-flight bundles can be non-consecutive (to cope with non-ordered delivery of BP). Moreover, DTNperf does not implement any retransmission mechanisms, as acknowledgments are used only to trigger the transmission of new bundles.

12.2.3.2 Rate-Based Congestion Control

In alternative to the window-based policy, useful to evaluate goodput (see sections about the DTNperf use), the user can select a rate-based mechanism, particularly suitable to emulate streaming services, or, more generally, to be used whenever the generated traffic should be independent of the actual delivery of previous bundles. The desired rate can be entered in a very flexible way, either in bundle/s or in Mbit/s or kbit/s. In the last two cases DTNperf converts automatically the bit rate into a bundle rate, given the dimension of the bundle, which proved very convenient. Note that in response of rate-based traffic the server does not generate any ACKs, like UDP. As a result, the client instead of calculating a goodput (data confirmed/interval), calculates a throughput (data sent/interval).

12.2.4 Collecting Information

In all operating modes, DTNperf_3 provides real time information on the console, whose amount can be increased in two steps by means of the debug option (--debug). In simple experiments, involving a pair of nodes, this is usually sufficient, but not in more complex ones, where it is essential to have the possibility of inspecting status reports generated by DTN nodes along the path to destination. The monitor task is to collect them in a .csv log file, for further analysis. Note that by contrast with ordinary tools, the DTNperf client always sets on "Delivered" and "deleted" status report request flags, while "custody" status report flag is on if the custody option is on. Only the setting of "Forwarded" and "received" is discretionary. An excerpt of the DTNperf_3 .csv log is given in Table 12.1, assuming one bundle transmission from vm1 to vm2 via vm6. The first row reports custody on vm1, then in order received and custody on vm6, finally received, custody and delivered on vm2. Note the distinction between fields that identify the report and fields that identify the bundle that triggered the report (alias bundle X, see [244]).

Table 12.1: Excerpt of a.csv Log File: Status Reports Triggered by bundle "2582.35" sent from vm1 to vm2 via vm6.

Report_SRC	Report TST	Rep. SQN	Bndl_SRC	Bndl TST	Bndl SQN	Bndl FO	Bndl FL	Dlv	Ct	Rcv	Fwd	Del
dtn://vm1.dtn	2582	36	dtn://vm1.dtn/dtnperf:/src_931	2582	35	0	0		2582			
dtn://vm6.dtn	2593	41	dtn://vm1.dtn/dtnperf:/src_931	2582	35	0	0			2594		
dtn://vm6.dtn	2594	43	dtn://vm1.dtn/dtnperf:/src_931	2582	35	0	0		2594			
dtn://vm2.dtn	2608	25	dtn://vm1.dtn/dtnperf:/src_931	2582	35	0	0			2608		
dtn://vm2.dtn	2608	27	dtn://vm1.dtn/dtnperf:/src_931	2582	35	0	0		2608			
dtn://vm2.dtn	2608	29	dtn://vm1.dtn/dtnperf:/src_931	2582	35	0	0	2608				

12.3   DTNperf_3 Client

The client must generate and send bundles to the server in a great variety of ways, to match the heterogeneity of DTN networks. To combine flexibility with user-friendliness, we kept as many inputs as possible optional, as described below.

12.3.1   *Syntax, Parameters and Options*

> *SYNTAX: dtnperf --client -d <dest_eid> <[-T <s> | -D <num> | -F <filename>]> [-W <size> | -R <rate>] [options]*

Required inputs and main options are given in Table 12.2. Compulsory inputs are just three, to keep the command interface as simple as possible: -d indicates the destination, -T, -D and -F specify the Tx mode (either time, or data, or file), -W and – R the congestion control (either window- or rate-based). Among the options, the most important are the payload dimension (-P), the custody option (-C) and the possibility of identifying an external monitor (-m). If the last is not specified, a DTNperf monitor is launched, in a transparent way, on the same node as the client. Other important options include lifetime (-l) priority (-p) and the CRC check of the bundle payload (--crc). The full list can be obtained with the command "dtnperf --client --help".

12.3.2   *Implementation Notes*

The most significant aspects of the implementation are highlighted below. The reader interested only in DTNperf_3 basic use can skip this sub-section (and the corresponding ones in the server and monitor sections). These notes are however essential for the expert user to attain a full comprehension of the tool.

12.3.2.1   *Initialization*

The most important task performed in the initialization phase is the registration of the client EID, which will contain the PID of the process. Assuming that the client is launched on vm1, the EID will be "dtn://vm1.dtn/dtnperf:/src_pid", where dtn://vm1.dtn is the address of the node and /dtnperf:/src_pid is the demux token that

Table 12.2: DTNperf_3 Client Required inputs and Main Options.

Required inputs	
-d, –destination <eid>	Destination EID (Endpint Identifier)
Tx modes (T|D|F)	
-T, –time <s>	Time-Mode: seconds of transmission
-D, –data <num[BkM]>|	Data-mode: bytes to transmit; B = Bytes, k = kBytes, M = MBytes. Default 'M' (MB). Note: following the SI and the IEEE standards 1 MB=10^6 bytes
-F, –file <filename>	File-mode: file to transfer
Congestion modes (W|R)	
-W, –window <size>	Window-based congestion control: size in bundle of transmission window (default 1).
-R, –rate<rate[k|M|b]>	Rate-based congestion control: bitrate of transmission. k = kbit/s, M = Mbit/s, b = bundles/s.Default is kbit/s
Options	
-P, –payload <size[BKM]>	Size in bytes of bundle payload; data unit default 'k' (kbytes). Note: following the SI and the IEEE standards 1 MB=10^6 bytes; default 50kB.
-C, –custody	Enable both Custody option and custody status reports. Default off.
-m, –monitor <eid>	Monitor EID. Default: internal monitor.
-M, –memory	Store the bundle into memory instead of file (if payload < 50KB).
-N, –nofragment	Disable bundle fragmentation. Default: off.
-l, –lifetime <time>	Bundles lifetime (s). Default is 60s.
-p, --priority <val>	Bundles priority [bulk|normal|expedited| reserved]. Default: normal.
-f, – forwarded	-f, –forwarded Enable request for status report forwarded. Default off.
-r –received	Enable request for status report received. Default off.
–crc	Enable a CRC check of the bundle payload. Default off.
-D, –debug <level>	Debug level [1-2]. Default off.
-L, –log <log_filename>	Create a log file. Default:dtnperf_client.log.
–force-eid <[DTN|IPN]>	Force the scheme of EID registration. Default: dtn on DTN2, ipn on ION.
–ipn-local <num>	Set ipn local number (Use only with –force-eid IPN on DTN2)
-h, –help	Help.

in BP performs the same demultiplexing task as TCP and UDP ports. ACKs sent back by the server will be addressed to this EID. The insertion of the PID allows the server to cope with multiple instances of the client running on the same node, sending ACKs to the right client. This may happen when traffic with different priorities must be generated on the same node, which requires a dtnperf client for each priority.

12.3.2.2 Threads

After initialization, three concurrent threads are launched. "Sender" is in charge of sending bundles to the server; "cong_ctrl" is in charge of the congestion control (in particular, it processes ACKs sent back by the server when in window mode, or manages a semaphore when in rate-based mode); the last, "wait_for_signal" is used to intercept a possible ctrl+c sent by the user to abort the client. This mechanism allows the client to send a "FORCE STOP" control bundle to an external monitor, before closing, to inform it that the client session has been aborted (if the monitor is internal the same information is sent directly). As a consequence, the monitor will close the corresponding .csv file.

12.3.2.3 Program Termination

Depending on the Tx mode selected, the sender thread terminates either after the specified amount of time (-T) has elapsed, or when the wanted data (-D) has been generated, or after having sent the entire file (-F). If the window-based congestion control is selected, the cong_ctrl thread cannot stop immediately, but has to wait until all bundles sent have been confirmed by server ACKs, which can take a while. To avoid deadlocks, a timeout is set to the lifetime value of bundles sent. After having received all ACKs (normal end), or the timeout has expired (abnormal end), the cong_ctrl terminates and the client closes. If the rate-based congestion control is selected, there are no ACKs to wait for and the cong_ctrl terminates as soon as the sender thread does. In both cases, the last operations performed are the calculation of goodput or throughput and the sending of a "STOP" bundle to the monitor (either internal or external) to inform it that the client session has terminated regularly after the transmission of a given amount of bundles.

12.4   DTNperf_3 Server

The server receives bundles, acknowledges them if sent in window-based mode, and reassembles bundle payloads if a file is transferred.

12.4.1   Syntax and Options

SYNTAX: dtnperf --server [options]

For the server, there are no compulsory inputs, but only options (Table 12.3). In

Table 12.3: DTNperf_3 Server Main Options.

Options	
-a, –daemon	Start the server as a daemon. Output is redirected to -
-o, –output <file>	Change the default output file (only with -a option).
-s, –stop.	Stop the server daemon.
-v, –verbose	Print some information message during the execution.
–debug[=level]	Debug level [1-2]. Default off.
–fdir <dir>	Destination directory of files transfered. Default is ˜/dtnperf/files .
-l, –lifetime <s>	Force minimum bundle-ACK lifetime (s). (60s if <s> is omitted). Default: =incoming bundle lifetime.
-p, –priority <val>	Force lowest bundle-ACK priority [bulk|normal| expedited|reserved]. Normal if <val> is omitted. Default: = incoming bundle priority.
–force-eid <[DTN|IPN]>	Force the scheme of EID registration. Default: dtn on DTN2, ipn on ION.
–ipn-local <num>	Set ipn local number (Use only with –force-eid IPN on DTN2)
–acks-to-mon	Send bundle ACKs in cc to the monitor
-h, –help	Help.

particular, it is possible to run the server in the background as a daemon (-a). The full list of options can be obtained with the command "dtnperf --server --help".

12.4.2   Implementation Notes

12.4.2.1   Initialization

In contrast to the client case, the registration of the server EID does not contain the PID of the process. Assuming that the client is launched on vm2, the complete EID is "dtn://vm2.dtn/dtnperf:/dest", where /dtnperf:/dest is the constant demux token. As only one server can be launched on the same DTN node, an initialization check based on the fixed demux string aborts the launch if another server has already registered.

12.4.2.2   Threads

After initialization, two threads start. The first receives and processes bundles, the second manages timers used in file transfers (see next subsection). The first thread performs a loop: when a bundle arrives, the bundle is processed and then the loop is re-entered. A small header in the payload of the incoming bundle indicates the Tx and the congestion modes set on the client, which allows the server to select the appropriate processing actions. If the Tx mode is time or data, payloads are dummy and data must be discarded; vice versa in file mode they must be saved and reassembled. With regard to congestion control modes, ACKs must be sent back to the client only in window mode. Each ACK contains the unique identifier of the bundle acknowledged, to avoid any ambiguity. If the CRC option is enabled on the client, the ACK also reports the success or the failure of the CRC check performed on the server. Finally, note that ACK priority and lifetime are automatically derived by the corresponding values of the ACKed bundle, also reported in the small payload header of the incoming bundle. However, minimum values can be forced on the server side by means of two specific options (-l, -p).

12.4.2.3   File Transfer

The file transfer mode is the most demanding on the server side, as it has to match three requirements: to manage multiple concurrent file transfers from multiple clients; to perform file segmentation and reassembly; and to cope with disordered bundle delivery. To this end, the server can work on parallel file transfer "sessions", identified by the EID of the client. A header in the payload of all bundles sent in file mode indicates the name of the file transferred, its dimension and the offset of the payload. For each session a dynamic list is maintained, to allow for reassembling the original file even in the presence of a disordered bundle delivery. The second thread manages one timer for each file transfer session. If a timeout (set to the incoming bundle lifetime) expires, the session structure is canceled. At the end, only files that have been completely reassembled are saved.

12.4.2.4   Program Termination

Unlike the client, the server is always on and must be terminated at the end of the experiments by pressing ctrl+c, which triggers a controlled closure. The same happens when a server launched as a daemon is stopped by means of the command "dtnperf --server –s".

12.5   DTNperf_3 Monitor

The monitor, added in this third major release, has been designed to receive and collect status reports. It can be launched either as an independent process on an external node (external monitor), in which case it can serve many concurrent clients, or as

Table 12.4: DTNperf_3 Monitor Specific Options.

Options	
-e, –session-expiration <s>	Max idle time of log files (s). Default: 120s.
–oneCSVonly	Generate a unique csv file
–rt-print[=filename]	Print realtime human readable status report information

a "child" process of a client (dedicated monitor), in which case it serves only its "parent" client.

12.5.1 Syntax and Options (external monitor only)

The external monitor syntax has not mandatory inputs, but only options, most of which are in common to the server. The specific options (see Table 12.4) are self-explanatory. Note that the "--rt-print[=filename]" option has been added in the latest DTNperf versions to allow the real-time use of the monitor. With this option, most of the collected information is printed on the standard output (optionally on file) in a format immediately readable by humans, as well as on the .csv file as usual. Again, the full list of options can be obtained with the command "dtnperf --monitor --help".

12.5.2 Implementation Notes

12.5.2.1 Initialization

If the monitor is external, the registration is like for the server. Assuming that the monitor is launched on vm3 the complete EID is "dtn://vm3.dtn/dtnperf:/mon", where /dtnperf:/mon is the constant demux token. With a dedicated monitor, however, the demux token must contain the PID of the client (the parent process), such as /dtnperf:/mon_pidclient, to allow as many dedicated monitors as clients on the same node.

12.5.2.2 Threads

As in the server, after initialization two threads start. The first receives and processes status reports and other DTNperf control bundles. The second manages timers used to force the closure of log files in case of deadlock. The first thread performs a loop: every time a bundle arrives, the bundle is processed and the information is saved in the appropriate log file, and so on until the monitor is closed. Bundle processing has many steps. First, the monitor has to distinguish if the bundle is a status report, generated by the BP, or a control bundle, such as a STOP or a FORCE STOP generated by the DTNperf client to indicate either regular or forced (ctrl+c) closure. DTNperf control bundles are easily distinguished by a small header in their payload.

12.5.2.3   Sessions and Log Files

If the monitor is dedicated, the task is easy as all status reports sent to it refer to the same client and must be saved in the same file. If the monitor is external, the task is more complex, as status reports refers to multiple clients (running on the same node or on multiple nodes) and must be saved in different files. To this end, the monitor must manage concurrent "sessions". A session is identified by the complete EID of the client process that has generated the bundles, which is also reported in the status reports and in the DTNperf control bundles directed to the monitor. Every time the monitor receives a bundle containing a new client EID, a new session is started and a new log file is created. If, vice versa, the client EID is that of an on-going session, the content of the status report or the informative bundle is appended to the corresponding log file. This default behavior can be overridden by means of the option –oneCSVonly, which force the collection of all status reports in a unique file. This may be convenient, although not necessary, when bundles with different priorities are generated by multiple clients on the same node, as in this case the bundle source is logically the same.

12.5.2.4   Closure of Log Files

When the client has completed the sender thread, it sends a STOP bundle to the monitor. The STOP bundle also contains the total number of bundles sent, which informs the monitor how many "delivered" status reports should be received to complete the corresponding session (delivered status report requests are always enabled in the client, as said). Once all the reports have arrived, the monitor closes the log file of the session. If the client is aborted with a ctrl+c, a FORCE STOP is sent to the external monitor to trigger the immediate closure of the corresponding log file (if the monitor is internal the same command is done directly). In the absence of any new status reports, or of a STOP or FORCE STOP, a log file is closed when a timeout expires (the default value can be overridden with the option –e).

12.5.2.5   Program Termination

As the server, an external monitor is always on and must be terminated at the end of the experiments by pressing ctrl+c, or, if launched as a daemon, by means of the command "dtnperf --server –s". By contrast, an internal monitor closes as soon as its sole log session concludes.

12.6   The Abstraction Layer

The "Abstraction Layer" is a library expressly designed to provide DTNperf (and possibly also other applications) a common interface towards the specific APIs (Application Programming Interface) of different BP implementations, such as DTN2, ION and IBR-DTN. By decoupling the DTNperf application from the BP implementation, the AL offers three advantages: first, features and syntax are the same; second, there is no need to develop and maintain a different DTNperf version for each BP implementation; third, interoperability tests are greatly facilitated by the fact that they involve just two instances of the same application and not two dedicated versions.

In the first versions of the AL, the support was limited to DTN2 and ION, but from AL 1.5.0 (bundled with DTNperf 3.5.0) it includes IBR-DTN too. To enable the support of a specific implementation, the user needs to add in the general (DTNperf and AL) "make" command the root directory of the implementation itself. One or more implementations can be enabled this way. In the former case, DTNperf can run on top of either DTN2, or ION or IBR-DTN only (Figure 12.2 a). In the latter, on whatever couple or even on all implementations, as the choice between DTN2, ION or IBR-DTN API procedures are performed at run time (Figure 12.2 b). To distinguish the application scope of each version, a postfix is added to the AL libraries and to the DTNperf application (_vDTN2, _vION, _vIBRDTN and their combinations). For the sake of simplicity, the lack of any postfix means that the scope is global.

The three BP implementations supported, although all compliant with RFC5050 [244], may offer different BP extensions. For this reason, a few DTNperf options are implementation specific, e.g. ECOS (Extended Class Of Service) features [59], which can be set only when DTNperf runs on top of ION. To avoid possible errors, a series of option compatibility checks are performed at run time, as soon as DTNperf becomes aware of the BP actual implementation on which is running.

12.6.1   Implementation Notes

The AL aims at providing the programmer with a unique programming interface, the AL API, abstracting the specific APIs of DTN2, ION and IBR-DTN. In brief, the AL consists of a set of abstracted APIs, a set of abstracted structures and a few auxiliary conversion functions, written in C (ION and DTN2) or in C++ (IBRDTN). Although the AL has been designed for DTNperf_3, it could be used by other DTN applications to make them independent of the BP implementation. However, as the present version has been created to support DTNperf, only the types and the functions necessary for

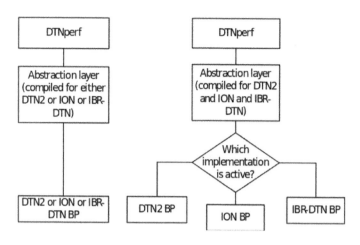

Figure 12.2: DTNperf scope. a) one implementation only, either DTN2, or ION, or IBR-DTN, b) multiple implementations with BP API selection at run time.

this purpose have been "abstracted". In other words, other DTN applications might require the abstraction of other elements as well, although this is not very likely. The programmer needing such extensions can autonomously build them, as all the code is free software.

12.6.1.1   Abstraction Layer Components: AL Types and AL API Procedures

Ideally, each AL procedure should be the abstracted version of one corresponding procedure of DTN2, ION and IBR-DTN. In practice this one-to-one correspondence is not always present, making the abstraction process much more complex. Moreover, data structures passed to specific API procedures are obviously different, and must thus be abstracted too. Another issue is given by the fact that AL, DTN2 and ION APIs are in C, while IBR-DTN is in C++ and object-oriented, making the correspondence only approximate. As a comprehensive description of the AL is out of the present scope, we will limit the description to a minimum.

12.6.1.2   AL Types

The AL Types are an abstraction of DTN2, ION, and IBR-DTN types. They are defined in file "al_bp_types.h" and are divided into four groups: general types, registration EID types, bundle types, status report types. A few type conversion functions from/to AL and DTN2 or ION types are available inside the AL API. There are no dedicated functions to convert between AL and IBR-DTN types. That is due to the lack of correspondence between most of the AL and IBR-DTN types, which results in conversions being performed directly when needed.

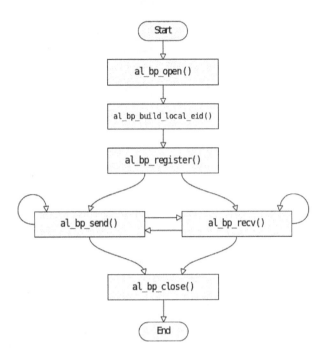

Figure 12.3: Typical flow of most significant AL functions.

12.6.1.3 AL API

The AL API procedures, defined in the file "al_bp_api.h", belong to three categories: principal, utility and high level procedures. The former two groups consist of the abstracted versions of corresponding DTN2, ION and IBR-DTN API procedures, while the latter contains procedure for error management or for bundle processing, which do not correspond to any specific APIs. The scheme below summarizes the typical flow of the most important AL functions.

12.6.1.4 AL API Files and Structure

The organization of AL files is the following:

- /al_bp/src: contains the declaration files and the implementation of the AL interface to the applications (al_bp_api.c);

- /al_bp/src/bp_implementations: contains the interfaces to DTN2, ION, and IBR-DTN APIs (al_bp_dtn.c, al_bp_ion.c, al_bp_ibr.cpp, etc.).

From the application to the API provided by the specific BP implementation, we have a chain of intermediate calls. Let us explain this with an example (see Figure 12.4). The AL procedure called by application is al_bp_send (contained in the file

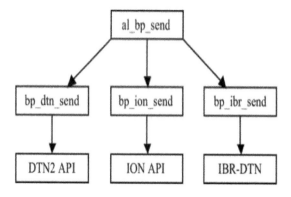

Figure 12.4: Example of relationship between the AL functions and the BP implementation API.

al_bp_api.c); if the DTN2 daemon is on, it calls bp_dtn_send (in al_bp_dtn.c). This procedure contains both a real and a dummy implementation. The former calls the corresponding DTN2 API procedure and is used when the AL is compiled to support DTN2; the latter is just inserted to avoid compilation errors (due to lack of DTN2 libraries) otherwise. Analogously for ION and IBR-DTN.

12.7   DTNperf_3 Use

DTNperf_3 is a general tool, with a wide application scope. The most important possible uses are discussed below, referring whenever possible to the authors' user-experience, with brief comments and increasing complexity, leaving, however, the most advanced use, such as interoperability tests, to the next section. The syntax is focused on the client, as both the server and the monitor can normally be launched without additional options as:

dtnperf - -server
dtnperf - -monitor

In the following, vm1, vm2 and vm3 will be the host names of DTN nodes used as source, destination or "report to". They correspond to nodes 1, 2 and 3 in the "ipn" scheme [244]. Debug options can be added at user choice if more verbosity is requested (e.g. - - debug=2).

12.7.1 Basic Applications

12.7.1.1 Ping

To ping vm2 from vm1 (if the server is registered with the "dtn" scheme)

dtnperf - -client -d dtn://vm2.dtn -T15 -W1

Alternatively, if the server is registered with the ipn scheme, the EID ipn:2.200 must replace dtn://vm2.dtn as destination (note that the demux token must be always specified in the ipn scheme).

With this command vm1 sends bundles of 50 kB (default) to vm2 for 15s (-T15), one by one (-W1 allows just one bundle in flight). The default lifetime (60s) is deliberately very short to force the deletion of undelivered bundles, which could interfere on subsequent experiments. As no external monitor is indicated, the .csv log file will be created on vm1 by a dedicated monitor. To avoid premature cancellation of bundles due to lifetime expiration, and also to collect consistent information on status reports, we recall that DTN nodes should be accurately synchronized. When synchronization is impossible, the lifetime should be enlarged to take synchronization errors into account.

12.7.1.2 Trace

To trace the route of a bundle (and also to check the custodial transfer):

dtnperf --client -d dtn://vm2.dtn -D100kB -P100kB -W1 -C - f -r

This command sends a single bundle of 100 kB as the total amount of data (-D100kB) coincides with the bundle payload (-P100kB). The custody option (-C) is necessary only if interested in the study of custodial transfer, while at least one of the additional status reports (forwarded, -f, received, -r) is necessary to trace the route of the bundle sent from the .csv file. Enabling all options allow the user a complete knowledge of the bundle transfer, a sort of "telemetry".

12.7.1.3 File Transfer

To transfer a file segmented into multiple bundles of desired dimension:

dtnperf - -client -d dtn://vm2.dtn -F picture.jpg -P100kB -W4

Here a file is sent (-F) instead of dummy data. The dimension of the bundle payload (-P100kB) is the dimension of segments into which the file is split. This feature is useful to match limited contact volumes as an alternative to proactive fragmentation. Note however that file segmentation is carried out at application layer (by DTNperf), while bundle fragmentation is performed at BP layer (by the BP daemon).

12.7.2 Performance Evaluation in Continuous and Disrupted Networks

12.7.2.1 Goodput (macro-analysis)

Goodput (data confirmed at application layer/interval) is a common figure of merit in networking. However, in a DTN environment it makes sense only if the DTN network is not partitioned or heavily disrupted (e.g. in GEO satellite communications, where the challenges are long propagation delay and losses). The following command could be suitable in the case of a hypothetical GEO satellite hop:

dtnperf --client -d dtn://vm2.dtn -T30 -P1MB -W4 -l 60

With this command vm1 sends bundles to vm2 for 30 s (-T30), i.e. for a much longer interval than the typical GEO RTT (600ms including processing time). To fill the (likely) large bandwidth-delay product and to reduce the impact of bundle overhead, bundles are large (-P1MB) and the congestion window W is greater than one (-W4). The same experiment should be repeated increasing W until the goodput reaches a maximum. Note that goodput evaluation should always be complemented by the analysis of status reports, collected in the example by the internal monitor (default), to control the regularity of the bundle flow and to recognize the reasons of the macro-results achieved.

12.7.2.2 Status Report Analysis (micro-analysis)

As the chances of disruption increase (e.g. in LEO satellites or Interplanetary communications), the study of individual bundles (i.e. micro-analysis), possible only from data collected by the monitor in the .csv file, becomes more important than goodput. A possible command in the presence of disruption is:

dtnperf --client -d dtn://vm2.dtn -D30MB -P 1 MB -W4 -l 200 -m dtn://vm3.dtn

Note that if server and/or monitor are registered with the ipn scheme; their EIDs, ipn:2.2000 and ipn:3.1000, should replace the dtn scheme EIDs.

This command dispatches 30 bundles of 1 MB each, with a limit of 4 bundles in flight. Status reports are collected (possibly in real time by means of dedicated links) by an external monitor (-m option). Note that the lifetime has been increased to 200s (-l200) to cope with disruption. Moreover, Data mode (-D30MB) has been selected instead of Time mode, because disruption makes uncertain the actual duration of bundle transfer.

To emulate a streaming source, the use of the rate based congestion control is preferable, as in the example below:

dtnperf --client -d dtn://vm2.dtn -T30 -P100kB -R2b -m dtn://vm3.dtn

This command generates a stream of 100 kB bundles for 30s, at 2 bundles per second (-R2b). No ACKs are generated by the server, as the congestion control is rate-based. As the client has not to wait for ACKs, the client terminates exactly after 30s once the last bundle has been generated and passed to BP.

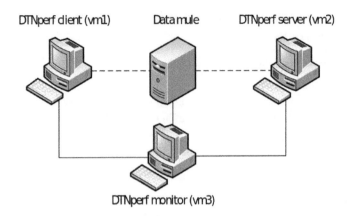

DTNperf client (vm1)   Data mule   DTNperf server (vm2)

DTNperf monitor (vm3)

Figure 12.5: DTNperf use in a data-mule testbed. Dotted lines denote intermittent data links; continuous lines denote continuous links dedicated to real time status report collection.

12.7.2.3 Performance Evaluation in Partitioned Networks: "Data Mule" Communications

In "data mule" communications, an intermediate node (the "mule", or "ferry") is alternatively connected, thanks to its movement, either to the sender or to the destination. In this case, the evaluation of goodput is useless, while the microanalysis is essential. A possible command is:

dtnperf --client -d dtn://vm2.dtn -D10MB -P1MB -W10 -l 200 -m dtn://vm3.dtn --debug=1

In a lab it is convenient to connect the monitor to all other nodes through dedicated links, as shown in Figure 12.5, since links used to transfer data are only occasionally active (dotted lines). Note that by setting the window to the total number of bundles (10 in the example) these are sent in one burst, which may be preferable in this kind of experiment, in order not to have to wait for ACKs (once the burst of data is sent, the user can interrupt the client by pressing ctrl+c). Alternatively, the user can take advantage of the rate-based congestion control, which does not imply any ACKs (e.g. by setting -R10b instead of -W10).

12.8 DTNperf_3 Advanced

In this section will be presented two advanced uses of DTNperf, interoperability tests, and independent use of client or monitor, intended for expert users only.

12.8.1 Interoperability Tests

As said in previous sections, thanks to the AL library, which now includes IBR-DTN support, DTNperf_3 (v3.5.0) can run on top of either DTN2, ION or IBR-DTN BP. If compiled for multiple implementations, the very same executable can be used, as the choice between the different implementation API is made at run time. This feature makes DTNperf_3 particularly suitable for carrying out interoperability tests among different implementations, which is an important feature to validate standards. To this end, a key requirement is the capability of coping with both the different EID URI schemes, "dtn" and "ipn", which actually required many efforts, but was also driven by the standardization needs of CCSDS for finalizing the BP standard [67]. DTN2 and IBR-DTN prefer the "dtn" scheme, ION the "ipn" one, but all of them support (with differences and limitations) also the alternate scheme. To facilitate experiments, in the latest versions of DTNperf_3 the registration is with the preferential scheme by default, for all modes (client, server and monitor); this behavior can be overridden, if necessary, by forcing the alternate URI scheme (dtn in ION, ipn in DTN2 and IBR-DTN), thus allowing full interoperability. In brief, DTNperf_3 can cope with every combination of DTN2, ION and IBR-DTN nodes. Mixed EIDs are allowed in the same command too, as in the following example (server and monitor registered with the dtn and ipn scheme, respectively):

dtnperf - -client -d dtn://vm2.dtn –m ipn:3.1000 –D30MB –P 1 MB –W4 –l 200

Of course, a prerequisite for DTNperf_3 interoperability is that DTN2, ION and IBR-DTN are correctly configured for supporting interoperability at bundle layer, which is far from being trivial. Not only configuration files must be accurately prepared, but also a few bugs can be fixed. For example, the "bleeding edge" version of DTN2, i.e. the development version, requires a patch developed by NASA for correctly supporting the ipn scheme. Analogously, a minor fix is necessary in IBR-DTN for allowing full interoperability with ION when the ipn scheme is used.

A few examples involving the use of the alternate scheme in DTN2, ION and IBR-DTN in the registration phase, are reported below.

12.8.1.1 DTNperf Running on Top of DTN2: Registration as "ipn" with Node Number Passed by the User

In DTN2 it is necessary to use two options: "force-eid" to override the dtn default and "ipn-local" to assign the node number (a DTN2 node has no notion of its "ipn" identity). For the client we have:

dtnperf --client dtn://vm2.dtn --D200k --R1b --force-eid IPN --ipn-local 1
--debug=2.

Analogously, for the server and the monitor.

12.8.1.2 DTNperf Running on Top of ION: Registration as "dtn"

In ION it is necessary to use only the "force-eid" option to override the ipn default, because the dtn node name is derived automatically by ION from the node hostname (no possibility of overriding this).

dtnperf --client ipn:5.2000 --D200k --R1b --force-eid DTN --debug=2

The DTNperf client on "vm1" (hostname) will register as dtn://vm1.dtn/x, being x the usual dtn demux token. Analogously, for server and monitor. Note that on the basis of author's experience, while ION is correctly case-sensitive concerning EIDs, DTN2 sometimes erroneously reports EIDs containing capital letters in lowercase. In brief, in dealing with heterogeneous ION-DTN2 networks, the use of capital letters in hostnames running ION should be avoided.

12.8.1.3 Dtnperf Running on Top of IBR-DTN: Registration as "ipn"

In IBR-DTN a node can have only one EID, either dtn or ipn, set in the configuration file. Running DTNperf on an IBR-DTN node with an ipn identity, it is necessary to use the "force-eid" option to override the dtn default. By contrast to DTN2, the ipn number must not be indicated by the user, as it is taken from the ipn EID of the node.

dtnperf --client dtn://vm2.dtn --D200k --R1b --force-eid IPN --debug=2

Running on ipn:1.0, the client will register as ipn:1.x, with x the ipn demux token number. Analogously for server and monitor.

12.8.2 Independent Use of Client and Monitor

DTNperf modes (client, server, monitor) were conceived to work together. However, it is perfectly possible, and sometimes also convenient, to use the client (rate based) or the monitor in a fully autonomous way.

12.8.2.1 Client

As the client in rate based congestion control does not need any feedback from the server, it can be used autonomously as a powerful and flexible source of bundles. The user can specify the bundle dimension, the total amount of data, the Tx rate (either in bundle/s or bit/s, at his choice), and all bundle flags (delivered is however always on). No more need of custom complex scripts!

12.8.2.2 Monitor

The monitor can be used in an independent way as well, to conveniently collect all status reports. To this end, it is enough to set the monitor EID as "report to" in the bundle source, which can be whatsoever program, not necessarily a DTNperf client. In case of multiple sources, the option "--oneCSVonly" can be useful to generate a unique .csv file. The option "--rt-print[=filename]", which prints on standard output the status report information could be useful too. It has been explicitly added in response to one user that wanted to monitor the traffic generated by his DTN applications in real time.

12.9 Summary

DTNperf was developed to assess goodput performance and collect logs in DTN bundle protocol architectures. Its third major release, DTNperf_3, consists of a unique application with three modes. The client can generate traffic with great flexibility by enforcing either a window-based or a rate based congestion control; the server can manage concurrent sessions with multiple clients, possibly running on the same source node; moreover, it provides the client with ACKs of incoming bundles, if sent in window-based congestion control; the monitor collects all status report in .csv files, ready to be analyzed by spreadsheets or can print this information on the screen, in a human readable format, possibly to be used in real time. Last but foremost, DTNperf_3 through the Abstraction Layer now supports all the most important free software BP implementations, such as DTN2, ION, and IBR-DTN. In brief, it is a powerful tool with a unique interface, independent of BP implementations, with a very wide range of possible applications, from basic to very advanced. It is released as free software and it is also included in ION.

Acknowledgments The author would like to thank his ex-students Piero Cornice, Marco Livini, Michele Rodolfi, Anna D'Amico, Davide Pallotti and his assistant Rosario Firrincieli for their essential contribution in the development of current and previous versions of DTNperf. A special thanks also to Scott Burleigh (NASA-JPL), David Zoller (NASA-MSFC), Robert Pitts (NASA-MSFC) and Keith Scott (MITRE) for their support and encouragement.

References

[1] Android service. http://developer.android.com/references/android/app/Service.html. Accessed in: February 4th, 2018.

[2] Contiki: The open source os for the internet of things. http://www.contiki-os.org/. Accessed in: January 28th, 2018.

[3] DTN2 code. http://sourceforge.net/projects/dtn/.

[4] The FreeRTOS Kernel. https://www.freertos.org/. Accessed in: January 28, 2018.

[5] IBR-DTN code. https://github.com/ibrdtn/ibrdtn/wiki.

[6] OpenSSL: Cryptography and SSL/TLS toolkit. https://www.openssl.org/. Accessed in: January 22nd, 2018.

[7] OpenWRT: Wireless freedom. https://openwrt.org/. Accessed in: January 23rd, 2018.

[8] OPNET tecnologies Ltd. http://www.opnet.com. Acessed in December 2011.

[9] Space packet protocol. Recommendation for Space Data Systems Standards (Blue Book) 133.0-B-1, September 2003.

[10] Routing in a Delay Tolerant Network. *ACM SIGCOMM Computer Communication Review*, 34(4):145, 2004.

[11] CCSDS File Delivery Protocol (CFDP), January 2007. CCSDS Blue Book 727.0-B-4.

[12] Encapsulation Service, October 2009. CCSDS Blue Book 133.1-B-2.

[13] Rationale, scenarios, and requirements for dtn in space. Report Concerning Space Data System Standards (Green Book) 734.0-G-1, August 2010.

[14] The network simulator ns2. Available in: `http://www.isi.edu/nsnam/ns/`, 2013. Accessed in March 23rd, 2013.

[15] Erasure correcting codes for use in near-earth and deep-space communications, November 2014. CCSDS Orange Book 131.5-O-1.

[16] CCSDS Bundle Protocol Specification. Recommendation for Space Data Systems Standards (Blue Book) 734.2-B-1, September 2015.

[17] Data distribution services specification v1.2. Object Manage. Group(OMG), 2015. Available: `http://www.omg.org/spec/DDS/1.2/`.

[18] Licklider Transmission Protocol (LTP) for CCSDS. Recommendation for Space Data Systems Standards (Blue Book) 734.1-B-1, May 2015.

[19] Stefan Achleitner, Thomas La Porta, Srikanth V. Krishnamurthy, and Victor S. Quizhpi. Network coding efficiency in the presence of an intermittent backhaul network. *2016 IEEE International Conference on Communications, ICC 2016*, 2016.

[20] F. Adelantado, X. Vilajosana, P. Tuset-Peiro, B. Martinez, J. Melia-Segui, and T. Watteyne. Understanding the limits of lorawan. *IEEE Commun. Mag.*, 55(9):34–40, 2017.

[21] Rudolf Ahlswede, Ning Cai, Shuo-Yen Robert Li, and Raymond W. Yeung. Network Information Flow. *IEEE Transactions on Information Theory*, 46(4):1204–1216, 2000.

[22] Shakeel Ahmad, Adli Mustafa, Bashir Ahmad, Arjamand Bano, and Al-Sammarraie Hosam. Comparative study of congestion control techniques in high speed networks. *International Journal of Computer Science and Information Security*, 6(2):222–231, 2009.

[23] Shabbir Ahmed and Salil S. Kanhere. HUBCODE: hub-based forwarding using network coding in delay tolerant networks. *Wireless Communications and Mobile Computing*, 13:828–846, 2013.

[24] Ian F. Akyildiz, Özgür B. Akan, Chao Chen, Jian Fang, and Weilian Su. Interplanetary internet: State-of-the-art and research challenges. *Comput. Netw.*, 43(2):75–112, October 2003.

[25] A. Al-Fuqaha, M. Guizani, M. Mohammadi, M. Aledhari, and M. Ayyash. Internet of things: A survey on enabling technologies, protocols, and applications. *IEEE Communications Surveys Tutorials*, 17(4):2347–2376, 2015.

[26] F. M. Al-Turjman, A. E. Al-Fagih, W. M. Alsalih, and H. S. Hassanein. A delay-tolerant framework for integrated rsns in iot. *Comput. Commun.*, 36(9):998–1010, 2013.

[27] Arshad Ali, Manoj Panda, Tijani Chahed, and Eitan Altman. Improving the transport performance in delay tolerant networks by random linear network coding and global acknowledgments. *Ad Hoc Networks*, 11(8):2567–2587, 2013.

[28] L. Alliance. Lpwa technologies unlock new iot market potential. *Machina Research*, 2015.

[29] N. Alon and M. G. Luby. A linear-time erasure-resilient code with nearly optimal recovery. *IEEE Transactions on Information Theory*, 42:1732–1736, November 1996.

[30] Ahmed B. Altamimi and T. Aaron Gulliver. On Network Coding in Intermittently Connected Networks. In *Vehicular Technology Conference (VTC Fall), 2014 IEEE 80th*, 2014.

[31] Ying An and Xi Luo. MACRE: A novel distributed congestion control algorithm in DTN. *Trans Tech Publications, Advanced Engineering Forum*, 1:71–75, 2011.

[32] Rabin Patra And. Dtnlite: A reliable data transfer architecture for sensor networks., 2003.

[33] G. Araniti, N. Bezirgiannidis, E. Birrane, I. Bisio, S. Burleigh, C. Caini, M. Feldmann, M. Marchese, J. Segui, and K. Suzuki. Contact graph routing in dtn space networks: Overview, enhancements and performance. *IEEE Communication Magazine*, 53(3):38–46, March 2015.

[34] Abdurrahman Arikan, Yuexin Mao, Xiaolan Zhang, Bing Wang, Shengli Zhou, and Song Han. Network Coding based Transmission Schemes in DTNs with Group Meetings. 2015.

[35] L. Atzori, A. Iera, and G. Morabito. The internet of things: A survey. *Computer Networks*, 54(15):2787–2805, 2010.

[36] Maël Auzias, Yves Mahéo, and Frédéric Raimbault. Coap over bp for a delay-tolerant internet of things. In *3rd International Conference on Future Internet of Things and Cloud (FiCloud 2015)*, 2015.

[37] Qaisar Ayub and Sulma Rashid. T-drop: An optimal buffer management policy to improve QoS in DTN routing protocols. *Journal of Computing*, 2(10):46–50, October 2010.

[38] Qaisar Ayub, Sulma Rashid, and M. Soperi Mohd Zahid. Buffer scheduling policy for opportunistic networks. *International Journal of Scientific & Engineering Research*, 2(7), July 2011.

[39] E. Baccelli, O. Hahm, M. Gunes, M. Wahlisch, and T. Schmidt. Riot os: Towards an os for the internet of things. In *2013 IEEE Conference on Computer Communications Workshops (INFOCOM WKSHPS)*, 2013.

[40] J. Banks. *RFID Applied*. John Wiley & Sons, 2007.

[41] F. Z. Benhamida, A. Bouabdellah, and Y. Challal. Using delay tolerant network for the internet of things: Opportunities and challenges. In *2017 8th International Conference on Information and Communication Systems (ICICS)*, 2017.

[42] E. Berlekamp. The technology of error-correcting codes. *Proceedings of the IEEE*, 68(5):564–593, May 1980.

[43] C. Berrou, A. Glavieux, and P. Thitimajshima. Near Shannon limit error-correcting coding and decoding: Turbo-codes. In *Proc. IEEE Int. Conf. on Communications*, Geneva, Switzerland, May 1993.

[44] N. Bezirgiannidis, C. Caini, and V. Tsaoussidis. Analysis of contact graph routing enhancements for dtn space. *International Journal of Satellite Communications and Networking*, 2015.

[45] E. J. Birrane. Congestion modeling in graph-routed delay tolerant networks with predictive capacity consumption. In *2013 IEEE Global Communications Conference (GLOBECOM)*, pages 3016–3022, Dec 2013.

[46] Edward Birrane. Contact graph routing extension block, 2013. draft-irtf-dtnrg-cgreb-00.

[47] Igor Bisio, Marco Cello, Tomasso de Cola, and Mario Marchese. Combined congestion control and link selection strategies for delay tolerant interplanetary networks. In *GLOBECOM 2009*.

[48] Andrey Bogdanov, Dmitry Khovratovich, and Christian Rechberger. Biclique cryptanalysis of the full aes. In *Proceedings of the 17th International Conference on the Theory and Application of Cryptology and Information Security*, ASIACRYPT'11, pages 344–371, Berlin, Heidelberg, 2011. Springer-Verlag.

[49] C. Bormanna, A. P. Castellani, and Z. Shelby. Coap: An application protocol for billions of tiny internet nodes. *IEEE Internet Comput.*, 16(2):62–67, 2012.

[50] L. Bruno, M. Franceschinis, C. Pastrone, R. Tomasi, and M. Spirito. 6LoWDTN: IPv6-enabled delay-tolerant WSNs for contiki. In *2011 International Conference on Distributed Computing in Sensor Systems and Workshops (DCOSS)*, pages 1–6, 2011.

[51] J. Burgess, B. Gallagher, D. Jensen, and B. N. Levine. Maxprop: Routing for vehicle-based disruption-tolerant networks. In *Proceedings IEEE INFOCOM 2006. 25TH IEEE International Conference on Computer Communications*, pages 1–11, April 2006.

[52] S. Burleigh. Interplanetary overlay network: An implementation of the dtn bundle protocol. In *2007 4th IEEE Consumer Communications and Networking Conference*, pages 222–226, 2007.

[53] S. Burleigh, C. Caini, J. J. Messina, and M. Rodolfi. Toward a Unified Routing Framework for Delay-Tolerant Networking. In *IEEE International Conference on Wireless for Space and Extreme Environments (WiSEE)*, pages 82–86, 2016.

[54] S. Burleigh, A. Hooke, L. Torgerson, K. Fall, V. Cerf, B. Durst, and K. Scott. Delay-tolerant networking: An approach to interplanetary internet. *IEEE Communication Magazine*, 41(6):128–136, June 2003.

[55] S. Burleigh, M. Ramadas, and S. Farrell. Licklider transmission protocol motivation. https://tools.ietf.org/html/rfc5325. Accessed September 12, 2016.

[56] Scott Burleigh. Interplanetary overlay network - design and operation. Technical report, Jet Propulsion Laboratory/ California Institute of Technology. Technical Report JPL D-48259. Jet Propulsion Laboratory/ California Institute of Technology, 2013.

[57] Scott Burleigh. Contact graph routing, 2009. IETF-draft draft-burleigh-dtnrg-cgr-01.

[58] Scott Burleigh. Compressed bundle header encoding (CBHE). Technical report, Jet Propulsion Laboratory - California Institute of Technology, 2011. IRTF RFC 6260.

[59] Scott Burleigh. Bundle protocol extended class of service (ECOS). Technical report, IETF Internet Draft, 2013.

[60] Scott Burleigh. ION-DTN. https://sourceforge.net/projects/ion-dtn/, 2016. Accessed October 14, 2016.

[61] Scott Burleigh, Esther Jennines, and Joshua Schoolcraft. Autonomous congestion control in delay tolerant network. Jet Propulsion Laboratory, American Institute of Aeronautics and Astronautic, 2007.

[62] Scott C. Burleigh and Edward J. Birrane. Toward a Communications Satellite Network for Humanitarian Relief. In *Proceedings of the 1st International Conference on Wireless Technologies for Humanitarian Relief - ACWR '11*, pages 219–224, 2011.

[63] B. Burns, O. Brock, and B. N. Levine. Mv routing and capacity building in disruption tolerant networks. In *Proceedings IEEE 24th Annual Joint Conference of the IEEE Computer and Communications Societies*, volume 1, pages 398–408, March 2005.

[64] D. Burshtein and G. Miller. An efficient maximum likelihood decoding of LDPC codes over the binary erasure channel. *IEEE Trans. Inf. Theory*, 50(11):2837–2844, November 2004.

[65] J.W. Byers, M. Luby, and M. Mitzenmacher. A digital fountain approach to reliable distribution of bulk data. *IEEE Journal on Selected Areas in Communications*, 20(8):1528–1540, October 2002.

[66] Ning Cai and R W Yeung. Secure Network Coding. In *Information Theory, 2002. Proceedings. 2002 IEEE International Symposium on*, 2002.

[67] C. Caini, P. Cornice, R. Firrincieli, and M. Livini. DTNperf_2: a performance evaluation tool for delay/disruption tolerant networking. In *ICUMT*, pages 1–6, October 2009.

[68] C. Caini, H. Cruickshank, S. Farrell, and M. Marchese. Delay and disruption-tolerant networking (DTN): An alternative solution for future satellite networking applications. *Proceedings of IEEE*, 99(11):1980–1997, November 2011.

[69] C. Caini, A. d'Amico, and M. Rodolfi. DTNperf_3: a further enhanced tool for delay-/disruption- tolerant networking performance evaluation. In *IEEE Globecom*, pages 3009–3015, 2013.

[70] C. Caini, R. Firrincieli, T. de Cola, I. Bisio, M. Cello, and G. Acar. Mars to earth communications through orbiters: DTN performance analysis. *International Journal of Satellite Communications and Networking*, pages 127–140, 2014.

[71] Q. Cao, T. Abdelzaher, J. Stankovic, and T. He. The liteos operating system: Towards unix-like abstractions for wireless sensor networks. In *2008 International Conference on Information Processing in Sensor Networks (ipsn 2008)*, 2008.

[72] Yue Cao and Zhili Sun. Routing in delay/disruption tolerant networks: A taxonomy, survey and challenges. *IEEE Communications Surveys and Tutorials*, 15(2):654–677, 2013.

[73] CCSDS. CCSDS file delivery protocol (CFDP). Technical Report CCSDS 727.0-B-4, Consultative Commitee for Space Data System CCSDS, 2007.

[74] V. Cerf, Scott Burleigh, R. Durst, K. Scott, E. Travis, and H. Weiss. Interplanetary internet (IPN): Architectural definition. Technical report, IETF Internet Draft, 2001. draft-irtf-ipnrg-arch-00.txt.

[75] V. Cerf, S.Burleigh, A. Hooke, L. Torgerson, R. Durst, K. Scott, K. Fall, and H. Weiss. Delay-tolerant networking architecture. Technical Report RFC4838, Internet RFC, 2007.

[76] Vint Cerf, Scott C. Burleigh, Adrian J. Hooke, Leigh Torgerson, Robert C. Durst, Keith L. Scott, Kevin Fall, and Howard S. Weiss. RFC 4838 - Delay-Tolerant Networking Architecture. *Internet Engineering Task Force (IETF)*, 2007.

[77] M. A. Chaqfeh and N. Mohamed. Challenges in middleware solutions for the internet of things. In *2012 International Conference on Collaboration Technologies and Systems (CTS)*, 2012.

[78] Ashutosh Chauhan, Sudhanhsun Kumar, Vivek Kumar, S. Mukherjee, and C. T. Bhunia. Implementing distributive congestion control in a delay tolerant network for different types of traffic. Available in: https://www.ietf.org/mail-archive/web/dtn.../docYrH2LrS75z.doc. Fremont, CA: IETF Trust, 2012. Accessed in June 18, 2012.

[79] H. C. H. Chen, Y. Hu, P. P. C. Lee, and Y. Tang. Nccloud: A network-coding-based storage system in a cloud-of-clouds. *IEEE Transactions on Computers*, 63(1):31–44, Jan 2014.

[80] Yuanzhu Chen, Xu Liu, Jiafen Liu, Walter Taylor, and Jason H. Moore. Delay-tolerant networks and network coding: Comparative studies on simulated and real-device experiments. *Computer Networks*, 83:349–362, 2015.

[81] Wang Cheng-jun, Gong Zheng-hu, Tao Yong, Zhang Zi-wen, and Zhao Bao-kang. CRSG: A congestion routing algorithm for security defense based on social psychology and game theory in DTN. *Journal of Central South University*, 20(2):440–450, February 2013. Springer.

[82] M. Chuah and P. Yang. Impact of selective dropping attacks on network coding performance in DTNs and a potential mitigation scheme. *Proceedings - International Conference on Computer Communications and Networks, IC-CCN*, 2009.

[83] M Chuah, P Yang, and Y Xi. How Mobility Models Affect the Design of Network Coding Schemes for Disruption Tolerant Networks. *IEEE International Conference on Distributed Computing Systems Workshops*, pages 172–177, 2009.

[84] Kun Cheng Chung, Yi Chin Li, and Wanjiun Liao. Exploiting network coding for data forwarding in delay tolerant networks. *IEEE Vehicular Technology Conference*, pages 0–4, 2010.

[85] Eric Coe and Cauligi Rachavendra. Token based congestion control for DTN. In *Aerospace Conference*. IEEE, 2010.

[86] T. D. Cola, H. Ernst, and M. Marchese. Joint application of CCSDS File Delivery Protocol and erasure coding schemes over space communications. In *2006 IEEE International Conference on Communications*, volume 4, pages 1909–1914, June 2006.

[87] T. D. Cola and M. Marchese. Reliable data delivery over deep space networks: Benefits of long erasure codes over ARQ strategies. *IEEE Wireless Communications*, 17(2):57–65, April 2010.

[88] T. De Cola and M. Marchese. Performance analysis of data transfer protocols over space communications. *IEEE Transactions on Aerospace and Electronic Systems*, 41(4):1200–1223, Oct 2005.

[89] W. Colitti, K. Steenhaut, N. De Caro, B. Buta, and V. Dobrota. Evaluation of constrained application protocol for wireless sensor networks. In *2011 18th IEEE Workshop on Local & Metropolitan Area Networks (LANMAN)*, 2011.

[90] D. J. Cook, A. S. Crandall, B. L. Thomas, and N. C. Krishnan. Casas: A smart home in a box. *Computer*, 46(7):62–69, 2013.

[91] D. J. Jr. Costello, J. Hagenauer, H. Imai, and S. B. Wicker. Applications of error-control coding. *IEEE Trans. Inf. Theory*, 44:2531–2560, October 1998.

[92] Laszló Czap and Istvan Vajda. Secure Network Coding in DTNs. *Communications Letters, IEEE*, 15(1):28–30, 2011.

[93] Aloizio Pereira da Silva. *A Novel Congestion Control Framework For Delay and Disruption Tolerant Networks*. PhD thesis, Instituto Tecnológico de Aeronáutica, 2015.

[94] F. A. Davis, J. K. Marquart, and G. Menke. Benefits of delay tolerant networking for earth science missions. Technical report, 2012.

[95] J. A. Davis, A. H. Fagg, and B. N. Levine. Wearable computers as packet transport mechanisms in highly-partitioned ad-hoc networks. In *Proceedings Fifth International Symposium on Wearable Computers*, pages 141–148, 2001.

[96] James A. Davis, Andrew H. Fagg, and Brian N. Levine. Wearable computers as packet transport mechanism in highly-partitioned ad-hoc networks. In *International Symposium on Wearable Computing*, October 2001. 5., Zurich. **Proceedings...** Piscataway: IEEE, 2001. p. 141–148.

[97] T. de Cola. A protocol design for incorporating erasure codes within CCSDS: The case of DTN protocol architecture. In *2010 5th Advanced Satellite Multimedia Systems Conference and the 11th Signal Processing for Space Communications Workshop*, pages 68–73, Sept 2010.

[98] Etienne C. R. de Oliveira and Célio V. N. de Albuquerque. Nectar: A dtn routing protocol based on neighborhood contact history. In *Proceedings of the 2009 ACM Symposium on Applied Computing*, SAC '09, pages 40–46, New York, NY, USA, 2009. ACM.

[99] M. Demmer and J. Ott. Delay-tolerant networking tcp convergence-layer protocol. Technical Report RFC 7242, Internet Research Task Force (IRTF), 2014.

[100] Michael Demmer and Kevin Fall. Dtlsr: Delay tolerant routing for developing regions. In *Proceedings of the 2007 Workshop on Networked Systems for Developing Regions*, NSDR '07, pages 5:1–5:6. ACM, 2007.

[101] Eletronic Design. Vxworks goes 64-bit. Technical report, 2011.

[102] Tim Dierks and Eric Rescorla. The transport layer security (tls) protocol. Technical Report RFC5246, IETF, 2008. version 1.2.

[103] Sebastian Domancich. Security in delay tolerant network for the android platform. Master's thesis, Aalto University - School of Science and Technology (TKK), 2010.

[104] W. Eddy and E. Davies. Using self-delimiting numeric values in protocols. RFC 6256, May.

[105] A. Elmangoush, A. Corici, M. Catalan, R. Steinke, T. Magedanz, and J. Oller. Interconnecting standard m2m platforms to delay tolerant networks. In *2014 International Conference on Future Internet of Things and Cloud*, 2014.

[106] D. Evans, 2011. CISCO, San Jose, CA, USA, White Paper.

[107] Kevin Fall. A delay tolerant network architecture for challenged internets. In *ACM Conference on Applications, Technologies, Architectures, and Protocols for Computer Communications*, 2003. Karlsruhe. **Proceedings...** New York: ACM, 2003. p. 27–34.

[108] S. Farrell, V. Cahill, D. Geraghty, and I. Humphreys. When TCP breaks: Delay and disruption-tolerant networking. *IEEE Internet Computing*, 10(4):72–78, 2006.

[109] S. Farrell, M. Ramadas, and S. Burleigh. Licklider transmission protocol security extensions, IRTF. https://tools.ietf.org/html/rfc5327, 2008. Accessed October 8, 2016.

[110] Stephen Farrell. *A Delay and Disruption Tolerant Transport Layer Protocol*. PhD thesis, University of Dublin, Trinity College, September 2008.

[111] C. Feng, R. Wang, Z. Bian, T. Doiron, and J. Hu. Memory dynamics and transmission performance of Bundle Protocol (BP) in deep-space communications. *IEEE Transactions on Wireless Communications*, 14(5):2802–2813, May 2015.

[112] Finkenzeller. *RFID Handbook - Radio-Frequency Identification Fundamentals and Applications*. Halsted Press, 2003.

[113] I. Florea, R. Rughinis, L. Ruse, and D. Dragomir. Survey of standardized protocols for the internet of things. In *2017 21st International Conference on Control Systems and Computer Science (CSCS)*, 2017.

[114] S. Floyd. Metrics for the evaluation of congestion control mechanism. Available in: http://www.ietf.org/mail-archive/web/ietf-announce/current/msg04676.html, 2006. Fremont, CA: IETF Trust, 2006. Acessed in May 13th, 2012.

[115] S. Floyd. Metrics for the evaluation of congestion control mechanisms. Available in: http://www.ietf.org/rfc/rfc5166.txt, March 2008. Fremont, CA: IETF Trust, 2008. RFC 5166. Network Working Group, Accessed in November 6th, 2014.

[116] Consultative Committee for Space Data Systems (CCSDS). Asynchronous message service draft recommended standard. Technical report, Red Book CCSDS 735.1-R-2, 2009.

[117] L. R. Ford, Jr. and D. R. Fulkerson. Maximal flow through a network. *Canadian Journal of Mathematics*, 8:399–404, 1956.

[118] C Fragouli, J Widmer, and J Y Le Boudec. On the Benefits of Network Coding for Wireless Applications. *2006 4th International Symposium on Modeling and Optimization in Mobile Ad Hoc and Wireless Networks*, pages 1–6, 2006.

[119] Christina Fragouli, Jean-Yves Le Boudec, and Jörg Widmer. Network Coding: An Instant Primer. *ACM SIGCOMM Computer Communication Review*, 36(1):63, 2006.

[120] Christina Fragouli, Jörg Widmer, and Jean Yves Le Boudec. Efficient broadcasting using network coding. *IEEE/ACM Transactions on Networking*, 16(2):450–463, 2008.

[121] J. A. Fraire, P. Madoery, J. M. Finochietto, and E. J. Birrane. Congestion modeling and management techniques for predictable disruption tolerant networks. In *2015 IEEE 40th Conference on Local Computer Networks (LCN)*, pages 544–551, Oct 2015.

[122] R. G. Gallager. *Low-Density Parity-Check Codes*. MIT Press, Cambridge, MA, 1963.

[123] J. Gantz and D. Reinsel. The digital universe in 2020: Big data, bigger digital shadows, and biggest growth in the far east. *IDC iView: IDC Anal. Future*, 2007:1–16, 2012.

[124] Nabil Ghanmy, Lamia Chaari Fourati, and Lotfi Kamoun. A comprehensive and comparative study of elliptic curve cryptography hardware implementations for wsn. *International Journal of RFID Security and Cryptography (IJRFIDSC)*, 3(1), 2015.

[125] M. Gigli and S. Koo. Internet of things: Services and applications categorization. *Advances in Internet of Things*, 01(02):27–31, 2011.

[126] D. Giusto, A. Iera, G. Morabito, and L. Atzori. The internet of things. In Springer Science & Business Media, editor, *20th Tyrrhenian Workshop on Digital Communications*, 2010.

[127] Christos Gkantsidis and Pablo Rodriguez. Cooperative security for network coding file distribution. In *Proceedings - IEEE INFOCOM*, 2006.

[128] Interagency Operations Advisory Group and Space Internetworking Strategy Group. Recommendation on a strategy for space internetworking. Technical Report IOAG.T.RC.002.V1, November 2008.

[129] Interagency Operations Advisory Group Space Internetworking Strategy Group. Recommendations on a strategy for space internetworking. Technical report, August 2010.

[130] Space Internetworking Strategy Group. Operations concept for a solar system internetwork. Technical Report IOAG.T.RC.001.V1, IOAG, 2010.

[131] Andrew Grundy and Milena RadenKovic. Promoting congestion control in opportunistic networks. In *IEEE 6th International Conference on Wireless and Mobile Computing, Networking and Communications*, pages 324–330, October 2010. Department of Computer Science, University of Nottingham.

[132] Andrew Michael Grundy. *Congestion Framework For Delay-Tolerant Communications*. PhD thesis, 145 p. Phd in Computer Science, University of Nottingham, Nottingham, July 2012.

[133] Mance E. Harmon and Stephanie S. Harmon. Reinforcement learning: a tutorial. Dayton, OH: US Air Force Office of Scientific Research, 1996.

[134] Khaled A. Harras, Kevin C. Almeroth, and Elizabeth M. Belding-Royer. *Delay Tolerant Mobile Networks (DTMNs): Controlled Flooding in Sparse Mobile Networks*, pages 1180–1192. Springer Berlin Heidelberg, 2005.

[135] Chun He, Ning Ning Huang, and Gang Feng. Network Coding Based Routing Scheme for Resource Constrained Delay Tolerate Networks. *International Conference on Computational Problem-Solving (ICCP)*, 2012.

[136] S. Hemminger. Network emulation with NetEm. National Linux Conference, April 2005. Canberra. **Proceedings...** Australia, 2005.

[137] Angela Hennessy, Alex Gladd, and Brenton Walker. Nullspace-based stopping conditions for network-coded transmissions in DTNs. *Proceedings of the seventh ACM international workshop on Challenged networks - CHANTS '12*, page 51, 2012.

[138] Fredrik Herbertsson. *Implementation of a Delay-Tolerant Routing Protocol in the Network Simulator NS-3*. PhD thesis, 54 p. Phd in Computer and Information Science, Linkopings Universitet, Linkopings, 2010.

[139] T. Ho, R. Koetter, M. Medard, D.R. Karger, and M. Effros. The benefits of coding over routing in a randomized setting. *IEEE International Symposium on Information Theory, 2003. Proceedings.*, page 7803, 2003.

[140] Tracey Ho and Desmond S. Lun. *Network Coding: An Introduction*. Cambridge University Press, 2008.

[141] Elizabeth Howell. Lagrange points: Parking places in space. `www.space.com`, October 2017.

[142] D. Hua, X. Du, Y. Qian, and S. Yan. A dtn routing protocol based on hierarchy forwarding and cluster control. In *2009 International Conference on Computational Intelligence and Security*, volume 2, pages 397–401, Dec 2009.

[143] Daowen Hua, Xuehvi Du, Lifeng Cao, Guoyu Xu, and Yanbin Qian. A DTN congestion avoidance strategy based on path avoidance. In *2nd International Conference on Future Computer and Communication, IEEE*, volume 1, pages 855–860, 2010.

[144] P. Hui, J. Crowcroft, and E. Yoneki. Bubble rap: Social-based forwarding in delay-tolerant networks. *IEEE Transactions on Mobile Computing*, 10(11):1576–1589, Nov 2011.

[145] Mouhamad Ibrahim. *Routing and Performance Evaluation of Disruption Tolerant Networks*. PhD thesis, Université de Nice - Informatique - Sophia Antípolis, November 2008.

[146] A. Indgren and K. S. Phanse. Evaluation of queuing policies and forwarding strategies for routing in intermittently connected networks. In *IEEE International Communication System Software and Middleware*, 2006. 1., **Proceedings...** Piscataway: IEEE, 2006. p. 1–10.

[147] Internet Research Task Force. Delay-tolerant networking research group (DTNRG). `https://irtf.org/concluded/dtnrg`. Accessed in March 2018.

[148] Internet Research Task Force. Interplanetary internet research group (IPNRG). `https://irtf.org/concluded/ipnrg`, 2018. Accessed in March 2018.

[149] D. Israel. Ladee presentation to ipnsig event. http://ipnsig.org/wp-content/uploads/2014/02/LLCD-DTN-Demonstration-IPNSIG-Final.pdf, 2014. Accessed: 2017-05-14.

[150] W. D. Ivancic. Security analysis of dtn architecture and bundle protocol specification for space-based networks. In *2010 IEEE Aerospace Conference*, 2010.

[151] V. Jacobson. Congestion avoidance and control. In *Computer Communication Review*, volume 18, pages 314–329, August 1988.

[152] Sushant Jain, Kevin Fall, and Rabin Patra. Routing in a delay tolerant network. In *Proceedings of the 2004 Conference on Applications, Technologies, Architectures, and Protocols for Computer Communications*, SIGCOMM '04, pages 145–158, 2004.

[153] A. Jenkins, S. Kuzminsky, K. K. Gifford, R. L. Pitts, and K. Nichols. Delay/disruption-tolerant networking: Flight test results from the international space station. In *IEEE Aerospace Conference*, 2010.

[154] Jet Propusion Laboratory. ION: Interplanetary overlay network. Athens: Ohio University, Available in: `https://ion.ocp.ohiou.edu/`, 2009. Accessed in January 7th, 2013.

[155] H. Jin, A. Khandekar, and R. McEliece. Irregular repeat-accumulate codes. In *Proc. of the 2nd Int. Symp. on Turbo codes & Related Topics*, pages 1–8, Brest, France, September 2000.

[156] J. Jin, J. Gubbi, S. Marusic, and M. Palaniswami. An information framework for creating a smart city through internet of things. *IEEE Internet of Things Journal*, 1(2):112–121, 2014.

[157] P. Juang, H. Oki, Y. Wang, M. Martonosi, L. S. Peh, and D. Rubenstein. Energy-efficient computing for wildlife tracking. In *the 10th international conference on architectural support for programming languages and operating systems (ASPLOS-X) - ASPLOS '02*, 2002.

[158] K. Bhasin and J. L. Hayden. Space internet architectures and technologies for NASA enterprises. In *2001 IEEE Aerospace Conference Proceedings (Cat. No.01TH8542)*, volume 2, pages 931–941, 2001.

[159] Ari Keränen. The ONE Simulator for DTN Protocol Evaluation. In *Proceedings of the Second International ICST Conference on Simulation Tools and Techniques*, number 55, 2009.

[160] Ari Keränen, Jörg Ott, and Teemu Kärkkäinen. The ONE Simulator for DTN Protocol Evaluation. In *International Conference on Simulation Tools and Techniques*, 2009. 2., Brussels. **Proceedings...** New York: ICST, 2009.

[161] Maurice J. Khabbaz, Chadi M. Assi, and Wissom F. Fawaz. Disruption-tolerant networking: A comprehensive survey on recent developments and persisting challenges. In *IEEE Communications Survey & Tutorials*, volume 14, pages 1–34. 2011.

[162] R. Khan, S. U. Khan, R. Zaheer, and S. Khan. Future internet: The internet of things architecture, possible applications and key challenges. In *2012 10th International Conference on Frontiers of Information Technology*, 2012.

[163] Dohyung Kim, Hanjin Park, and Ikjun Yeom. Minimize the impact of buffer overflow in DTN. In *International Conference on Future Internet Technologies*, 2008. Seoul. **Proceedings...** [S.1.:s.n], 2008.

[164] E. Konler, M. Handley, and S. Foyd. Datagram congestion control protocol (DCCP). Technical report, Network Working Group Request for Comments: 4340, 2006. Fremont, CA: IETF Trust, 2006. RFC4340.

[165] S. Krco, B. Pokric, and F. Carrez. Designing iot architecture(s): A European perspective. In *IEEE World Forum on Internet of Things (WF-IoT)*, 2014.

[166] Amir Krifa, Chadi Barakat, and Thrasyvoulos Spyropoulos. Optimal buffer management policies for delay tolerant networks. In *Annual IEEE Communications Society Conference on Sensor, Mesh and Ad Hoc Communications and Networks*, June 2008. 5., San Francisco, CA. **Proceedings...** Piscataway: IEEE, 2008.

[167] H. Kruse and S. Jero. Datagram convergence layers for the delay- and disruption-tolerant networking (dtn) bundle protocol and licklider transmission protocol (ltp). Technical Report RFC 7122, Internet Research Task Force (IRTF), 2014.

[168] N. Kushalnagar, G. Montenegro, and C. Schumacher. Ipv6 over low-power wireless personal area networks (6lowpans): Overview, assumptions, problem statement, and goals. Technical report, RFC 4919 (Informational), Internet Engineering Task Force Std. 4919, 2007.

[169] Jani Lakkakorpi, Mikko Pitkanen, and Jorg Ott. Using buffer space advertisements to avoid congestion in mobile opportunistic DTNs. In *9th IFIP TC 6 International Conference on Wired/Wireless Internet Communications*, pages 386–397, 2011.

[170] B. A. LaMacchia and A. M. Odlyzko. Solving Large Sparse Linear Systems over Finite Fields. *Lecture Notes in Computer Science*, 537:109–133, 1991.

[171] C. Lanczos. Solution of systems of linear equations by minimized iterations. *J. Res. Nat. Bureau of Standards*, 49:33–53, 1952.

[172] Lertluck Leela-amornsin and Hiroshi Esaki. Heuristic congestion control for massage deletion in delay tolerant network. In *Smart Spaces and Next Generation Wired/Wireless Networking*, August 2010. Third Conference on Smart Spaces and 10th International Conference.

[173] Y. Leng and L. Zhao. Novel design of intelligent internet-of-vehicles management system based on cloud-computing and internet-of-things. In *2011 International Conference on Electronic & Mechanical Engineering and Information Technology*, 2011.

[174] P. Levis, S. Madden, J. Polastre, R. Szewczyk, K. Whitehouse, A. Woo, D. Gay, J. Hill, M. Welsh, E. Brewer, and D. Culler. *TinyOS: An Operating System for Sensor Networks*, pages 115–148. Springer Berlin Heidelberg, 2005.

[175] Lingzhi Li, Shukui Zhang, Zhe Yang, and Yanqin Zhu. Data dissemination in mobile wireless sensor network using trajectory-based network coding. *International Journal of Distributed Sensor Networks*, 2013.

[176] Yong Li, Mengjiong Qian, Depeng Jin, Li Su, and Lieguang Zeng. Adaptive optimal buffer management policies for realistic DTN. In *IEEE Global Telecommunications Conference*, 2009. Honolulu. **Proceedings...** Piscataway: IEEE, 2009.

[177] Rudolf Lidl and Harald Niederreiter. *Finite Fields*. Cambridge University Press, 1997.

[178] Yunfeng Lin, Baochun Li, and Ben Liang. Efficient network coded data transmissions in disruption tolerant networks. *Proceedings - IEEE INFOCOM*, pages 2180–2188, 2008.

[179] Yunfeng Lin, Ben Liang, and Baochun Li. Performance Modeling of Network Coding in Epidemic Routing. *Proceedings of the 1st international MobiSys workshop on Mobile opportunistic networking - MobiOpp '07*, pages 67–74, 2007.

[180] A. Lindgren, C. Mascolo, M. Lonergan, and B. McConnell. Seal-2-seal: A delay-tolerant protocol for contact logging in wildlife monitoring sensor networks. In *2008 5th IEEE International Conference on Mobile Ad Hoc and Sensor Systems*, 2008.

[181] Anders Lindgren, Avri Doria, and Olov Schelén. Probabilistic routing in intermittently connected networks. *SIGMOBILE Mob. Comput. Commun. Rev.*, 7(3):19–20, July 2003.

[182] G. Liva, E. Paolini, and M. Chiani. Bounds on the error probability of block codes over the q-ary erasure channel. *IEEE Trans. Commun.*, 61(6):2156–2165, June 2013.

[183] G. Liva, P. Pulini, and M. Chiani. On-line construction of irregular repeat accumulate codes for packet erasure channels. *IEEE Trans. Wireless Commun.*, 12(2):680–689, February 2013.

[184] Shou Chih Lo and Chuan Lung Lu. A dynamic congestion control based routing for delay tolerant network. In *International Conference on Fuzzy Systems and Knowledge Discovery*, 2012. 9., Sichuan. **Proceedings...** Piscataway: IEEE, 2012. p. 2047–2051.

[185] D. Locke. Mq telemetry transport (mqtt) v3. 1 protocol specification. Markham, 2010.

[186] Max Loubser. Delay tolerant networking for sensor networks. Technical report, Swedish Institute of Computer Science, 2006. SICS Technical Report T2006:01.

[187] M. Luby, T. Gasiba, T. Stockhammer, and M. Watson. Reliable multimedia download delivery in cellular broadcast networks. *IEEE Transactions on Broadcasting*, 53:235–246, March 2007.

[188] M. G. Luby, M. Mitzenmacher, M. A. Shokrollahi, and D. A. Spielman. Improved low-density parity-check codes using irregular graphs. *IEEE Trans. Inf. Theory*, 47(2):585–598, 2001.

[189] J. E. Luzuriaga, M. Zennaro, J. C. Cano, C. Calafate, and P. Manzoni. A disruption tolerant architecture based on mqtt for iot applications. In *2017 14th IEEE Annual Consumer Communications & Networking Conference (CCNC)*, 2017.

[190] P. Madoery, J. Fraire, and J. Finochietto. Congestion management techniques for disruption tolerant satellite networks. *Wiley International Journal of Satellite Communications and Networking*, 2017.

[191] Mallick and P. Kumar. *Research Advances in the Integration of Big Data and Smart Computing*. IGI Global, 2015.

[192] Lefteris Mamatas, Tobias Harks, and Vassilis Tsaoussidis. Approaches to congestion control in packet networks. *Journal of Internet Engineering*, 1(1), 2007.

[193] J. Manyika, M. Chui, J. Bughin, R. Dobbs, and P. Bisson. Disruptive technologies: Advances that will transform life, business, and the global economy, 2013.

[194] T. Matsuda and T. Takine. (p,q)-epidemic routing for sparsely populated mobile ad hoc networks. *Selected Areas in Communications, IEEE*, pages 783–793, 2008.

[195] A. McMahon and S. Farrell. Delay and disruption-tolerant networking. *IEEE Internet Computing*, 13(6):82–87, 2009.

[196] Preechai Mekbungwan, Bidur Devkota, Sarita Gurung, Apinun Tunpan, and Kanchana Kanchanasut. Network Coding Based Bulk Data Synchronization in Mobile Ad Hoc Networks. *Proceedings of the 9th Asian Internet Engineering Conference*, pages 17–24, 2013.

[197] Preechai Mekbungwan, Apinun Tunpan, and Kanchana Kanchanasut. An NC-DTN framework for many-to-many bulk data dissemination in OLSR MANET. *IWCMC 2015 - 11th International Wireless Communications and Mobile Computing Conference*, pages 964–969, 2015.

[198] K. Michaelsen, J. L. Sanders, S. M. Zimmer, and G. M. Bump. Overcoming patient barriers to discussing physician hand hygiene: Do patients prefer electronic reminders to other methods? *Infect. Control Hosp. Epidemiol.*, 34(09):929–934, 2013.

[199] G. Montenegro, N. Kushalnagar, J. Hui, and D. Culler. Transmission of ipv6 packets over ieee 802.15.4 networks. Technical report, RFC 4944 (Proposed Standard), Internet Engineering Task Force Std. 4944, 2007.

[200] J. Morgenroth, S. Schildt, and L. Wolf. Ibr-dtn – the versatile, efficient open-source framework for a bundle protocol ecosystem. *PIK - Praxis der Informationsverarbeitung und Kommunikation*, 36(1), 2013.

[201] Johannes Morgenroth, Sebastian Schildt, and Lars Wolf. A bundle protocol implementation for android devices. In *Proceedings of the 18th Annual International Conference on Mobile Computing and Networking*, Mobicom'12, pages 443–446, 2012.

[202] I. Peña López. Itu internet report 2005: The internet of things, 2005.

[203] Piyush Naik and Krishna M. Sivalingam. *A Survey of MAC Protocols for Sensor Networks*, pages 93–107. Springer US, 2004.

[204] P. Narendra, S. Duquennoy, and T. Voigt. Ble and ieee 802.15.4 in the iot: Evaluation and interoperability considerations. Lecture Notes of the Institute for Computer Sciences, Social Informatics and Telecommunications Engineering, 2016.

[205] National Security Agency. NSA suite b cryptography. Technical report, National Security Agency, 2015.

[206] Sergiu Nedevschi and Rabin Patra. DTNlite: A reliable data transfer architecture for sensor network. https://people.eecs.berkeley.edu/~culler/cs294-f03/finalpapers/dtnlite.pdf. Accessed in: February 1st, 2018.

[207] Bradford Nichols, Dick Buttlar, and Jacqueline Proulx Farrell. *Pthreads Programming A POSIX Standard for Better Multiprocessing*. O'Reilly, 1996.

[208] Hervé Ntareme, Marco Zennaro, and Björn Pehrson. Delay tolerant network on smartphones: Applications for communication challenged areas. In *ExtremeCom 2011*, 2011.

[209] National Institute of Standards and Technology (NIST). Federal information processing standards publication 197. Technical report, United States National Institute of Standards and Technology (NIST), 2001.

[210] [Committee on National Security Systems]. National policy on the use of the advanced encryption standard aes to protect national security systems and national security information. Technical report, National Security Agency (NSA), 2003.

[211] E. Paolini, G. Liva, B. Matuz, and M. Chiani. Generalized ira erasure correcting codes for hybrid iterative/maximum likelihood decoding. *IEEE Commun. Lett.*, 12(6):450–452, June 2008.

[212] E. Paolini, G. Liva, B. Matuz, and M. Chiani. Maximum likelihood erasure decoding of ldpc codes: Pivoting algorithms and code design. *IEEE Trans. Commun.*, 60(11):3209–3220, November 2012.

[213] M. Petrova, J. Riihijarvi, P. Mahonen, and S. Labella. Performance study of ieee 802.15.4 using measurements and simulations. In *IEEE Wireless Communications and Networking Conference*, 2006.

[214] Agoston Petz, Chien-Ling Fok, Christine Julien, Brenton D. Walker, and Calvin Ardi. Network coded routing in delay tolerant networks: An experience report. *ExtremeCom '11: Proceedings of the 3rd Extreme Conference on Communication: The Amazon Expedition*, pages 1–6, 2011.

[215] H. Pishro-Nik and F. Fekri. On decoding of low-density parity-check codes over the binary erasure channel. *IEEE Trans. Inf. Theory*, 50(3):439–454, March 2004.

[216] Robert D. Poor. Gradient routing in ad hoc networks. https://www.media.mit.edu/pia/Research/ESP/texts/poorieeepaper.pdf.

[217] W. B. Pottner, F. Busching, G. von Zengen, and L. Wolf. Data elevators: Applying the bundle protocol in delay tolerant wireless sensor networks. In *2012 IEEE 9th International Conference on Mobile Ad-Hoc and Sensor Systems (MASS 2012)*, pages 218–226, 2012.

[218] Juhua Pu, Xingwu Liu, Nima Torabkhani, Faramarz Fekri, and Zhang Xiong. Delay analysis of two-hop network-coded delay-tolerant networks. *Wireless Communications and Mobile Computing*, 15:742–754, 2015.

[219] Shuang Qin and Gang Feng. Performance Modeling of Network Coding Based Epidemic Routing in DTNs. In *Wireless Communications and Networking Conference (WCNC)*, 2013.

[220] Shuang Qin and Gang Feng. Performance modeling of data transmission by using random linear network coding in DTNs. *International Journal of Communication Systems*, 28:2275–2288, 2015.

[221] R. C. Singleton. Maximum distance q-nary codes. *IEEE Trans. Inf. Theory*, 10(2):116–118, April 1964.

[222] Milena Radenkovic and Andrew Grundy. Congestion aware forwarding in delay and social opportunistic networks. In *Eight International Conference on Wireless on Demand Network Systems and Services*, pages 60–67, January 2011.

[223] Milena Radenkovic and Sameh Zakhary. Flexible and Dynamic Network Coding for Adaptive Data Transmission in DTNs. In *Wireless Communications and Mobile Computing Conference (IWCMC), 2012 8th International*, pages 567–573, 2012.

[224] M. Ramadas, S. Burleigh, and S. Farrell. Licklider transmission protocol-specification. RFC 5326, September 2008.

[225] Fioriano De Rango, Mauro Tropea, Giovanni Battista Laratta, and Salvatore Marano. Hop-by-hop local flow control over interplanetary networks based on DTN architecture. In *Communications, 2008. ICC'08. IEEE International Conference on*, pages 1920–1924, May 2008.

[226] Sulma Rashid, A. H. Abdullah, M. Soperi Mohd Zahid, and Qaisar Ayub. Mean drop and effectual buffer management policy for delay tolerant network. *European Journal of Scientific Research*, 70(3):396–407, 2012.

[227] Sulma Rashid and Qaisar Ayub. Effective buffer management policy DLA for DTN routing protocols under congestion. *International Journal of Computer and Network Security*, 2(9):118–121, 2010.

[228] Sulma Rashid, Qaisar Ayub, M. Soperi Mohd Zahid, and S. Hanan Abdullah. E-DROP: An effective drop buffer management policy for DTN routing protocols. *International Journal of Computer Applications*, 13(7), January 2011.

[229] P. Raveneau, E. Chaput, R. Dhaou, and A.-L. Beylot. Dtn for wsn: Freak, implementations and study. In *2015 12th Annual IEEE Consumer Communications and Networking Conference (CCNC)*, 2015.

[230] U. Raza, P. Kulkarni, and M. Sooriyabandara. Low power wide area networks: An overview. *IEEE Communications Surveys & Tutorials*, 19(2):855–873, 2017.

[231] RedHat. Common criteria: Security objectives. `https://access.redhat.com/documentation/en-US/Red_Hat_Certificate_System_Common_Criteria_Certification/8.1/html/Deploy_and_Install_Guide/security-objectives.html`, Accessed: July 2017.

[232] E. Rescorla and B. Korver. Guidelines for writing rfc text on security considerations. Technical Report RFC 3552, IETF.

[233] T. Richardson and R. Urbanke. The capacity of low-density parity-check codes under message-passing decoding. *IEEE Trans. Inf. Theory*, 47, 2001.

[234] T. Richardson and R. Urbanke. Efficient encoding of low-density parity-check codes. *IEEE Trans. Inf. Theory*, 47:638–656, February 2001.

[235] T. J. Richardson, M. A. Shokrollahi, and R. L. Urbanke. Design of capacity-approaching irregular low-density parity-check codes. *IEEE Trans. Inf. Theory*, 47(2):619–637, February 2001.

[236] S. Rottmann, R. Hartung, J. Käberich, and L. Wolf. Amphisbaena: A two-platform DTN node. In *2016 IEEE 13th International Conference on Mobile Ad Hoc and Sensor Systems (MASS)*, pages 246–254, Oct 2016.

[237] S. Tennina, A. Koubâa, R. Daidone, M. Alves, P. Jurcík, R. Severino, M. Tiloca, J.-H. Hauer, N. Pereira, G. Dini, E. Bouroche and M. Tovar. *IEEE 802.15.4 and ZigBee as Enabling Technologies for Low-Power Wireless Systems with Quality-of-Service Constraints*. Springer Science & Business Media, 2013.

[238] P. Saint-Andre. Extensible messaging and presence protocol (xmpp). Technical report, IETF RFC 6120, 2011.

[239] S. M. Sajjad and M. Yousaf. Security analysis of ieee 802.15.4 mac in the context of internet of things (iot). In *2014 Conference on Information Assurance and Cyber Security (CIACS)*, 2014.

[240] Lucile Sassatelli and Muriel Medard. Network Coding for Delay Tolerant Networks with Byzantine Adversaries.

[241] Lucile Sassatelli and Muriel Medard. Inter-session Network Coding in Delay-Tolerant Networks Under Spray-and-Wait Routing. In *10th International Symposium on Modeling and Optimization in Mobile, Ad Hoc and Wireless Networks (WiOpt)*, May, pages 14–18, 2012.

[242] Sebastian Schildt, Johannes Morgenroth, Wolf-Basian Pottner, and Lars Wolf. IBR-DTN:a lightweight, modular and highly portable bundle protocol implementation. In *WowKiVS 2011*, volume 37, 2011.

[243] D. Schürmann, G. V. Zengen, M. Priedigkeit, and L. Wolf. Dtnsec: a security layer for disruption-tolerant networks on microcontrollers. In *2017 16th Annual Mediterranean Ad Hoc Networking Workshop (Med-Hoc-Net)*, pages 1–7, June 2017.

[244] K. Scott and S. Burleigh. Bundle protocol specification. Technical Report RFC5050, Internet RFC, 2007.

[245] K. Scott and S. Burleigh. RFC 5050 - Bundle Protocol Specification. *Internet Engineering Task Force (IETF)*, 2007.

[246] L. Selavo, A. Wood, Q. Cao, T. Sookoor, H. Liu, A. Srinivasan, Y. Wu, W. Kang, J. Stankovic, D. Young, and J. Porter. Luster: Wireless sensor network for environmental research. In *Proceedings of the 5th International Conference on Embedded Networked Sensor Systems*, SenSys '07, pages 103–116, 2007.

[247] Matthew Seligman, Kevin Fall, and Padma Mundur. Alternative custodians for congestion in delay tolerant networks. In *SIGCOMM'06 Workshops*, 2006.

[248] Matthew Seligman, Kevin Fall, and Padma Mundur. Storage routing for DTN congestion control. In *Wireless Communications and Mobile Computing*, volume 7, pages 1183–1196. 2007.

[249] M. Z. Shafiq, A. X. Liu, L. Ji, J. Pang, and J. Wang. A first look at cellular machine-to-machine traffic. In *ACM SIGMETRICS Performance Evaluation Review*, volume 40, 2012.

[250] Z. Shelby, K. Hartke, and C. Bormann. The constrained application protocol (CoAP). Technical report, IETF RFC 7252, 2014.

[251] Z. Shelby, K. Hartke, and C. Bormann. The constrained application protocol (coap). Technical report, IETF RFC7252, 2014.

[252] Zhang Shengli, Soung-Chang Liew, and Patrick PK Lam. Physical layer network coding. *arXiv preprint arXiv:0704.2475*, 2007.

[253] Jang Ping Sheu, Chih-Yin Lee, and Chuang Ma. An Efficient Transmission Protocol Based on Network Coding in Delay Tolerant Networks. *2013 Seventh International Conference on Innovative Mobile and Internet Services in Ubiquitous Computing*, pages 399–404, 2013.

[254] A. Shokrollahi and M. Luby. Raptor codes. *Found. and Trends on Commun. and Inf. Theory*, 6:213–322, March 2009.

[255] Neetya Shrestha and Lucile Sassatelli. On Control of Inter-session Network Coding in Delay-Tolerant Mobile Social Networks. In *17th ACM international conference on Modeling, analysis and simulation of wireless and mobile systems*, pages 231–239, 2014.

[256] Aloizio P. Silva, Scott Burleigh, Celso M. Hirata, and Katia Obraczka. A survey on congestion control for delay and disruption tolerant networks. *Ad Hoc Netw.*, 25(PB):480–494, February 2015.

[257] R. S. Sinha, Y. Wei, and S.-H. Hwang. A survey on lpwa technology: Lora and nb-iot. *ICT Express*, 3(1):14–21, 2017.

[258] Tara Small and Zygmunt J. Haas. Resource and performance tradeoffs in delay-tolerant wireless networks. In *Proceedings of the 2005 ACM SIGCOMM Workshop on Delay-tolerant Networking*, pages 260–267. ACM, 2005.

[259] Tara Small and Zygmunt J. Haas. Resource and performance tradeoffs in delay-tolerant wireless networks. In *Proceedings of the 2005 ACM SIGCOMM Workshop on Delay-tolerant Networking*, WDTN '05, pages 260–267, New York, NY, USA, 2005. ACM.

[260] T. Spyropoulos, K. Psounis, and C. S. Raghavendra. Single-copy routing in intermittently connected mobile networks. In *2004 First Annual IEEE Communications Society Conference on Sensor and Ad Hoc Communications and Networks, 2004. IEEE SECON 2004*, pages 235–244, Oct 2004.

[261] T. Spyropoulos, K. Psounis, and C. S. Raghavendra. Spray and focus: Efficient mobility-assisted routing for heterogeneous and correlated mobility. In *Pervasive Computing and Communications Workshops, 2007. PerCom Workshops '07. Fifth Annual IEEE International Conference on*, pages 79–85, March 2007.

[262] T. Spyropoulos, R. N. Bin Rais, T. Turletti, Katia Obraczka, and A. Vasilakas. *DTNs: Protocols and Applications*, chapter DTN Routing: Taxonomy and Design. CRC Press, 2011.

[263] Thrasyvoulos Spyropoulos, Konstantinos Psounis, and Cauligi S. Raghavendra. Spray and wait: An efficient routing scheme for intermittently connected

mobile networks. In *Proceedings of the 2005 ACM SIGCOMM Workshop on Delay-tolerant Networking*, WDTN '05, pages 252–259. ACM, 2005.

[264] R. Srikant. *The Mathematics of Internet Congestion Control*. Boston: Birkhauser Basel, 2004.

[265] Ramanan Subramanian and Faramarz Fekri. Throughput Performance of Network-Coded Multicast in an Intermittently-Connected Network. In *WiOpt*, pages 212–221, 2010.

[266] Wei Sun, Qingbo Liu, and Kangkang Li. Research on congestion management in delay tolerant networks. In *2011 International Conference on Computer Science and Information Technology - ICCSIT 2011*, volume 51, pages 612–618, 2012.

[267] Richard S. Sutton and Andrew G. Barto. *Introduction to Reinforcement Learning*. MIT Press, Cambridge, MA, 1998.

[268] S. Symington, S. Farrell, H. Weiss, and P. Lovell. Bundle security protocol specification. https://tools.ietf.org/html/rfc6257, 2011.

[269] C. Talcott. Cyber-physical systems and events. In Lecture Notes in Computer Science, 2008. pages 101–115.

[270] Kun Tan, Qian Zhang, and Wenwu Zhu. Shortest path routing in partially connected ad hoc networks. In *Global Telecommunications Conference, 2003. GLOBECOM '03. IEEE*, volume 2, pages 1038–1042, December 2003.

[271] Lu Tan, L. Tan, and N. Wang. Future internet: The internet of things. In *2010 3rd International Conference on Advanced Computer Theory and Engineering(ICACTE)*, 2010.

[272] Biaoshuai Tao and Hongjun Wu. Improving the biclique cryptanalysis of aes. In *Information Security and Privacy: 20th Australasian Conference on Information Security and Privacy*, ACISP'2015, pages 39–56, Brisbane, QLD, Australia, 2015. Springer International Publishing.

[273] Nathanael Thompson and Robin Kravets. Understanding and controlling congestion in DTNs. In *Mobile Computing and Communications*, volume 13. 2010.

[274] Nathanael Thompson, Samuel C. Nelson, Mehedi Baknt, Tarek Abdelzaher, and Robin Kravets. Retiring replicants: Congestion control for intermittently - connected networks. In *INFOCOM'2010, IEEE*, pages 1–9, March 2010. University of Illinois at Urbana-Champaign, Department of Computer Science.

[275] J. Tripathi, J. C. de Oliveira, and J. P. Vasseur. A performance evaluation study of rpl: Routing protocol for low power and lossy networks. In *2010 44th Annual Conference on Information Sciences and Systems (CISS)*, pages 1–6, 2010.

[276] N. Tsiftes, A. Dunkels, Z. He, and T. Voigt. Enabling large-scale storage in sensor networks with the coffee file system. In *the 8th ACM/IEEE International Conference on Information Processing in Sensor Networks (IPSN 2009)*, 2009.

[277] Pierre ugo Tournoux and Jeremie Leguay. The accordion phenomenon: Analysis, characterization and impact on DTN routing. In *INFOCOM 2010*, pages 1116–1124, 2010.

[278] Amin Vahdat and David Becker. Epidemic routing for partially-connected ad hoc networks. Technical report, Duke University, 2000.

[279] Amin Vahdat and David Becker. Epidemic routing for partially connected ad hoc networks. Technical Report CS-200006, 2000.

[280] Anna Vazintari, Christina Vlachou, and Panayotis G. Cottis. Network coding for overhead reduction in delay tolerant networks. *Wireless Personal Communications*, 72(4):2653–2671, 2013.

[281] O. Vermesan and P. Friess. *Finkenzeller, RFID Handbook - Radio-Frequency Identification Fundamentals and Applications*. Halsted Press, 2003. River Publishers, 2016.

[282] S. Vinoski. Advanced message queuing protocol. *IEEE Internet Comput.*, 10(6):87–89, 2006.

[283] Georg von Zengen, Felix Büsching, Wolf-Bastian Pöttner, and Lars C. Wolf. An overview of DTN: Unifying DTNs and WSNs. https://www.ibr.cs.tu-bs.de/papers/buesching-fgsn2012.pdf, 2012. Accessed in: January 28th, 2018.

[284] Artemios G. Voyiatzis. A survey of delay- and disruption-tolerant networking applications. *Journal of Internet Engineering*, 5(1), 2012.

[285] Brenton Walker, Calvin Ardi, Agoston Petz, Jung Ryu, and Christine Julien. Experiments on the spatial distribution of network code diversity in segmented DTNs. In *17th Annual International Conference on Mobile Computing and Networking, MobiCom'11 and Co-Located Workshops - 6th ACM Workshop on Challenged Networks, CHANTS'11*, pages 15–20, 2011.

[286] Chengen Wang, Z. Bia, C. Wang, and L. Da Xu. Iot and cloud computing in automation of assembly modeling systems. *IEEE Transactions on Industrial Informatics*, 10(2):1426–1434, 2014.

[287] Chengjun Wang, Baokang Zhao, Wei Peng, Chunqing Wu, and Zhenghu Gong. Following routing: An active congestion control approach for delay tolerant networks. In *International Conference on Network-Based Information Systems*, 2012. 15., Melbourne. **Proceedings...** Piscataway: IEEE, 2012. p. 727–732.

[288] Chengjun Wang, Baokang Zhao, Wanrong Yu, Chunqing Wu, and Zhenghu Gong. SARM: An congestion control algorithm for dtn. In *International Conference on Ubiquitous Intelligence and Computing and International Conference on Autonomic and Trusted Computing*, 2012. 9., Fukuoka. **Proceedings...** Piscataway: IEEE, 2012. p. 869–875.

[289] R. Wang, S. C. Burleigh, P. Parikh, C. J. Lin, and B. Sun. Licklider Transmission Protocol (LTP)-based DTN for cislunar communications. *IEEE/ACM Transactions on Networking*, 19(2):359–368, April 2011.

[290] R. Want. An introduction to rfid technology. *IEEE Pervasive Comput.*, 5(1):25–33, 2006.

[291] Jörg Widmer and Jean-Yves Le Boudec. Network coding for efficient communication in extreme networks. *Proceeding of the 2005 ACM SIGCOMM workshop on Delay-tolerant networking - WDTN '05*, pages 284–291, 2005.

[292] Keith Winstein and Hari Balakrishnan. End-to-end transmission control by modeling uncertainty about the network state. In *Tenth ACM Workshop on Hot Topics in Networks*, Cambridge, MA, Nov. 2011.

[293] Keith Winstein and Hari Balakrishnan. Tcp ex machina: Computer-generated congestion control. In *ACM Special Interest Group on Data Communication*, SIGCOMM '13, 2013. Hong Kong. **Proceedings...** Piscataway: ACM, 2013. p. 123–134.

[294] Hanno Wirtz, Jan Rüth, Martin Serror, Jó Ágila Bitsch Link, and Klaus Wehrle. Opportunistic interaction in the challenged internet of things. In *Proceedings of the 9th ACM MobiCom Workshop on Challenged Networks*, CHANTS '14, pages 7–12, 2014.

[295] Yunnan Wu, Philip A. Chou, and Kamal Jain. A Comparison of Network Coding and Tree Packing. In *Proceedings. International Symposium on Information Theory*, 2004.

[296] J. Wyatt, S. Burleigh, R. Jones, and L. Torgerson. Disruption tolerant networking flight validation experiment on NASA's epoxi mission. *First International Conference on Advances in Satellite and Space Communications*, 2009.

[297] Bartek Peter Wydrowski. *Techniques in Internet Congestion Control*. PhD thesis, Electrical and Electronic Engineering Department, The University of Melbourne, February 2003.

[298] B. Xu, D. Zhang, and W. Yang. Research on architecture of the internet of things for grain monitoring in storage. *Communications in Computer and Information Science*, page 431–438, 2012.

[299] Y. Yan, Y. Qian, H. Sharif, and D. Tipper. A survey on smart grid communication infrastructures: Motivations, requirements and challenges. *IEEE Communications Surveys Tutorials*, 15(1):5–20, 2013.

[300] M. Yang and Y. Yang. Peer-to-peer file sharing based on network coding. In *2008 The 28th International Conference on Distributed Computing Systems*, pages 168–175, June 2008.

[301] Z. Yang, R. Wang, Q. Yu, X. Sun, M. De Sanctis, Q. Zhang, J. Hu, and K. Zhao. Analytical characterization of Licklider Transmission Protocol (LTP) in cislunar communications, July 2014.

[302] Zhihong Yang. Study and application on the architecture and key technologies for iot. In *2011 International Conference on Multimedia Technology*, 2011.

[303] Lei Yin, Hui-mei Lu, Yuan da Cao, and Jian min Gao. An incentive congestion control strategy for DTNs with mal-behaving nodes. In *International Conference on Networks Security, Wireless Communications and Trusted Computing*, 2010. Wuhan. **Proceedings...** Piscataway: IEEE, 2010. p. 91–94.

[304] Seung-Keun Yoon and Zygmunt J. Haas. Application of Linear Network Coding in Delay Tolerant Networks. In *The Second International Conference on Ubiquitous and Future Networks (ICUFN 2010)*, 2010.

[305] Q. Yu, X. Sun, R. Wang, Q. Zhang, J. Hu, and Z. Wei. The effect of DTN custody transfer in deep-space communications. *IEEE Wireless Communications*, 20(5):169–176, October 2013.

[306] Li Yun, Cheng Xinjian, Liu Qilie, and You Xianohu. A novel congestion control strategy in delay tolerant networks. In *Second International Conference on Future Networks*, 2010.

[307] Deze Zeng, Song Guo, Hai Jin, and Victor Leung. Dynamic segmented network coding for reliable data dissemination in delay tolerant networks. *2012 IEEE International Conference on Communications (ICC)*, pages 63–67, 2012.

[308] Deze Zeng, Song Guo, Zhuo Li, and Sanglu Lu. Performance Evaluation of Network Coding in Disruption Tolerant Networks. 2010.

[309] G. Von Zengen, F Busching, WB Pottner, and L Wolf. An overview of dtn: Unifying dtns and wsns. In *11th GI/ITG KuVS Fachgespräch Drahtlose Sensornetze (FGSN)*, 2012.

[310] Gengxin Zhang, Zhidong Xie, Dongming Bian, and Qian Sun. A survey of deep space communications. *Journal of Electronics (China)*, 28(2):145, 2011.

[311] Guohua Zhang and Yonghe Liu. Congestion management in delay tolerant networks. In *WICON*, pages 1–9, 2008.

[312] Qian Zhang, Zhigang Jin, Zhenjing Zhang, and Yantai Shu. Network coding for applications in the Delay Tolerant Network (DTN). *MSN 2009 - 5th International Conference on Mobile Ad-hoc and Sensor Networks*, pages 376–380, 2009.

[313] Xiaolan Zhang, Giovanni Neglia, and Jim Kurose. Chapter 10 - Network Coding in Disruption Tolerant Networks. *Network Coding*, pages 267–308, 2012.

[314] Xiaolan Zhang, Giovanni Neglia, and Jim Kurose. *Network Coding in Disruption Tolerant Networks*. Elsevier Inc., 2012.

[315] Xiaolan Zhang, Giovanni Neglia, Jim Kurose, and Don Towsley. On benefits of random linear coding for unicast applications in disruption tolerant networks. *2006 4th International Symposium on Modeling and Optimization in Mobile, Ad Hoc and Wireless Networks, WiOpt 2006*, 2006.

[316] Xiaolan Zhang, Giovanni Neglia, Jim Kurose, Don Towsley, and Haixiang Wang. Benefits of network coding for unicast application in disruption-tolerant networks. *IEEE/ACM Transactions on Networking*, 21(5):1407–1420, 2013.

[317] Baokang Zhao, Wei Peng, Ziming Song, Jinshu Su, Chunqing Wu, Wanrong Yu, and Qiaolin Hu. Towards efficient and practical network coding in delay tolerant networks. *Computers and Mathematics with Applications*, 63(2):588–600, 2012.

[318] Fang Zhao, Ton Kalker, Muriel Medard, and Keesook J Han. Signatures for Content Distribution with Network Coding. In *IEEE International Symposium on Information Theory*, pages 556–560, 2007.

[319] K. Zhao, R. Wang, S. C. Burleigh, A. Sabbagh, W. Wu, and M. D. Sanctis. Performance of Bundle Protocol for deep-space communications. *IEEE Transactions on Aerospace and Electronic Systems*, 52(5):2347–2361, October 2016.

Index